KNIGHTLY PIETY
AND THE LAY RESPONSE
TO THE FIRST CRUSADE

Knightly Piety
and the Lay Response
to the First Crusade

The Limousin and Gascony,
*c.*970–*c.*1130

MARCUS BULL

CLARENDON PRESS · OXFORD
1993

Oxford University Press, Walton Street, Oxford OX2 6DP
Oxford New York Toronto
Delhi Bombay Calcutta Madras Karachi
Kuala Lumpur Singapore Hong Kong Tokyo
Nairobi Dar es Salaam Cape Town
Melbourne Auckland Madrid
and associated companies in
Berlin Ibadan

Oxford is a trade mark of Oxford University Press

Published in the United States
by Oxford University Press Inc., New York

British Library Cataloguing in Publication Data
Data available

Library of Congress Cataloging in Publication Data
Bull, Marcus Graham.
Knightly piety and the lay response to the First Crusade:
the Limousin and Gascony, c. 970–c. 1130 / Marcus Graham Bull.
p.   cm.
Includes bibliographical references and index.
ISBN 0-19-820354-3
1. Limousin (France)—Religious life and customs.   2. Gascony
(France)—Religious life and customs.   3. Crusades—First,
1096–1099.   I. Title.
BR844.B85   1993
282'.4466'09021—dc20   92-21403

1 3 5 7 9 10 8 6 4 2

Typeset by Best-set Typesetter Ltd., Hong Kong

Printed in Great Britain
on acid-free paper by
Biddles Ltd., Guildford and King's Lynn

For my family

Preface

IN what follows, most references to a text found in edited cartularies and collections of documents are to the text's number, which is rendered in arabic numerals whatever the convention used by the editor. In those cases where I wish to draw attention to a specific passage within a long document, or whenever there is scope for ambiguity, the text's number is introduced by 'no.' and/or the page reference is supplied. Documents' dates are usually given in the footnote if this is not apparent from the main text. In some instances, however, fixing upon a date must involve speculation, and in those cases in which too much guesswork appeared necessary it has seemed preferable not to volunteer suggestions. The dates given for certain documents depart from those suggested by the editors; but when this work was in preparation it became obvious that to justify every proposed correction in a consistent manner would require an unwieldy, and frequently superfluous, critical apparatus. I have therefore elaborated on my reasons for revising dates only when this appeared integral to the argument. Dates are rendered using the following conventions: 1068–96 = a continuous period between the two years; 1068 × 1096 = an unspecifiable year within the two years inclusive. Narrative sources which exist in standard and commonly available editions are not cited by book and/or chapter. In the case of old editions which may soon be superseded, however, it has seemed useful to supply book and/or chapter number in parentheses; the same applies to recent editions of works which readers may wish to consult in an older version. Biblical quotations are from the Authorized Version or the Revised Standard Version, Catholic Edition.

It is a pleasure to acknowledge the many debts which I have incurred. The Institute of Historical Research and the British Academy elected me to Fellowships which afforded the opportunity to complete this work. The librarians and staff of the Institute of Historical Research, the British Library, and the Bibliothèque Municipale, Bordeaux, provided invaluable assistance in finding materials and answering my many queries. Requests for microfilms, photocopies, information, and advice were kindly met by the Bibliothèque Nationale, the Société Archéologique et Historique du Limousin, and the Archives Départementales of Creuse, Gers, Gironde,

Hautes-Pyrénées, and Landes. The Department of History at Royal Holloway and Bedford New College made available funds for me to attend a conference in the United States, where I was able to sound out some of the ideas contained in the Introduction. My indebtedness to a number of individual friends and scholars is very great. They have saved me from many errors; those that remain are, of course, entirely my own. My interest in Medieval History was first aroused by the inspirational teaching of David Jones. Professor Daniel Callahan generously provided me with a volume which did much to aid my understanding of the Peace of God. Justin Jacyno kindly drew drafts of the two maps. Dr Nigel Saul read an early draft of two chapters and suggested improvements, and Professor Norman Housley read the whole work in a later version. Professor Christopher Brooke and Brenda Bolton, who examined the thesis on which this book is based, made many pertinent observations and offered useful advice, for which I am very grateful. I have benefited a great deal from many discussions with Dr Jonathan Phillips and Dr Christoph Maier. John Doran has been an unfailing source of interesting insights into the life of the church, medieval and modern. In the later stages of this work's preparation Lucy Pratt gave much valued support. Vera and Eric Burden graciously submitted themselves to the toil of reading many drafts and never flagged in their encouragement. My warm thanks also go to the Delegates and staff of the Oxford University Press, especially Sophie MacCallum, who addressed my often trifling queries with patience, scholarship, and good sense. To my former supervisor, Professor Jonathan Riley-Smith, I owe an enormous debt. It has been a privilege to benefit from his kind generosity, enthusiasm, and erudition. Finally, I must thank the members of my family, without whose emotional and material support this project could never have reached completion. The dedication of this book can be no more than a gesture of my deep appreciation.

M.G.B.

Contents

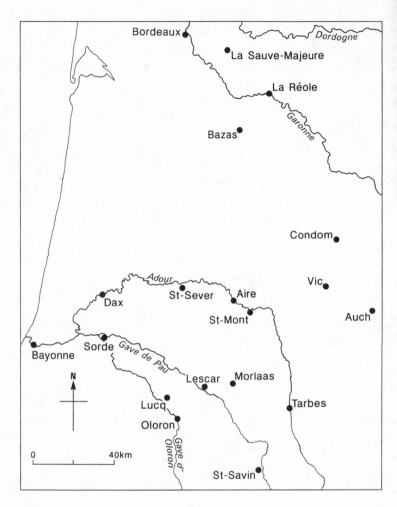

1. Gascony

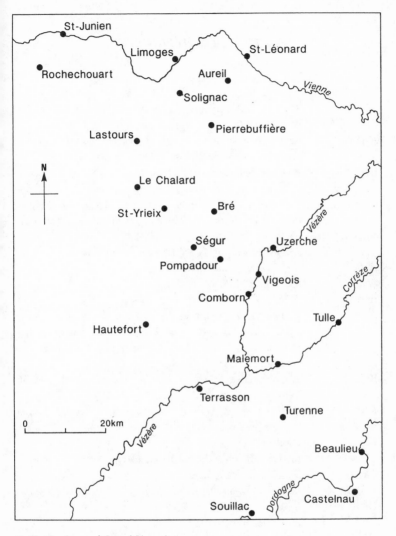

2. The Southern and Central Limousin

List of Abbreviations

A.   'Cartulaire du prieuré d'Aureil', ed. G. de Senneville, *BSAHL* 48 (1900), pp. 1–289.

AASS   *Acta Sanctorum Bollandiana*, ed. Société des Bollandistes, 3rd edn., 62 vols. (Brussels, 1863–1925).

AB   *Analecta Bollandiana*.

Adhemar,   Adhemar of Chabannes, *Chronique*, ed. J. Chavanon
Chronique   (Collection de textes pour servir à l'étude et à l'enseignement de l'histoire, 20; Paris, 1897); all references are from book III.

AHG   *Archives historiques du Département de la Gironde*.

AHR   *American Historical Review*.

AM   *Annales du Midi*.

AESC   *Annales: économies, sociétés, civilisations*.

Auch   *Cartulaires du chapitre de l'église métropolitaine Sainte-Marie d'Auch*, ed. C. Lacave La Plagne Barris (Archives historiques de la Gascogne², 3; Paris, 1899); all references are from the *Cartulaire noir*.

B.   *Cartulaire de l'abbaye de Beaulieu (en Limousin)*, ed. M. Deloche (Paris, 1859).

BHL   *Bibliotheca hagiographica latina antiquae et mediae aetatis*, ed. Société des Bollandistes, 2 vols. in 3 (Brussels, 1898–1911).

BL Add. MS   British Library, Additional Manuscript.

BRAH   *Boletín de la Real Academia de la Historia*.

BSAHL   *Bulletin de la Société Archéologique et Historique du Limousin*.

C.   *Recueil des chartes de l'abbaye de Cluny*, ed. A. Bernard and A. Bruel, 6 vols. (Paris, 1876–1903).

CCM   *Cahiers de civilisation médiévale*.

DDC   *Dictionnaire de droit canonique*, ed. R. Naz, 7 vols. (Paris, 1935–65).

DTC   *Dictionnaire de théologie catholique*, ed. A. Vacant, E. Mangenot, and É. Amann, 15 vols. (Paris, 1908–50).

EEMCA   *Estudios de Edad Media de la Corona de Aragón*.

EHR   *English Historical Review*.

ES   *España Sagrada*, ed. E. Flórez and M. Risco, 2nd edn., 51 vols. (Madrid, 1754–1879).

GC   *Gallia Christiana in provincias ecclesiasticas distributa*, ed. D. Sammarthanus, P. Piolin, and B. Hauréau, 16 vols. (Paris, 1739–1877).

Glaber   Ralph Glaber, 'Historiarum Libri Quinque', ed. and trans. J. France, in *Rodulfus Glaber Opera*, ed. J. France, N. Bulst, and P. Reynolds (Oxford, 1989), 1–253.

Guide   *Le Guide du pèlerin de Saint-Jacques de Compostelle*, ed. and trans. J. Vielliard, 3rd edn. (Mâcon, 1963).

HGL   *Histoire générale de Languedoc*, ed. C. Devic and J. Vaissete, rev. E. Roschach, A. Molinier, *et al.*, 16 vols. (Toulouse, 1872–1904).

HR   *Historical Reflections*.

JEH   *Journal of Ecclesiastical History*.

JL   *Regesta Pontificum Romanorum*, ed. P. Jaffé, rev. S. Loewenfeld, W. Wattenbach, F. Kaltenbrunner, and P. Ewald, 2 vols. (Leipzig, 1885–8).

Lair   J. Lair, *Études critiques sur divers textes des Xe et XIe siècles*, ii. *Historia d'Adémar de Chabannes* (Paris, 1899).

LR   'Cartulaire du prieuré de Saint-Pierre de la Réole, ed. C. Grellet-Balguerie, *AHG* 5 (1863), pp. 99–186.

LSM   'Grand Cartulaire de la Sauve-Majeure', Bibliothèque Municipale, Bordeaux, MS 769.

Mansi   *Sacrorum conciliorum nova et amplissima collectio*, ed. J. D. Mansi, 55 vols. (Florence, 1759–1962).

Mart. I.   'Premier cartulaire de l'aumônerie de S. Martial', ed. A. Leroux, E. Molinier, and A. Thomas, *Documents historiques bas-latins, provençaux et français concernant principalement la Marche et le Limousin*, 2 vols. (Limoges, 1883–5), ii, pp. 1–17.

MGH   *Monumenta Germaniae historica inde ab anno Christi quingentesimo usque ad annum millesimum et quingentesimum auspiciis societatis aperiendis fontibus rerum germanicarum medii aevi*, ed. G. H. Pertz *et al.* (Hanover, 1826–).

MGH SS   *MGH Scriptores in Folio et Quarto*.

OV   Orderic Vitalis, *The Ecclesiastical History*, ed. and trans. M. Chibnall, 6 vols. (Oxford, 1969–80).

PL   *Patrologiae cursus completus, series Latina*, ed. J.-P. Migne, 221 vols. (Paris, 1844–64).

RB — *Revue bénédictine.*

RHC — *Recueil des historiens des croisades*, ed. Académie des Inscriptions et Belles-Lettres, 16 vols. in 17 (Paris, 1841–1906).

RHC Occ. — *RHC historiens occidentaux*, 5 vols. (Paris, 1844–95).

RHGF — *Recueil des historiens des Gaules et de la France*, ed. M. Bouquet, rev. L. Delisle, 24 vols. in 25 (Paris, 1840–1904).

RM — *Revue Mabillon.*

RS — Rolls Series: Chronicles and memorials of Great Britain and Ireland during the Middle Ages, published under the direction of the Master of the Rolls (London, 1838–96).

S. — *Cartulaire de l'abbaye de Saint-Jean de Sorde*, ed. P. Raymond (Paris, 1873).

SEL — 'Sancti Stephani Lemovicensis Cartularium', ed. J. de Font-Réaulx, *BSAHL* 69 (1922), pp. 5–258.

Sol. — 'Cartulaire de Solignac', Bibliothèque Nationale, MS lat. 18363.

Ste-C. — 'Cartulaire de l'abbaye Sainte-Croix de Bordeaux', ed. A. Ducaunnès-Duval, *AHG* 27 (1892), pp. 1–157.

St-M. — *Cartulaire du prieuré de Saint-Mont*, ed. J. de Jaurgain (Archives historiques de la Gascogne², 7; Paris, 1904).

St-M(Sam). — 'Le Plus Ancien Cartulaire de Saint-Mont (Gers) (XIᵉ–XIIIᵉ siècles)', ed. C. Samaran, *Bibliothèque de l'École des Chartes*, 110 (1952), pp. 5–56.

St-Seurin — *Cartulaire de l'église collégiale Saint-Seurin de Bordeaux*, ed. J.-A. Brutails (Bordeaux, 1897).

T. — *Cartulaire des abbayes de Tulle et de Roc-Amadour*, ed. J.-B. Champeval (Brive, 1903).

TRHS — *Transactions of the Royal Historical Society.*

Uz. — *Cartulaire de l'abbaye d'Uzerche*, ed. J.-B. Champeval (Paris, 1901).

V. — *Cartulaire de l'abbaye de Vigeois en Limousin (954–1167)*, ed. M. de Montégut (Limoges, 1907).

Introduction

THIS book examines the religious ideas of nobles and knights, with
particular reference to the reasons why men went on the First Crusade
(1095–1101). There have been many valuable studies of aristocratic piety
in the Central Middle Ages, but few have concentrated specifically upon
the response to crusading.[1] It would be impracticable to study the whole
of Latin Europe, or indeed all of France. Consequently what follows
concentrates upon the aristocracies of three parts of south-western France:
the southern and central (or Bas-) Limousin; the Bordelais and Bazadais;
and southern Gascony, an area roughly bounded by the upper reaches of
the River Adour, the Landes, and the Pyrenees. (The last two areas will
sometimes be collectively termed 'Gascony'.)[2] To this extent the present
study is inspired by the methods of a popular product of French medieval
scholarship in recent years, the regional monograph. It departs from most
of the works of this genre, however, in that they seldom devote much
attention to the significance of crusading apart from its influence on a
small number of themes, such as patterns of land-tenure and mortality
rates within a few well-documented families.[3] This study's terminal dates
are c.970–c.1130. They are intended to be approximate only, and have
been chosen for two reasons: it is from around the third quarter of the

---

[1] e.g. M. Castaing-Sicard, 'Donations toulousaines du Xᵉ au XIIIᵉ siècle', *AM* 70 (1958),
27–64; C. Blanc, 'Les Pratiques de piété des laïcs dans les pays du Bas-Rhône aux XIᵉ et
XIIᵉ siècles', *AM* 72 (1960), 137–47; É. Delaruelle, 'La Culture religieuse des laïcs en
France aux XIᵉ et XIIᵉ siècles', in *I laici nella 'Societas Christiana' dei secoli XI e XII: atti della
terza settimana internazionale di studio Mendola* (Miscellanea del Centro di Studi Medioevali, 5;
Milan, 1968), 548–81; R. Mortimer, 'Religious and Secular Motives for some English
Monastic Foundations', in D. Baker (ed.), *Religious Motivation: Biographical and Sociological
Problems for the Church Historian* (Studies in Church History, 15; Oxford, 1978), 77–85; C.
Harper-Bill, 'The Piety of the Anglo-Norman Knightly Class', in R. A. Brown (ed.),
Proceedings of the Battle Conference on Anglo-Norman Studies, ii. 1979 (Woodbridge, 1980),
63–77; V. Chandler, 'Politics and Piety: Influences on Charitable Donations during the
Anglo-Norman Period', *RB* 90 (1980), 63–71; J. Paul, *L'Église et la culture en Occident
IXᵉ–XIIᵉ siècles* (Nouvelle Clio, 15; Paris, 1986), ii. 645–702.

[2] But for evidence that the inhabitants of the Bordelais and Bazadais distinguished
between themselves and Gascons, see St-Seurin, 16, and LSM, p. 6.

[3] e.g. G. Duby, *La Société aux XIᵉ et XIIᵉ siècles dans la région mâconnaise*, 2nd edn. (Paris,
1971), 283–4, 334–5; G. Devailly, *Le Berry du Xᵉ siècle au milieu du XIIIᵉ: étude politique,
religieuse, sociale et économique* (Civilisations et sociétés, 19; Paris, 1973), 382–4, 524–6.

tenth century that documentary evidence for these areas becomes available in significant quantities (though there is some earlier Limousin material); and by continuing for about three decades after the First Crusade it is possible to follow the careers of early crusaders to their conclusion.

The crusade proved to be one of the most important and enduring features of medieval civilization. On 27 November 1095, in a meadow near Clermont in the Auvergne, Pope Urban II made a speech which formally proclaimed an armed expedition to the Holy Land. Within four years forces from western Europe were in control of Jerusalem and pockets of land in Palestine and Syria. Within six years large expeditions which had set out from the West to support the fledgling Latin territories had been crushed by Turkish armies in Asia Minor. Despite this and other early setbacks, however, the Latin presence in the Levant survived. With the benefit of hindsight, the expedition which Pope Urban launched has become known as the First Crusade. Historians can now see that the events of 1095–1101 marked the beginnings of a distinctive institution which was to spread to such diverse regions as the Iberian peninsula, the shores of the Baltic, Italy, and Languedoc, and which was to occupy the energies of many thousands of people throughout the Middle Ages and into the modern period.

Given the popularity and longevity of crusade ideas, it can sometimes require an effort of the imagination to appreciate that Pope Urban's plan in 1095 was remarkably innovative. Urban did not devise the crusade *in vacuo*, of course, and it is possible to identify precedents which almost certainly influenced his ideas. Particularly relevant was a plan for a Western expedition to the East which had been mooted by Pope Gregory VII in 1074. In the event, Gregory's scheme came to nothing, partly because it was overtaken by the papacy's quarrels with the German monarchy, and partly because Gregory placed too great an emphasis on what the papacy stood to gain from the expedition at the expense of clear statements of the material or spiritual benefits which the participants might enjoy. Nevertheless, the scheme of 1074 provides valuable pointers to some of the themes which Urban developed twenty-one years later.[4] Other precedents were also significant. For example, in 1089 Urban himself had proposed to the faithful of Catalonia that its support of the frontier church of Tarragona would merit the same spiritual rewards

---

[4] H. E. J. Cowdrey, 'Pope Gregory VII's "Crusading" Plan of 1074', in B. Z. Kedar, H. E. Mayer, and R. C. Smail (edd.), *Outremer: Studies in the History of the Crusading Kingdom of Jerusalem Presented to Joshua Prawer* (Jerusalem, 1982), 27–40.

which an individual might achieve by means of a penitential pilgrimage to Jerusalem or elsewhere. Urban's readiness to link ideas of Christian expansion, pilgrimage, the Holy Land, and penance anticipated some of the most important elements of the crusade appeal. As in the case of the abortive scheme of 1074, the papacy's concern for Tarragona demonstrates that Urban's crusade plans had some form of prehistory.[5]

It would be wrong to conclude, however, that, because Urban's crusade scheme was not a bolt from the blue, Latin Christian society saw, or should have seen, the crusade coming. It is salutary to read the words of two observers who were deeply impressed by the manner in which the crusade appeared to the faithful as an exciting and new opportunity. One, Guibert of Nogent, described the crusade's impact upon Christians generally:

in our time God instituted holy warfare, so that the arms-bearers [*ordo equestris*] and the wandering populace [*vulgus oberrans*], who after the fashion of the ancient pagans were engaged in mutual slaughters, should find a new way of attaining salvation; so that they might not be obliged to abandon the world completely, as used to be the case, by adopting the monastic way of life or any other form of professed calling, but might obtain God's grace to some extent while enjoying their accustomed freedom and dress, and in a way consistent with their own station.[6]

The other observer, Ralph of Caen, condensed the appeal of the crusade to its effect upon his hero, the Norman lord from southern Italy Tancred fitz Marquis, who was described as a conspicuous model of lay probity and a 'diligent listener to God's precepts':

day after day his prudent mind was in turmoil, and he burned with anxiety all the more because he saw that the warfare which flowed from his position of authority [*militiae suae certamina*] obstructed the Lord's commands. For the Lord enjoins that the struck cheek and the other one be offered to the striker, whereas secular authority [*militia saecularis*] requires that not even relatives' blood be spared. The Lord warns that one's tunic, and one's cloak too, must be given to the man intending to take them away; but the imperatives of authority [*militiae necessitas*] demand that a man who has been deprived of both should have whatever else

---

[5] *La documentación pontificia hasta Inocencio III (965–1216)*, ed. D. Mansilla (Monumenta Hispaniae Vaticana, Sección Registros, 1; Rome, 1955), no. 29, pp. 46–7; J. S. C. Riley-Smith, *The First Crusade and the Idea of Crusading* (London, 1986), 18–20. See also *Papsturkunden in Spanien*, i. *Katalanien*, ed. P. Kehr (Abhandlungen der Gesellschaft der Wissenschaften zu Göttingen, Phil.-hist. Kl., NS 18²; Berlin, 1926), no. 23, pp. 287–8 [*recte* 1096 × 1099].

[6] 'Gesta Dei per Francos' (I, 1), *RHC Occ.* 4. 124.

remains taken from him. Thus, this incompatibility dampened the courage of the wise man whenever he was allowed an opportunity for quiet reflection. But after the judgement of Pope Urban granted a remission of sins to every Christian setting out to overcome the Gentiles, then at last the man's energies were aroused, as though he had earlier been asleep; his strength was renewed, his eyes opened, and his courage was redoubled. For until then . . . his mind was torn two ways, uncertain which path to follow, that of the Gospel, or that of the world.[7]

Despite their very different perspectives—Guibert, a learned and intro-spective abbot, was one of the finest products of the Benedictine culture of his day,[8] Ralph a Norman knight who did not go on the First Crusade but who settled in the Latin East some years later[9]—these two com-mentators saw the impact of the crusade in remarkably similar terms. Both men understood that the crusade provided a novel opportunity for laymen to be saved. Both appreciated that before 1095–6 the lay con-dition had inevitably impeded men's chances of salvation. Both realized that the crusade vocation addressed, if only in part, pious impulses which had only been met previously by entry into professed religion. It is perhaps unremarkable that Guibert should have placed a monastic slant on his interpretation of crusading; but it is significant that Ralph, whose convoluted and allusive Latin shows that he was impressively educated (he had once been intended for a career in the secular church), did so too. Ralph's passage is similar to contemporary hagiographical texts describing the mental agonies of men who converted to religion as adults;[10] and the necessary implication of Tancred's desire to follow the 'path of the Gospel' was that he would have had to enter a religious order had not the crusade opportunity presented itself.

The following study may be treated as an extended commentary upon Guibert's and Ralph's observations, in particular the symbiosis they believed existed between crusading and professed religion. It is reasonable to suppose that the two commentators' analysis of the crusade appeal can be applied to parts of western Europe other than those regions where they lived or knew well (Ralph, after all, had not witnessed Tancred's

[7] 'Gesta Tancredi in expeditione Hierosolymitana' (ch. 1), *RHC Occ.* 3. 605–6.

[8] See Guibert of Nogent, *Autobiographie*, ed. and trans. E.-R. Labande (Les classiques de l'histoire de France au Moyen Âge, 34; Paris, 1981), pp. ix–xv, xix–xxi.

[9] See the ed.'s comments in *RHC Occ.* 3, pp. xxxviii–xxxix; L. Boehm, 'Die "Gesta Tancredi" des Radulf von Caen: Ein Beitrag zur Geschichtsschreibung der Normannen um 1100', *Historisches Jahrbuch*, 75 (1956), 49–51.

[10] e.g. Gilbert Crispin, 'Vita Domni Herluini Abbatis Beccensis', ed. J. A. Robinson, *Gilbert Crispin Abbot of Westminster* (Notes and Documents relating to Westminster Abbey, 3; Cambridge, 1911), 90.

'conversion'). On the basis of what Guibert and Ralph believed were the decisive elements of Pope Urban's appeal, it is a valuable exercise to put oneself in the position of laymen from an area such as south-western France hearing the crusade message for the first time in 1095–6, and to examine the influences which might have operated on men's minds when they decided to take the cross. The body of ideas which came to make up the received orthodoxy of the First Crusade by about 1110 had three principal sources: the original crusade message preached by Pope Urban and other ecclesiastics in the West; the beliefs which developed within the crusade army between 1097 and 1099, as it became convinced that it had become the chosen instrument of God's will; and the subsequent analysis by western writers, such as Guibert, of how the literally miraculous success of the expedition fitted into God's providential scheme for mankind.[11] The last two sources of ideas were influenced by the experience of crusading itself; the first, patently, was not. This means that it is necessary to sift through the features of Latin Christian society before 1095 in order to isolate factors which would have reified the novel crusade appeal for the faithful.

This study proceeds from the basic premiss that the reasons why men went on the First Crusade were overwhelmingly ideological. Pope Urban expressly forbade men to join the expedition for pay or honours, but it seems that in so doing he was simply being cautious.[12] The crusaders had to finance themselves, although the humbler among them might have anticipated becoming the clients of the richer lords taking part. It has been estimated that a lord or knight intending to fund his journey to the East would have had to liquidize assets to the value of four or five times his annual income.[13] These are, if anything, conservative figures. According to more than one version of Urban II's speech at Clermont, the pope appealed to the fact that many Westerners were acquainted with the eastern Mediterranean world by reason of pilgrimages to Jerusalem.[14] Thus it is reasonable to suppose that, on the basis of direct experience or stories related by others, the intelligent crusader would have had few

---

[11] Riley-Smith, *First Crusade*, 15–30, 91–119, 138–52. For the crusade as an occasion for miracles, see also B. Ward, *Miracles and the Medieval Mind: Theory, Record and Event 1000–1215*, rev. edn. (Aldershot, 1987), 203–4.

[12] *The Councils of Urban II*, i. *Decreta Claromontensia*, ed. R. Somerville (Annuarium Historiae Conciliorum, Suppl., 1; Amsterdam, 1972), 74. See also *Die Kreuzzugsbriefe aus den Jahren 1088–1100*, ed. H. Hagenmeyer (Innsbruck, 1901), no. 3, p. 137.

[13] Riley-Smith, *First Crusade*, 43.

[14] Guibert of Nogent, 'Gesta Dei per Francos' (II. 4), 139–40; Baldric of Bourgueil, 'Historia Jerosolimitana', (I. 4), *RHC Occ.* 4. 13.

illusions about any supposed treasures of the Orient. What is more, the crusaders were not driven by land-hunger. It is likely that the pope anticipated the creation of some permanent Latin presence in the East under the authority of the Byzantine emperor, and many more departing crusaders may have intended to settle in Palestine than in fact did so.[15] But, as the Cistercians and countless assarters were to prove in the twelfth century, western Europe had far from exhausted its capacity for internal territorial expansion by the 1090s; and the crusade seems an almost perversely esoteric way to have satisfied any powerful colonial instincts. In the same way that crusaders were prepared to pay heavily for their experience, the intention of some westerners to stay in the East should be treated as an expression of crusade ideology.

This crusade ideology was predominantly religious in its inspiration. A caveat needs to be entered immediately, however, for other, overtly secular, values were also important. In particular two overlapping groups of ideas contributed to crusaders' enthusiasm. One set of ideas was expressed through appeals to the faithful's sense of national or racial pride, especially (because the crusade was pitched mainly at Frenchmen)[16] the emotional associations of Frankishness. One crusade chronicler, Robert the Monk, believed that at Clermont Urban II evoked the Franks' historical self-awareness and pride with great force;[17] other writers, with varying emphasis, exploited honorific connotations of Franc-stem words in their accounts of the crusade's origins and progress.[18]

The impact of Frankish sentiment in 1095–6, however, needs to be set in its broad context. In the late eleventh century France was politically fragmented, and its king, who was the natural focus of national feeling, had little residual authority in temporal affairs beyond the Île-de-France. Furthermore, there is evidence that the participants on the First Crusade retained a strong sense of identity based on their attachment to localities or regions back in the West.[19] When the army was subjected to great

[15] Riley-Smith, *First Crusade*, 40, 42–3.

[16] Guibert of Nogent, 'Gesta Dei per Francos' (I. 1), 125. See J. S. C. Riley-Smith (ed.), *The Atlas of the Crusades* (London, 1991), 29.

[17] 'Historia Iherosolimitana' (I. 1), *RHC Occ.* 3. 727, 728. See also ibid. (III. 17), 765.

[18] e.g. Guibert of Nogent, 'Gesta Dei per Francos' (II. 1; IV. 3), 135–6, 169; Baldric of Bourgueil, 'Historia Jerosolimitana' (II. 5; IV. 3), 38, 91. See J. S. C. Riley-Smith, 'The First Crusade and St. Peter', in Kedar, Mayer, and Smail (edd.), *Outremer*, 45–9.

[19] See the use of Franc-terms specifically of crusaders from the Île-de-France associated with Hugh of Vermandois, the brother of the French king: *Die Kreuzzugsbriefe*, no. 15, p. 160; Baldric of Bourgueil, 'Historia Jerosolimitana' (III. 16), 75; Ralph of Caen, 'Gesta Tancredi' (ch. 85), 666. For other instances of regional or racial distinctiveness, see Peter Tudebode, *Historia de Hierosolymitano Itinere*, ed. J. H. and L. L. Hill (Documents relatifs à

stresses, linguistic and cultural differences became exacerbated—between northern and southern French in particular—and could sometimes erupt into violence.[20] Yet, in spite of these antagonisms, the crusade forces usually retained a remarkable level of cohesion. Thus, for observers such as the southern Italian Norman author of the *Gesta Francorum*, who went on the crusade and identified strongly with the world of northern France, Franc-stem terms expressed well enough the unifying features of an army which came up against non-Latin races such as Greeks, Turks, and Arabs, and in which men from Gaul in general, and speakers of mutually intelligible northern *langue d'oïl* dialects in particular, formed the dominant element.[21] This unifying process was accelerated within the crusade army by the Latins' perception that their Muslim adversaries treated them as a single entity.[22] As the ability of the crusaders to weld polyglot forces together became perceived as one sign of God's support for the expedition, so a unitary sense of Frankishness became an element of crusade ideology.[23] (It is significant that the work of ideological justification was mainly the work of writers with roots in northern France.) There are interesting parallels here with the manner in which the conquest of England was retrospectively explained and justified by notions of Norman prowess and readiness to fight for the church.[24] Given that in some contexts Franc-stem vocabulary excluded men from southern France, and that some crusaders—Anglo-Saxons and non-Catalan Spaniards, for

l'histoire des croisades, 12; Paris, 1977), 110 ('gente Pictavensis comitis'); Raymond of Aguilers, *Le 'Liber'*, ed. J. H. and L. L. Hill (Documents relatifs à l'histoire des croisades, 9; Paris, 1969), 40 (Raymond of St-Gilles's battle cry of 'Tolosa'), 52, 60, 64, 121 (a *miles* 'genere Biterensis' (Béziers)), 153 (Provençals); Fulcher of Chartres, *Historia Hierosolymitana (1095–1127)* (I. 6, 23), ed. H. Hagenmeyer (Heidelberg, 1913), 157–8, 255 (Normans, Provençals/Goths, Gascons); Robert the Monk, 'Historia Iherosolimitana' (VII. 9), 828 ('Aquitanicum quemdam, quem nos [northern Frenchmen] Provincialem dicimus'); Ralph of Caen, 'Gesta Tancredi' (chs. 8, 79, 92), 610, 662, 672 (Normans); 'Fragment d'une *Chanson d'Antioche* en provençal', v. 677, ed. and trans. P. Meyer, *Archives de l'Orient Latin*, 2 (1884), 493 ('nostre Lemosi').

[20] Ralph of Caen, 'Gesta Tancredi' (chs. 15, 61, 98–103, 109), 617 (southern Frenchmen not behaving 'ut Franci'), 651, 675–9, 682–3. See also L. Boehm, 'Gedanken zum Frankreich-Bewußtsein im frühen 12. Jahrhundert', *Historisches Jahrbuch*, 74 (1955), 684–5.

[21] See B. Schneidmüller, *Nomen Patriae: Die Entstehung Frankreichs in der politisch-geographischen Terminologie (10.–13. Jahrhundert)* (Nationes, 7; Sigmaringen, 1987), 106–22.

[22] Raymond of Aguilers, '*Liber*', 52.

[23] Fulcher of Chartres, *Historia Hierosolymitana* (I. 6, 13), 161, 202–3. See also Ekkehard of Aura, 'Hierosolymita' (ch. 6), *RHC Occ.* 5. 16.

[24] D. C. Douglas, *The Norman Achievement 1050–1100* (London, 1969), 102–3, 105–6; R. H. C. Davis, *The Normans and their Myth* (London, 1976), 65–7, 103–4.

example—came from groups to which even the widest acceptation of Frankishness was inapplicable before the crusade,[25] it is likely that appeals to racial or patriotic self-esteem in 1095–6 were, more often than not, pitched at a regional level. A closely contemporary account suggests that Pope Urban himself did precisely this in Anjou in February 1096;[26] and the technique was most probably repeated elsewhere.[27]

The other group of secular values which contributed to crusade enthusiasm comprised ideas which were forerunners of the chivalric ethos, particularly, in this context, notions of military prestige and family honour. It is significant, for example, that the crusade vocation very soon came to be interpreted as a form of vendetta.[28] As in the case of the appeal to Frankishness, however, the original call to crusaders' militaristic and honorific ideology was amplified and distorted by subsequent events. Moreover, as a self-sustaining value-system which could motivate action for its own sake, chivalry did not exist in anything more than an embryonic form before the second half of the twelfth century.[29] Before then—as a broad generalization embracing regional variations—men of the various social and juridical classes who fought on horseback were bonded by their shared experiences of training and war, by family and feudal ties, and by a horse- and weapons-centred martial culture. As yet, however, there was no coherent ethos to lend all mounted warriors a caste-identity with its own distinctive duties and patterns of behaviour. It is noteworthy that recent research has established that dubbing, which eventually became the formal rite of passage into the chivalric caste, had not fully developed this precise function by *c.*1100.[30] As Pope Urban

[25] See M. Lugge, *'Gallia' und 'Francia' im Mittelalter: Untersuchungen über den Zusammenhang zwischen geographisch-historischer Terminologie und politischem Denken vom 6.–15. Jahrhundert* (Bonner Historische Forschungen, 15; Bonn, 1960), 160–80; Schneidmüller, *Nomen Patriae*, 62–103. See also K. F. Werner, 'Les Nations et le sentiment national dans l'Europe médiévale', *Revue historique*, 244 (1970), 292–4.

[26] Fulk IV of Anjou, 'Fragmentum Historiae Andegavensis', ed. P. Marchegay and A. Salmon, *Chroniques des comtes d'Anjou* (Paris, 1871), 380.

[27] For an instance of military pride operating on a regional level, see *Guide* (ch. 7), 16–18.

[28] Riley-Smith, *First Crusade*, 9, 48–9, 54–7.

[29] The literature on chivalry is vast. For two excellent treatments, see M. H. Keen, *Chivalry* (New Haven, Conn., 1984), and J. Flori, *L'Essor de la chevalerie XIᵉ–XIIᵉ siècles* (Travaux d'histoire éthico-politique, 46; Geneva, 1986).

[30] J. Flori, 'Sémantique et société médiévale: le verbe adouber et son évolution au XIIᵉ siècle', *AESC* 31 (1976), 915–40; id., 'Les Origines de l'adoubement chevaleresque: étude des remises d'armes et du vocabulaire qui les exprime dans les sources historiques latines jusqu'au début du XIIIᵉ siècle', *Traditio*, 35 (1979), 215–18, 222–49. For a slightly different view, see J. M. van Winter, '*Cingulum militiae*, Schwertleite en *miles*-terminologie als spiegel van veranderend menselijk gedrag', *Tijdschrift voor Rechtsgeschiedenis*, 44 (1976), 1–92, esp. 13–16, 17–19, 24, 41–6.

toured France in 1095–6 preaching the cross, he passed through areas which differed in terms of the social status conveyed by the word *miles*, and the social exclusivity, authority, and size of the senior aristocratic élite identified by such terms as *principes* and *nobiles*.[31] Those who preached the cross on the pope's behalf elsewhere would have found the same.[32] Extraordinary levels of advance intelligence and precise attention to semantic subtleties would have been required for the crusade message to have been pitched differently from area to area according to regional variations in social structure. In fact, this was not necessary, for, when the pope encouraged *milites* to go on the crusade, he was above all making a practical point about restricting participation to the men best suited for the task in hand.[33] Only indirectly, and differently from region to region, could his words be construed as a coded appeal to what, with the benefit of hindsight, might be termed proto-chivalric values. All that was necessary for Urban's appeal to have relevance in a given region was that the upper social levels of that locality should comprise men who professed expertise in the use of arms, especially on horseback. This very straightforward precondition was satisfied throughout France and in virtually all parts of Latin Europe.

Like sentiments of regional or national pride, embryonic chivalric values may be treated as secondary reasons why men went on the crusade— but not in the sense that they simply 'topped up' other motives. One way to approach the significance of racial–national and militaristic ideas is to imagine that they operated on the level of what in psychoanalytical parlance are termed 'cognitive assumptions': sets of values (drawn from everyday life and so self-justifying that they seldom need articulating) by which an individual understands an experience and judges his or her response to it. In other words, patriotic and militaristic enthusiasms might have influenced the way in which an arms-bearer interpreted the crusade appeal: they cannot adequately explain why he should have been

---

[31] e.g. Saintonge (Apr. 1096): A. Debord, *La Société laïque dans les pays de la Charente Xᵉ–XIIᵉ siècles* (Paris, 1984), 189–207; Bordelais (May 1096): C. Higounet, 'En Bordelais: "Principes castella tenentes"', in P. Contamine (ed.), *La Noblesse au Moyen Âge XIᵉ–XVᵉ siècles: essais à la mémoire de Robert Boutruche* (Paris, 1976), 97–104.

[32] e.g. in Normandy: D. Bates, *Normandy before 1066* (London, 1982), 106–11. For Germany see J. Johrendt, '"Milites" und "Militia" im 11. Jahrhundert in Deutschland', in A. Borst (ed.), *Das Rittertum im Mittelalter* (Wege der Forschung, 349; Darmstadt, 1976), 419–36; B. Arnold, *German Knighthood 1050–1300* (Oxford, 1985), esp. 23–52.

[33] 'Papsturkunden in Florenz', ed. W. Wiederhold, *Nachrichten der K. Gesellschaft der Wissenschaften zu Göttingen, Phil.-hist. Kl.* (1901), no. 6, p. 313; *Papsturkunden*, i. *Katalanien*, no. 23, p. 287.

thinking about it in the first place. At the heart of the crusade message lay an appeal to piety.

Most treatments of the origins of the First Crusade have not distinguished clearly between the roots of the religious values of the laymen who went on the expedition and the ideas of the clerical intellectuals who developed the canonical and theological theories which made the crusade concept possible. An emphasis upon the crusade's intellectual origins is methodologically sound, for the historian can draw upon the products of an impressive literary culture—chronicles, collections of canon law, polemical tractates, papal letters—through which arguments and counter-arguments can be seen to develop and ideas mature or perish. In a sense, research along these lines has a satisfying focus, for it all converges upon Pope Urban's mind at the very moment he began to speak at Clermont.

The greatest exponent of this type of approach was the German scholar Carl Erdmann, whose *Die Entstehung des Kreuzzugsgedankens* appeared in 1935.[34] Erdmann traced the idea of crusading through the medieval church's shifting attitudes to violence, and in particular to two issues: whether the church had the powers to authorize war; and whether there were circumstances in which soldiers, the Fifth Commandment notwithstanding, could treat violence as spiritually beneficial or at least morally neutral. Erdmann's arguments have been extensively revised by recent scholarship. In particular, there is now a broad consensus that he underestimated the extent to which the crusade was conceived as a form of pilgrimage as well as a holy war;[35] and it has recently been argued that the canonical position on just violence in the later eleventh and early twelfth centuries was not as consistent as he supposed.[36] But Erdmann's contribution to crusade studies has been so valuable that he does not deserve to become a soft target for modern researchers. Some of the following chapters open with a discussion of Erdmann's arguments, not in order to score easy points, but as a recognition of his great influence on all subsequent work in the field. However much Erdmann has been criticized in matters of detail, his overall vision has remained compelling. Studies

[34] Eng. trans. by M. W. Baldwin and W. Goffart, *The Origin of the Idea of Crusade* (Princeton, NJ, 1977).

[35] See H. E. J. Cowdrey, 'The Origin of the Idea of Crusade', *International History Review*, 1 (1979), 121–5. For a recent restatement that pilgrimage was not an important element of Urban II's crusade plan, see A. Becker, *Papst Urban II. (1088–1099)* (MGH Schriften, 19; Stuttgart, 1964–88), ii. 396–8.

[36] J. T. Gilchrist, 'The Erdmann Thesis and the Canon Law 1083–1141', in P. W. Edbury (ed.), *Crusade and Settlement: Papers read at the First Conference of the Society for the Study of the Crusades and the Latin East and presented to R. C. Smail* (Cardiff, 1985), 37–41.

by Delaruelle,[37] Villey,[38] Rousset,[39] Mayer,[40] Brundage,[41] and Becker[42] have examined the crusade's causes wholly or in substantial part through theological and canonical texts produced by the church's educated élite.

This study does not question the validity of the techniques of Erdmann and his successors, for unquestionably Urban II's ideas in 1095 had an intellectual pedigree. Rather, it is worth challenging an assumption often made on the basis of the Erdmann-type perspective: that the intellectual origins of the crusade idea largely subsumed the reasons why at least the more discriminating members of the rank and file took part in the expedition. Such an assumption requires the belief that, for every important stand which the eleventh-century church took on such issues as just war and meritorious violence, there was an equivalent and concordant lay position, simplified but not fundamentally different in character. The notion of clerical–lay parallelism is not inherently implausible, for it is important to remember that at most times and in most circumstances the church has been a largely reactive body, accommodating its ideas and practices to the faithful's expectations. The two generations before the First Crusade were, however, an exception to the general rule. The church had become an active proponent of reform, challenging old habits of mind. More particularly, the papacy had placed itself at the forefront of reforming efforts (perhaps uniquely in its history). This great shift in the attitudes of the later eleventh-century church (a movement embraced by the terms 'Gregorian Reform' and 'Investiture Contest') means that the notion of a lay culture in tune with all ecclesiastics' ideas needs to be tested carefully against specific problems concerning the crusade.

The first two chapters of this book examine the two most commonly cited arguments in support of clerical–lay parallelism: the Peace and Truce of God, church-inspired peace movements which touched some parts of Latin Europe from the late tenth century; and the eleventh-century Spanish *Reconquista*, the process of Christian expansion at the expense of the Moors of southern Iberia. It is argued here that the

---

[37] 'Essai sur la formation de l'idée de croisade', *Bulletin de littérature ecclésiastique*, 42 (1941), 24–45, 86–103; 45 (1944), 13–46, 73–90; 54 (1953), 226–39; 55 (1954), 50–63.

[38] *La Croisade: essai sur la formation d'une théorie juridique* (L'Église et l'État au Moyen Âge, 6; Paris, 1942), esp. 21–73.

[39] *Les Origines et les caractères de la Première Croisade* (Neuchâtel, 1945), esp. 43–62. See also id., *Histoire d'une idéologie: la Croisade* (Lausanne, 1983), 29–31, 33–40.

[40] *The Crusades*, trans. J. Gillingham, 2nd edn. (Oxford, 1988), 14–20.

[41] *Medieval Canon Law and the Crusader* (Madison, Wis., 1969), 19–29.

[42] *Papst Urban II.*, ii. 272 ff.

influence of the Peace of God on the laity's reaction to the crusade message has been exaggerated. The fact that Peace ideas appear in early crusade texts has no deep significance, because the expedition's organizers simply wanted to promote domestic stability in order to stimulate recruitment. The eleventh-century Peace was sporadic and geographically uneven in its impact. In Aquitaine, for example, it was fitfully important between *c*.990 and *c*.1035, and even at its height it did not involve appeals to Christian morality or the supposed ethics of a warrior caste as much as an accommodation between two small and closely related groups, the ecclesiastical hierarchy and the successors of the Carolingian high aristocracy, who could no longer take for granted strong public authority predicated on effective royal power.

Turning to Spain, it is often supposed that the eleventh-century church by turns stimulated and was influenced by a nascent ethos of holy war directed against Islam and bred by the Christian expansion which is now known by the loaded term *Reconquista*. Chapter 2 argues that Frenchmen—to whom the First Crusade message was mainly directed—had had minimal experience of war in Spain by 1095–6. Before then French campaigns in the peninsula had been few in number, on a much smaller scale than the First Crusade, firmly controlled by peninsular rulers (whose domestic policies and rivalries were more complex than a straightforward notion of holy war could embrace), and motivated predominantly by a desire for material gain. Furthermore, crusading in Spain was not wholly an autochthonous phenomenon for some years after the First Crusade. It rather drew much of its inspiration and support from external forces such as the papacy and former first crusaders from France. Influences of this sort, and the fact that it was about two decades before crusading values had a significant impact on peninsular affairs, demonstrate that Spain was not the breeding-ground for a holy war ethos which only needed fine tuning to become crusade enthusiasm.

The Peace of God and Spain before 1095 do not exhaust the possible parallelisms between lay values and ecclesiastical ideas about violence— the Christian expansion into Sicily in the later eleventh century is another potential area of enquiry—but these two topics have occupied a dominant place in scholarly debate.[43] They are particularly relevant within the scope of this study because south-western France was the region where the Peace originated, and where links to Spain west of Catalonia might be expected to have been strong. Chapters 1 and 2 necessarily differ in

[43] See now I. S. Robinson, *The Papacy 1073–1198: Continuity and Innovation* (Cambridge, 1990), 323–7.

emphasis from the rest of this book in that they attempt to establish a negative. Even so, it is a valuable exercise to discuss the Peace and Spain at some length, for in this way the ground can be cleared satisfactorily for an analysis of the positive roots of first crusaders' ideas.

Only a minority of crusade studies have treated the ideas of the laity as a discrete subject. What attention has been given to the topic has relied largely on vernacular literature, especially epic songs.[44] Valuable as vernacular works are as sources for laymen's ideas, they must be used with great caution. No extant text of a *chanson de geste* can be confidently dated to before the First Crusade. What is generally considered the oldest text, the Bodleian version of the *Chanson de Roland*,[45] was almost certainly produced in the twelfth century.[46] Of course, it is a matter of debate how far the surviving *chansons* had an oral or textual prehistory before the twelfth century.[47] But even the earliest known version of the *Roland* cannot embody a tradition which had passed unaltered through the experience of the First Crusade: the influence of crusade values and ideas is simply too strong.[48] Thus too great a reliance on vernacular texts can produce a picture of lay ideas distorted by a process similar to the ideological reworking which influenced the early western chroniclers of the First Crusade: what may appear to be antecedent, formative factors contributing to the crusade's origins prove to be, in fact, the creations of the crusading experience itself.

One of the most extended treatments of the laity's contribution to First Crusade ideas, by Alphandéry and Dupront, appeared in 1954.[49] While the authors stressed the importance of popular religious sentiment, particularly as it was manifested by pilgrimage traditions and western interest in Jerusalem, they did so in general terms and by focusing upon eschatological responses which, they believed, sufficed to explain the crusade's attraction. It has more recently been argued that eschatological fears were only a minor element in the crusaders' motives.[50] In 1956

---

[44] e.g. Rousset, *Origines*, 110–33. See now D. A. Trotter, *Medieval French Literature and the Crusades (1100–1300)* (Histoire des idées et critique littéraire, 256; Geneva, 1987), esp. 35–70, 77–89, 94–9.

[45] MS Digby 23.

[46] See *The Song of Roland*, ed. and trans. G. J. Brault, (Phil., 1978), i. 5, 342 n. 27.

[47] Ibid. 4, 341 n. 19.

[48] e.g. vv. 1129–41, 2253–8, 2383–8, 2503–6, ed. Brault, ii. 72, 138, 146, 152.

[49] *La Chrétienté et l'idée de croisade* (L'évolution de l'humanité, 38; Paris, 1954–9), i. 10 ff.

[50] B. McGinn, '*Iter Sancti Sepulchri*: The Piety of the First Crusaders', in B. K. Lackner and K. R. Philp (edd.), *Essays on Medieval Civilization* (The Walter Prescott Memorial Lectures, 12; Austin, Tex., 1978), 47–8, 66–7 n. 91.

Adolf Waas made the clearest statement to date that the ideas of ecclesiastics and lay crusaders were not coextensive.[51] He maintained that the
ideas of the laity stemmed from a self-contained value-system which was
distinct from the preoccupations of the church, though borrowings were
possible in both directions. Waas expressed the lay value-system as
Ritterfrömmigkeit, the piety of knights. His thesis is weakened by his
extensive use of vernacular texts which post-date the First Crusade, and
by his belief that the code of chivalry was reaching maturity by 1095.
Nevertheless, his broad concept of *Ritterfrömmigkeit* is valuable and
deserves to be explored more fully.

Contrary to Waas's position, this book argues that the piety of nobles
and knights was profoundly influenced by the church, and in fact needed
the church for its inspiration—but not where Erdmann and his successors
were looking. This study puts itself in the position of an arms-bearer
from south-western France whose mental map was most likely very
localized (apart from possible experiences of some distant places through
pilgrimage) and who could not have had the modern historian's perspective on contemporary intellectual currents. Where did his religious values
come from? The answer is, in large measure, from local religious communities staffed by monks or canons. These institutions, it is true, did
not have a complete monopoly of religious influence over the arms-
bearing aristocracy. There is evidence that it became more common from
the latter half of the eleventh century for lords to retain the services of
chaplains;[52] and there are a few indications of laymen attracted to the
spirituality of hermits.[53] But the main potential rivals to religious
communities, the parish clergy, were generally poorly educated and often
of a lower social status than arms-bearers. In terms of learning, wealth,
and prestige, most regional churches were dominated by monks, to whom
may be added bishops, and the canons, secular or regular, who were based
in cathedrals or separate communities.

A good deal of useful information about laymen's religious practices
and ideas can be gleaned from narrative sources such as hagiographical
texts and chronicles. Gascony was not a notable centre of history-writing
in this period, but the study of the Limousin is enriched by the work of
two writers: Adhemar of Chabannes, a Limousin who maintained a keen
interest in his homeland after he had become a monk in Angoulême,

---

[51] *Geschichte der Kreuzzüge*, (Freiburg, 1956), i. 1–53.
[52] e.g. T. 412 (1087 or 1088); Uz. 1148 (mid-11th C.); V. 156 (1100 × 1104), 206
(1111 × 1124), 234 (1111 × 1124).
[53] See LSM, p. 18 (*c.*1110).

producing three recensions of a valuable chronicle between *c.*1025 and *c.*1032 as well as many other works;[54] and Geoffrey of Vigeois, who wrote his chronicle after our period, in 1183, but recorded a good deal of important information about the eleventh century.[55]

Valuable as narratives are, however, the principal sources for the impact of religious communities upon their local aristocracies are charters: documents which record laymen's grants to religious bodies by way of gift, sale, or pledge, as well as confirmations of existing rights. Chapters 3–6 contain treatments of particular features of the charters, but it is worth making some general points here about how these documents have been used. In the regions of south-western France covered by this study (and in many other places) very few original charters from before the twelfth century are extant. Documents earlier than this typically survive in cartularies, compendia of a religious community's muniments which were drawn up by the monks or canons for ease of reference.[56] Some of the cartularies were produced long after the date of the earliest material they contain; and some are later medieval copies of earlier versions. Others survive only as copies or extracts made by antiquaries in the seventeenth and eighteenth centuries, before the turmoil of the French Revolution swept the originals away. It is fortunate, for example, that the indefatigable copier of medieval texts Étienne Baluze (1630–1718) was a native of the Limousin. He retained an interest in the region of his birth and recorded many of its old documents.[57] It is largely thanks to him and other copyists that many of the records of the important abbeys of Uzerche and Tulle, to name but two, have survived. The layered history of copying and recopying means that most eleventh-century documents are now in some state of textual decomposition, so that, for example, many can be dated only very approximately, if at all.

Enough charters survive in a tolerably intact state, however, to establish an important point: that it is a valid exercise to examine *en bloc* charters from different types of religious community, from different parts of south-western France, and from throughout the period covered by this

---

[54] See R. Landes, 'The Dynamics of Heresy and Reform in Limoges: A Study of Popular Participation in the "Peace of God" (994–1033)', *HR* 14 (1987), 470 n. 11.

[55] See M. Aubrun 'Le Prieur Geoffroy du Vigeois et sa chronique', *RM* 58 (1974), 313–26.

[56] See generally D. Walker, 'The Organization of Material in Medieval Cartularies', in D. A. Bullough and R. L. Storey (edd.), *The Study of Medieval Records: Essays in Honour of Kathleen Major* (Oxford, 1971), 132–50.

[57] Baluze's researches into Limousin materials bore fruit in his *Historiae Tutelensis Libri Tres* (Paris, 1717).

study. Religious communities differed in their spirituality, so that the doctrinal and devotional statements made in their charters might vary subtly. To pursue this question in detail, however, would require a book in its own right. The absence here of this sort of detailed analysis can be justified on two grounds. First, the influence of using earlier documents as models, and the constant applicability to acts of benefaction of a limited range of scriptural or doctrinal statements, resulted in a notable level of homogeneity among charters from different places and times. Secondly, whatever fine distinctions might be detected between the diplomatic utterances of individual religious communities would be impossible to correlate with what is known of local levels of response to the First Crusade appeal. Consider the case of the great abbey of Cluny in southern Burgundy. It was the most prestigious monastery in eleventh-century Europe, and its charters contain some of the most elaborate and varied preambles to be found in this period. Furthermore, Cluny's documents contain references to more early crusaders than do the records of most contemporary institutions.[58] Once we discount, however, the abbey's wealth and prestige, the fact that its documents survive in unusual abundance, and Pope Urban II's close personal ties to the community (he had been prior there), it would be impossible to state with confidence that the level of response to the crusade appeal in Cluny's part of Burgundy was directly linked to particular nuances in the religious statements of its charters. Thus it is licit to examine together the records of Benedictine (including Cluniac) communities, canons regular, and secular cathedral chapters to the extent that the differences between their vocations did not impinge appreciably on their charters. (Variations in such documents as privileges and custumals were more pronounced, of course.) Similarly, the charters' religious ideas were remarkably consistent between the terminal dates of this study. Given the broadly homogenous characteristics of the documentary evidence, it is possible to identify the broad themes behind the ideas which were transmitted between the laity and religious institutions. If this approach requires some generalization, it is at least hoped that it is sufficiently grounded in the pious careers of identifiable individuals and families, and in the histories of specific religious communities, to avoid a criticism which can be levelled at some treatments of the lay response to crusading: that an emphasis upon sources such as *chansons* or canonical collections depersonalizes what happened at the grass-roots level when men heard the cross preached.

---

[58] See H. E. J. Cowdrey, 'Cluny and the First Crusade', *RB* 83 (1973), 291–4, 302–3.

This study mostly uses the term 'arms-bearer' to denote individuals within a broad social spectrum ranging from the great territorial princes to humble mounted soldiers, men based in castles as dependants of their lords or living from allods or fiefs. This usage is intended to avoid the ambiguity of the usual translation of *miles* as 'knight', a loaded term which bears many chivalric connotations inapplicable to the eleventh century. It is unfortunate that there is in English no convenient way to express the subtle but significant nuances conveyed in French by the cognates *chevalier/cavalier* or in German by *Ritter/Reiter*. Arms-bearers, thus defined, dominate the charters. Even when due allowance is made for the fact that humbler persons were far less likely to generate written records, it is clear that by the time of the First Crusade arms-bearers and their families were the lay group closest to religious communities and thus most receptive to the ideas which were associated with such practices as conversion and benefaction. Senior princes and castellans had the most resources from which to endow religious bodies and, generally speaking, long family traditions of support of the church. But by the later eleventh century lesser *milites* were also often in close contact with monks and canons. This is particularly demonstrated by cases in which lords consented to alienations of property rights which had earlier been made by their *fideles*.[59]

In this context, it is worth explaining why this study concentrates upon the religious ideas of men characterized by the use of the horse in combat. When Pope Urban planned the crusade, he was in part reacting to appeals for military help from the Byzantine empire, which was on the defensive against Turkish forces. The pope would have been aware that there were recent precedents for western mounted troops serving (at least notionally) under Byzantine command.[60] Any type of force might have been welcome to the Greeks; but Latin Europe's pre-eminent and unique contribution to warfare in this period was the corps of armoured, mounted shock troops.[61] These were the men whom Emperor Alexius I Comnenus wanted most and the pope understood best.

---

[59] Uz. 66 (1091); V. 80 (1092 × 1110); Ste-C. 92 (1122 × 1126); St-Seurin, 89 (1120 × 1143).

[60] M. Angold, *The Byzantine Empire 1025–1204: A Political History* (London, 1984), 93–4, 127, 136–7.

[61] See R. C. Smail, *Crusading Warfare (1097–1193)* (Cambridge, 1956), 106–15; I. Peirce, 'The Knight, his Arms and Armour in the Eleventh and Twelfth Centuries', in C. Harper-Bill and R. Harvey (edd.), *The Ideals and Practice of Medieval Knighthood: Papers from the First and Second Strawberry Hill Conferences* (Woodbridge, 1986), 152–64; J. Flori, 'Encore l'usage de la lance . . . la technique du combat chevaleresque vers l'an 1100', *CCM*

In the event, the response to the crusade appeal embraced many levels of lay society, for Urban II discovered that he could not restrict what was essentially a pastoral message to only a minority of the faithful. Apart from clerics and women of all classes who went on crusade, those who travelled east with the mounted forces can be placed in two broad categories: the poor; and the foot-soldiers. The poor, urban and rural, were generally non-combatants. During the expedition they assumed some useful functions as auxiliary personnnel, and their numbers meant that they were not without influence in the direction of certain stages of the campaign. Generally, however, they were a logistical and financial burden on the army.[62] The foot-soldiers (*pedites*) were in a more ambivalent position. Once Pope Urban realized, such was the enthusiasm for his crusade message, that the armies he had launched were larger than anticipated, he would have accepted that large-scale western forces often included archers and men equipped to fight hand-to-hand.[63] The *pedites* proved to be a significant element of the crusade army, for example by strengthening the Christian battle-lines between cavalry engagements.[64] They also shared to some extent the devotional self-awareness of the whole 'army of God', although the blend of literal and vocational connotations expressed in the term *milites Christi* could not apply to them.[65]

During the crusade, as most of the mounted troops' horses died and all the Latins suffered great hardships, the dividing lines between the three main classes of crusaders became very blurred, making it difficult by this stage to isolate a distinctive arms-bearers' response to crusading. On the other hand, the blurring process largely coincided with the second and third phases in the formation of crusade ideas which were noted earlier. In 1095–6 there still existed something approaching a pristine crusade appeal directed principally at mounted soldiers. This does not resolve the problem why the poor and the *pedites* went east, and it is beyond the scope of this study to attempt a full answer. It can only be tentatively suggested here that, although educated observers were sometimes mortified

---

31 (1988), 213–40; R. H. C. Davis, *The Medieval Warhorse: Origin, Development and Redevelopment* (London, 1989), 15–22, 23, 25–6.

[62] W. Porges, 'The Clergy, the Poor, and the Non-combatants on the First Crusade', *Speculum*, 21 (1946), 1–4, 9–12. The problem proved to be persistent: see E. Siberry, *Criticism of Crusading 1095–1274* (Oxford, 1985), 25–8.

[63] See P. Contamine, *War in the Middle Ages*, trans. M. Jones (Oxford, 1984), 49.

[64] e.g. *Gesta Francorum et aliorum Hierosolimitanorum*, ed. and trans. R. Hill (London, 1962), 19. See generally Smail, *Crusading Warfare*, 115–20.

[65] But see *Chartes et documents pour servir à l'histoire de l'abbaye de Charroux*, ed. P. de Monsabert (Archives historiques du Poitou, 39; Poitiers, 1910), 22.

by the naïvety of humble crusaders' ideas,[66] it would be unwise to make a rigid distinction between an 'orthodox', church-inspired motivation on the part of the mounted élite, and the beliefs of the 'wandering populace', millenarian, simplistic, and ill-informed. There is more likely to have been a sliding scale of responses corresponding to the extent to which different social groups had been exposed to the ecclesiastical hierarchy's teachings in the years before 1095. In addition, it is worth noting that the religious values of arms-bearers are an area of study in their own right simply because they are far better documented than the beliefs of other lay groups in this period.

Chapter 3 establishes the mechanics of how religious ideas passed between monasteries and local arms-bearing families. It argues in particular that the variety of forms of entry into professed religion created a network of family ties between churches and their neighbours. Given the great significance which arms-bearers attached to notions of kinship, these networks expressed the intimacy of ecclesiastical–lay contacts and created the channels through which religious ideas could flow back and forth. Chapter 4 turns to those religious ideas themselves, chiefly as they are revealed by the charters, it being argued that these documents can profitably be used to reconstruct lay pious beliefs. A discussion of the notions of the afterlife which inform the charters is supplemented by an analysis of some literary accounts of the next world. The charters and narratives demonstrate that arms-bearers had some quite sophisticated ideas about penance and a 'middle place' between Heaven and Hell: ideas which have a direct bearing upon the response to the First Crusade appeal. Chapter 5 argues that of the traditional trinity of crusade origins— holy war (particularly in Spain), the Peace of God, and pilgrimage—the last element needs to be retained in a discussion of laymen's ideas. This is so for two reasons. First, the crusade was a species of pilgrimage, and pilgrimage was a common expression of the cultic popular religion of the years before (and after) 1095. Second, pilgrimage was not a discrete, self-contained form of devotion, but rather fitted easily into the broader patterns of pious behaviour revealed by laymen's relations with local religious institutions. Pilgrimage thus provides further evidence of the value-system discussed in Chapters 3 and 4. Finally, Chapter 6 discusses how arms-bearers heard and responded to the preaching of the cross in 1095–1101, and links the impact of the crusade message to themes contained in Chapters 3–5.

---

[66] e.g. Bernold of St Blasien, 'Chronicon', *MGH SS* 5. 464.

Reduced to a single statement, this study argues that the reasons why arms-bearers from certain parts of south-western France (and very possibly from elsewhere) went on the First Crusade can be traced in patterns of behaviour and sets of ideas which were principally moulded by contacts with professed religion. To this extent it is unnecessary to hunt for grand themes or movements to explain the crusade, extraordinary as it was. The crusaders' ideas were rooted in the commonplace and unexceptional. This is not to argue that the crusade was a necessary consequence of the nature of Latin Christian society at the end of the eleventh century; it would be mistaken to imagine the West in the years before 1095 consciously preparing itself for the crusade message. Nevertheless, the success of Urban's appeal deserves to be explained on a deeper level than that of historical accident. It is in this context that the comments of Guibert of Nogent and Ralph of Caen are important clues, for when these observers attempted to explain the crusade's attraction they chose to emphasize a range of ideas associated with 'the monastic way of life or any other form of professed calling'.

I

The Peace of God and the First Crusade

HISTORIANS have often linked the origins of crusading to the tenth- and eleventh-century movements for the Peace and the Truce of God, movements which attempted to limit respectively the objects and persons against whom, and the times during which, arms-bearers could indulge in violence.[1] In his famous work *The Origin of the Idea of Crusade* Carl Erdmann devoted a chapter to the Peace, which, he argued, was important on two levels.[2] Broadly, it represented the first medieval religious movement in which the mass of the people was involved, and as such demonstrated the influence which the eleventh-century church was able to exert upon the laity.[3] More specifically, the Peace movement was one of the principal means by which ecclesiastics attempted to Christianize the military profession, thereby providing knighthood with a religious duty to act in the church's interest and stimulating the emergence of an ethos of holy war.[4] Pursuing the origins of the crusade, like Erdmann, through the development of ideas of holy war, Étienne Delaruelle expressed less confidence that the Peace was of seminal importance, but he too argued that it fostered among arms-bearers ideas of responsibility for the church's safety and the value of fighting non-Christians.[5] In his study of the

[1] Whenever the context allows, the term 'Peace' will be used of both the Peace and Truce of God. By the 2nd half of the 11th C. most Peace legislation was a hybrid of the two institutions.

[2] Trans. M. W. Baldwin and W. Goffart (Princeton, NJ, 1977), 57–94. The German original also contained an app. on the dating of certain Peace councils: *Die Entstehung des Kreuzzugsgedankens* (Forschungen zur Kirchen- und Geistesgeschichte, 6; Stuttgart, 1935), 335–8.

[3] *Origin*, 75–6. See also L. C. Mackinney, 'The People and Public Opinion in the Eleventh-Century Peace Movement', *Speculum*, 5 (1930), 181–206, esp. 200–4.

[4] *Origin*, 57–8, 62–4, 92. See also H. Hoffmann, *Gottesfriede und Treuga Dei* (MGH Schriften, 20; Stuttgart, 1964), 249–50.

[5] 'Essai sur la formation de l'idée de Croisade', *Bulletin de littérature ecclésiastique*, 45 (1944), 37–42. Delaruelle's study of the crusade was contemporary with Erdmann's and independent of it, but he generously allowed the *Origin* prominence, held back publication of his own study, and confined it to a journal. This has tended to obscure Delaruelle's own contribution, which was at some points better argued than Erdmann's. For example, his observation (ibid. 41) that knighthood developed first in northern France, whereas the Peace was originally a southern institution, has not been fully addressed by later research.

impact upon the laity of the Peace movement, Georges Duby concluded that, following the emergence of a distinct *ordo* of warriors, a Christian-ized ethic developed which lent the prescriptions of the Peace, and especially those of the Truce, a moral force. As the shedding of Christian blood became progressively more stigmatized—this process reaching its zenith at the Council of Narbonne in 1054—war against non-believers became the only licit act of violence open to nobles and knights. The crusade was thus the natural consequence of the Peace movement.[6] Duby's conclusions are weakened by his misunderstanding of the nature of the First Crusade: he describes the crusaders as 'a flood of confident, pacific and unarmed people that was to carry all before it', marching under the protection of an armed cordon of knights.[7] But his argument about the emergence of an ethos for *milites*, in so far as it complements Erdmann's thesis, remains very influential.[8] Modern studies of the early crusade movement have included passages on the Peace in order to establish that the eleventh-century church was concerned to find some-thing positive for warriors to do.[9] The possible existence of a Christian warrior ethic, especially one which was supposedly stimulated by the direct initiative of the church, is clearly an important problem in a discussion of knightly piety on the eve of the First Crusade. It is the subject of this chapter.

Of the three areas covered by this study, the Limousin made much the most important contribution to the Peace of God movement. Our sources for the Peace in south-western France largely stem from the busy quill of Adhemar of Chabannes, which means that there is a danger of ex-aggerating Limoges's role in the Peace movement relative to other parts of Aquitaine. None the less, it is clear that Limoges was a major centre of the Aquitanian Peace, particularly during two peaks of activity, one in the 990s and the other between *c*.1025 and *c*.1033. It might therefore seem an attractive proposition that the Limousin's importance in the

---

[6] 'Laity and the Peace of God', *The Chivalrous Society*, trans. C. Postan (London, 1977), 131–2.

[7] Ibid. 132. See also J. Paul, *L'Église et la culture en Occident IXᵉ–XIIIᵉ siècles* (Nouvelle Clio, 15; Paris, 1986), ii. 583–4, 585–7.

[8] See e.g. I. S. Robinson, *The Papacy 1073–1198: Continuity and Innovation* (Cambridge, 1990), 325–7. See also, J.-P. Poly, *La Provence et la société féodale (879–1166): contribution à l'étude des structures dites féodales dans le Midi* (Paris, 1976), 266–8.

[9] J. S. C. Riley-Smith, *The First Crusade and the Idea of Crusading* (London, 1986), 3–4; H. E. Mayer, *The Crusades*, trans. J. Gillingham, 2nd edn. (Oxford, 1988), 8, 15–16, 36. See also A. Becker, *Papst Urban II. (1088–1099)* (MGH Schriften, 19; Stuttgart, 1964–88), ii. 279–84.

Peace movement is linked to the fact that (as will be seen in Chapter 6) many first crusaders came from that region. It is not necessary to rehearse the history of the Peace in great detail, for it has been the subject of many recent studies.[10] The following discussion will limit itself to a consideration of factors which suggest that the superficially attractive link between the Peace and the crusade is a chimera. Several points lead towards this conclusion: the social position of those laymen to whom the Peace message was directed; the duration and effectiveness of the Peace in Aquitaine; and the significance of Peace ideas in the sources for the First Crusade. First it is necessary to set the Peace in its political and social context.

The Background of Violence in Aquitaine *c.*950–*c.*1030

It is conventional wisdom that the Peace movement grew out of the decline in public authority in France, especially south of the Loire, in the latter half of the tenth century.[11] Faced by mounting political and social insecurity, the church, led by bishops but with monastic communities in close support, instituted a series of conciliar decrees and oaths which were intended to protect certain classes of persons (clergy, the poor, merchants, women) and property from violence.[12] Why the church went about its peacemaking mission is the subject of some debate. A traditional interpretation is that the church, being the most durable institution of Carolingian governance and finding itself threatened by endemic anarchy, assumed the royal function of guarantor of peace. It thus filled a power vacuum, almost by default, until such time as effective secular authority could reassert itself. (An important extension of this line of argument is that the church transferred an ethic of protection of clerics and the vulnerable from kings to much lower levels of lay society, thereby

---

[10] The essential guide is Hoffmann, *Gottesfriede*. See also B. Töpfer, *Volk und Kirche zur Zeit der beginnenden Gottesfriedensbewegung in Frankreich* (Berlin, 1957). The best survey in English is H. E. J. Cowdrey, 'The Peace and the Truce of God in the Eleventh Century', *Past and Present*, 46 (1970), 42–67. F. S. Paxton, 'The Peace of God in Modern Historiography: Perspectives and Trends', *HR* 14 (1987), 385–404, reviews work done on the Peace since the 19th C.; see also H.-W. Goetz, 'Kirchenschutz, Rechtswahrung und Reform: Zu den Zielen und zum Wesen der frühen Gottesfriedensbewegung in Frankreich', *Francia*, 11 (1983), 202–8.

[11] See e.g. T. N. Bisson, 'The Organized Peace in Southern France and Catalonia, ca.1140–ca.1233', *AHR* 82 (1977), 292–3.

[12] Cowdrey, 'Peace', 46–8.

stimulating the emergence of chivalric values and preparing knights' minds for crusading ideas.)[13] More recently scholars influenced by comparisons with the German institution of *Landfrieden* (territorial peaces) have argued that the Peace did not involve the church simply acting as a king-substitute but rather trimming to local conditions, co-operating with the secular survivals of Carolingian authority, and strengthening its lordships just as the lay nobility was trying to do.[14] The evidence for the Peace in Aquitaine suggests that this more subtle argument has great merit, with important implications for the supposed ethical antecedents of the crusade vocation.

The background to the origins of the Aquitanian Peace movement, in particular that part of it which turned upon an axis comprising Poitiers, Charroux, and Limoges, was the warfare in northern and western Aquitaine between *c*.950 and *c*.1030. The most important sources are narratives such as Adhemar's chronicle and the *Miracles of St Benedict*, which means that chronological precision is often impossible. Even so, a broad outline of some significant events may be traced. In 963, for example, Viscount Ebles of Thouars went to war against his wife's brother, the viscount of Limoges. It is probable that the *casus belli* was Ebles's claims to property by virtue of his marriage. A late source favourable to the viscount of Thouars stated that Ebles would have wreaked great damage upon his Limousin enemies had not the death of Count William III Tête d'Étoupe of Poitou, Ebles's supporter, persuaded the warring parties to make peace.[15]

Further instances of violence, as revealed by the narrative sources, appear as a succession of bitter princely struggles for resources, military power, authority, and prestige, compounded by vendettas and family rivalries. At some point between *c*.963 and *c*.975, for example, Boso I of La Marche[16] and Viscount Gerald of Limoges went to war. Possibly in anticipation of help from Duke William IV Fier-à-Bras of Aquitaine,

---

[13] See now G. Althoff, 'Nunc fiant Christi milites, qui dudum extiterunt raptores: Zur Entstehung von Rittertum und Ritterethos', *Saeculum*, 32 (1981), 317–18, 329–30.

[14] O. Engels, 'Vorstufen der Staatwerdung im Hochmittelalter: Zum Kontext der Gottesfriedensbewegung', *Historisches Jahrbuch*, 97/8 (1978), 71–86, esp. 83–4; Goetz, 'Kirchenschutz', 206–8, 221–2, 223–4, 226–7, 238.

[15] 'Fragmenta chronicorum comitum Pictaviae, ducum Aquitaniae', ed. E. Martène and U. Durand, *Veterum Scriptorum et Monumentorum, Historicorum, Dogmaticorum, Moralium, Amplissima Collectio* (Paris, 1724–33), v. 1147–8.

[16] A. Debord, *La Société laïque dans les pays de la Charente X^e–XIIe siècles* (Paris, 1984), 73–4, observes that Boso's son Helias was the first of his line to assume the title of Count of La Marche.

Boso's forces laid siege to Gerald's fortress of La Brosse on the borders of Berry and the Limousin. A Limousin relief force broke the siege.[17] Boso's son Helias blinded Benedict, the choir-bishop of Limoges, who had influenced William IV to favour Viscount Gerald, whereupon war broke out between Helias on the one side and Gerald of Limoges and his son Guy on the other. Although victorious at first, Helias and his brother Hildebert were captured by Guy. Narrowly avoiding the blinding which Duke William had requested, Helias escaped. Hildebert, however, was held captive at Limoges 'for a long time' before agreeing to marry Guy's sister.[18] This bitter dispute spilled over into another feud. Helias of La Marche's brother Gauzbert took refuge with his cousin Ranulf (the son of Count Bernard of Angoulême) who was contesting Angoulême with Count William II Taillefer's bastard son Arnald Manzer. Gauzbert was captured, handed over to Duke William IV, and blinded in order to avenge Benedict the choir-bishop. These events probably took place around the time that Ranulf died at the hands of Arnald Manzer (975).[19] More turbulence was to follow in the next generation. In 1000 Viscount Guy of Limoges's son Adhemar disputed La Brosse with a local lord, Hugh of Gargilesse. A coalition of castellans defeated Adhemar, took him prisoner, and seized the castle.[20]

This last case is a clear example of an important characteristic of the warfare of the period: that it was largely fuelled by competition for castles.[21] Competition of this sort was itself generated by potentially divergent appeals to the senior princes' banal control of fortifications, the feudal obligations of lords and their castellans, and hereditary claims.[22]

---

[17] *Les Miracles de saint Benoît* (II. 16), ed. E. de Certain (Paris, 1858), 118–20.

[18] Adhemar, *Chronique* (ch. 25), 147–8. The dating of these connected disputes raises problems: see Lair, 145 [version *C* of the *Historia*, i.e. Adhemar's chronicle]; A. Richard, *Histoire des comtes de Poitou 778–1204* (Paris, 1903), i. 110 and n. 4; J. Becquet, 'Les Évêques de Limoges aux xe, xie et xiie siècles', *BSAHL* 104 (1977), 86; Debord, *Société laïque*, 74 n. 101. To add to the difficulties, it seems that Adhemar confused the siege of La Brosse with a second siege in 999 or 1000: see G. Thomas, 'Les Comtes de la Marche de la Maison de Charroux (xe siècle–1177)', *Mémoires de la Société des Sciences Naturelles et Archéologiques de la Creuse*, 23 (1925–7), 572–3, corrected by Debord, *Société laïque*, 74 n. 100.

[19] Adhemar, *Chronique* (ch. 28), 149–50; Lair, 149–50 [versions *H* and *C* of Adhemar's chronicle]; 'Annales Engolismenses', *MGH SS* 16. 487; Thomas 'Comtes', 573; Debord, *Société laïque*, 75–6, 102–3.

[20] *Miracles de saint Benoît* (III. 5, 6), 136–41, 142–3.

[21] See e.g. J. Martindale, 'Conventum inter Guillelmum Aquitanorum comes et Hugonem Chiliarchum', *EHR* 84 (1969), 533–4, 535–6, 543–8; *Cartulaire de l'abbaye de Saint-Amant-de-Boixe*, ed. A. Debord (Poitiers, 1982), 2 (1020 × 1028).

[22] For the origins of castles in this period, and the problems associated with their control, see G. Fournier, *Le Peuplement rural en Basse Auvergne durant le Haut Moyen Âge*

Aimo of Fleury, who favoured Hugh of Gargilesse, observed that Adhemar of Limoges's ambitions towards La Brosse were the result of disaffected youthfulness which predisposed him to steal others' property in order to compensate for the meagre portion of the family lands granted him by Viscount Guy.[23] Adhemar, however, seems to have had a very solid claim in this instance, for his grandfather Gerald had married a woman named Rothildis of La Brosse. Aimo himself observed that land near La Brosse belonged to Gerald's *parentes*,[24] and the lords of La Brosse would seem to have assumed the title of viscount in the early eleventh century on the basis of their kinship to the viscounts of Limoges.[25] Aimo also implicitly recognized that hereditary rights in castles were important when he stressed that this was the basis of Hugh of Gargilesse's own claim to La Brosse.[26]

Given the importance of castles and competing rights in fomenting war, it is important to emphasize the role in most of the recorded disputes of the senior princes of Aquitaine, in particular the count-dukes of Poitou-Aquitaine. According to Aimo of Fleury, Helias of La Marche approached Duke William IV for help with the first siege of La Brosse because William was his *dominus*.[27] It was in support of Hugh of Gargilesse that Duke William V and Count Boso II of La Marche besieged La Brosse in 999 or 1000.[28] A little earlier, in 996, Count Hildebert of La Marche had launched a vigorous attack upon Poitiers itself in alliance with Count Fulk Nerra of Anjou; war ranged as far north as the Touraine. William V's castle at Gençay near Poitiers was destroyed by Hildebert but then quickly rebuilt; besieging the castle a second time Hildebert was killed, probably in 997. Hildebert's successor Boso II continued the war against William V, who attacked Boso's fortress at Bellac. An agreement was eventually reached whereby William married

---

(Publications de la Faculté des Lettres et Sciences Humaines de Clermont-Ferrand[2], 12; Aurillac, 1962), 365–77, 383–7, 577–603; M. Garaud, *Les Châtelains de Poitou et l'avènement du régime féodal XIe et XIIe siècles* (Mémoires de la Société des Antiquaires de l'Ouest[4], 8; Poitiers, 1964), 17–20, 22–7, 29–35; G. Devailly, *Le Berry du Xe siècle au milieu du XIIIe: étude politique, religieuse, sociale et économique* (Civilisations et sociétés, 19; Paris, 1973), 168–76; Debord, *Société laïque*, 125–30, 139–49, 151–9.

[23] *Miracles de saint Benoît* (III. 5), 136.
[24] Ibid. (II. 16), 119.
[25] See Devailly, *Le Berry*, 371, 378.
[26] *Miracles de saint Benoît* (III. 5), 136. Hugh's genealogy is obscure.
[27] Ibid. (II. 16), 119.
[28] Ibid. (III. 5), 136. Adhemar, *Chronique* (ch. 34), 156–7 probably refers to the same event.

Hildebert's widow Almodis.[29] William V's alliance with Boso II was not to last, however. While besieging Rochemaux, near Charroux, William inflicted a heavy defeat upon a relief force led by Boso and was able to storm the castle.[30]

Other cases further demonstrate the significance which the senior princes of Aquitaine attached to castles as means to enhance their authority. For example, Count William IV of Angoulême took control of Blaye by force with the help of Duke William V, receiving it from him as a benefice.[31] It was with William V's assistance that Bishop Hilduin of Limoges (990–1014) built the castle of Beaujeu near Rochechouart in order to contain the ambitions of a local lord, Jordan II of Chabanais; when William withdrew, Hilduin was unable to withstand Jordan's attack in spite of an alliance with his brother, Viscount Guy I of Limoges. A bitter feud between Hilduin's and Jordan's kindreds developed thereafter and persisted as long as Beaujeu stood.[32] When Boso II of La Marche died (*c.*1014), William V moved to capture Périgueux and become guardian to Boso's sons and nephew. The subsequent separation of Boso's two counties of La Marche and Périgueux reveal the wisdom of William's policy and the fundamental importance of the control of fortified centres as the precondition for, and effective expression of, political power.[33] As Count Bernard of La Marche (one of William V's wards) approached adulthood, he and the duke broke the power of Bernard's tutor, Abbot Peter of Le Dorat. Peter had, for reasons which are unclear, burned the castle of Mortemart, which had been built by Peter's father Abbo on behalf of Bernard's uncle Count Hildebert.[34] When Count William IV of Angoulême was absent on pilgrimage, probably in 1024, Aymeric of Rancon erected the castle of Fractabotum in the Saintonge as a direct challenge to his authority. Aymeric was defeated and killed by William's son Geoffrey. On his return, William and his other son Hilduin destroyed the castle, then rebuilt it so that it might be held by

[29] *Miracles de saint Benoît* (III. 7), 147–8; Adhemar, *Chronique* (ch. 34), 156; Lair, 164–5 [C]; Peter of Maillezais, 'De antiquitate et commutatione in melius Malleacensis insulae' (I. 5–6), ed. P. Labbe, *Novae Bibliothecae Manuscriptorum Librorum* (Paris, 1657), ii. 227–8 (in which Hildebert and Boso are confused); Thomas, 'Comtes', 578–82.

[30] Adhemar, *Chronique* (ch. 41), 165. Versions *C* and *H* of Adhemar's chronicle differ as to Boso's fate in the action: Lair, 183. Cf. Peter of Maillezais, 'De antiquitate' (I. 6), 228.

[31] Adhemar, *Chronique* (ch. 41), 165; Lair, 184 [C]; *Historia Pontificum et Comitum Engolismensium* (ch. 24), ed. J. Boussard (Paris, 1957), 16–17.

[32] Adhemar, *Chronique* (ch. 42), 165–6.

[33] Ibid. (ch. 45), 167; Debord, *Société laïque*, 105–6.

[34] Adhemar, *Chronique* (ch. 45), 167; Lair, 187 [C]; Thomas, 'Comtes', 587.

Geoffrey as a fief.[35] The same Count William, allied with William V of Aquitaine, besieged and burned Marcillac in *c.*1020 in support of Hilduin, the brother of Viscount William of Marcillac and his rival for possession of the castle of Ruffec. (Ruffec had been a ducal possession originally granted by William V to Count William IV.) William of Marcillac and a third brother, Odolric, had broken a sworn agreement arranged by Count William, and had had Hilduin blinded. Years later Hilduin's son, also named Hilduin, was allowed to refortify and hold Marcillac.[36] Shortly after the death of Count William in 1028, his successsor Hilduin had to fight his brother Geoffrey for possession of Blaye; in time an agreement was reached involving the control of this and two other castles in the Saintonge.[37]

This last case is a good example of how the causes of wars and the means used to prosecute them might become detached from the operation of ducal authority. Blaye had been granted by William V to Count William of Angoulême as a fief, but the agreement reached between Hilduin and Geoffrey, which probably took place in 1029 as the ageing William V was preparing to become a monk, turned upon the rights as feudal lord of Count Hilduin alone. From *c.*1020 onwards there are also signs of second-rank lords waging war on their own account without reference to the major princes. In 1012 × 1025, for example, Odo I of Déols seized Argenton from Viscount Guy I of Limoges and his kinsmen the viscounts of La Brosse, thus ensuring control of the upper valley of the Creuse.[38] The value which a successful and expansionist lord such as Odo placed upon a network of fortresses is further demonstrated by his construction of a castle near Massay to protect his lordship at Issoudun.[39] In a rare excursion into the Bas-Limousin, Adhemar of Chabannes records that a war was fought *c.*1010 × 1020 between Viscount Ebles I of Comborn and Gauzbert of Malemort. During the hostilities the castle of Merle was stormed and destroyed by a band of peasants acting on Gauzbert's behalf.[40]

---

[35] Adhemar, *Chronique* (ch. 60), 185–6; *Historia Pontificum* (ch. 25), 17. For *Fractabotum*, see Debord, *Société laïque*, 112, 137, 141, 459–60.

[36] Adhemar, *Chronique* (chs. 41, 60), 165, 186; *Historia Pontificum* (ch. 25), 17–18; Debord, *Société laïque*, 79, 112, 210 and n. 130. A. Richard, *Histoire des comtes de Poitou 778–1204* (Paris, 1903), i. 157 dates the Marcillac war to 1021 but adduces no firm evidence.

[37] Adhemar, *Chronique* (ch. 67), 193; *Historia Pontificum* (ch. 27), 22–3; Debord, *Société laïque*, 168.

[38] Adhemar, *Chronique* (ch. 51), 174; Devailly, *Le Berry*, 371.

[39] Adhemar, *Chronique* (ch. 51), 174; Devailly, *Le Berry*, 372.

[40] Adhemar, *Chronique* (ch. 48), 171; Lair, 194–5 [C]. See M. Aubrun, *L'Ancien Diocèse*

Aimo of Fleury provides evidence of how castles could stimulate violence and competition for resources on a very local level. He records that in the area around Fleury's dependency at Sault, in southern Berry, an alarming number of fortresses had sprung up in recent times (Aimo was writing in the first years of the eleventh century). From one fortress—it is not named—armed men sallied forth to seize supplies (*praedae*) of draught animals and livestock. In time the raiders were defeated by forces from the castle of Argenton in retaliation for the pillaging of their lands. Although Aimo's purpose in telling this story was to extol the merits of St Benedict, through which victory had been granted, he still noted that the victors themselves brought back captives to ransom as well as booty, only part of which was given in alms.[41] Elsewhere Aimo reports that Adhemar of Limoges preyed upon the lands of St-Benoît-du-Sault in order to resupply the recently besieged inhabitants of La Brosse.[42] In yet another incident reflecting the impact of castles on their immediate vicinity, Abbot Aymeric of St-Martial ordered that a fortification be destroyed because it threatened the monks of Ste-Valery, Chambon and their peasants.[43]

However much castles and their occupants created trouble for their neighbours on a local level, however, it is also important to note that in the more small-scale conflicts reported by Adhemar and Aimo the status of the protagonists was only slightly lower than that of those who featured in the wars of the dukes of Aquitaine. Jordan of Chabanais, Guazbert of Malemort, and Odo of Déols, for example, were all described as *principes*.[44] Hugh of Gargilesse was a *fidelis* of the counts of La Marche.[45] Abbot Aymeric had vassals as important as Boso I of La Marche and Viscount Gerald of Limoges.[46] If the duke of Aquitaine himself was not directly involved in some of these conflicts, his *fideles* such as the count of Angoulême or the viscount of Limoges often were.

A region's stability was particularly vulnerable to disruption when

de Limoges des origines au milieu du XI* *siècle* (Publication de l'Institut d'Études du Massif Central, 21; Clermont-Ferrand, 1981), 199.

[41] *Miracles de saint Benoît* (II. 15), 117–18. Devailly, *Le Berry*, 193 argues that the incident is legendary, but its incidental details do seem plausible. He dates the incident to *c*.950 (ibid. and n. 7), but Aimo (*Miracles* (II. 15, 16), 118, 119) is clear that it took place shortly before the war between Gerald of Limoges and Boso I of La Marche (*c*.963 × *c*.975).

[42] *Miracles de saint Benoît* (III. 5), 136–7.

[43] Adhemar, *Chronique* (ch. 29), 150; Lair, 151 [C].

[44] Adhemar, *Chronique* (chs. 42, 48, 51), 166, 171, 174.

[45] *Miracles de saint Benoît* (III. 7), 147.

[46] Adhemar, *Chronique* (ch. 29), 150.

rivalries surfaced within its political élites. Looking back more than a century later, for example, Geoffrey of Vigeois firmly believed that the early years of the eleventh century had been very turbulent in the Limousin. In a rather cryptic passage he noted that in the years around 1020 violent warfare had raged 'amongst the viscounts' on account of a viscountess named Ava/Blanche.[47] The identity of this viscountess is uncertain. She is possibly to be identified with the Ava, wife of Viscount Aymeric I Ostafrancs of Rochechouart, who is recorded subscribing one of her husband's charters in 1019; it is probable that this woman was the daughter of Fulcald of La Roche.[48] She is almost certainly the 'Aduis, cognomento Blancha' who appears as the wife of Viscount Guy II of Limoges (Aymeric Ostafrancs's great-nephew) in a document from 1036.[49] If the two charters refer to the same woman, her career in the intervening years is unknown. A charter of 1025 which records a gathering of the vicecomital family of Limoges makes no reference to the future Guy II (1036–52) or any spouse, although Guy's parents Adhemar I and Senegundis are recorded.[50] It may be that sometime between 1025 and 1036 the future Guy II married his great-aunt as part of a power struggle within the vicecomital kindred. This can only be conjecture.

The identity of the other warring viscounts referred to by Geoffrey of Vigeois is also uncertain, but it is recorded that Guy Niger of Lastours was one of the protagonists in the hostilities.[51] It is possible that rivalries within the vicecomital kindred of Limoges heightened after Guy I's death in 1025,[52] and that these were exploited by Guy Niger—who appears second in a list of Guy I's *fideles* from that year[53]—to consolidate his own

[47] Geoffrey of Vigeois, 'Chronica' (ch. 3), ed. P. Labbe, *Novae Bibliothecae Manuscriptorum Librorum* (Paris, 1657), ii. 281.

[48] Uz. 53; Debord, *Société laïque*, 220–1.

[49] Uz. 315.

[50] Ibid. 47.

[51] Geoffrey of Vigeois, 'Chronica' (ch. 6), 281. Cf. 'Historia monasterii Usercensis', extr. *RHGF* 14. 337, which was probably Geoffrey's source and places the war during the time of Abbot Richard (1002–36). Geoffrey dated the war by the foundation of Arnac by Guy Niger of Lastours, which probably took place in the 1010s: A. Sohn, *Der Abbatiat Ademars von Saint-Martial de Limoges (1063–1114): Ein Beitrag zur Geschichte des cluniacensischen Klösterverbandes* (Beiträge zur Geschichte des alten Mönchtums und des Benediktinertums, 37; Münster, 1989), pp. 100–3, and nos. 10–11, pp. 326–30. Arnac was consecrated in 1028: Geoffrey of Vigeois, 'Chronica' (chs. 4, 6), 281; Bernard Itier, 'Chronicon', ed. H. Duplès-Agier, *Chroniques de Saint-Martial de Limoges* (Paris, 1874), 47.

[52] 'Chronicon Aquitanicum', *MGH SS* 2. 253. The account of Adhemar, *Chronique* (ch. 62), 188 suggests that Guy's son Adhemar needed the support of Count William of Angoulême to secure his succession.

[53] Uz. 47.

territorial power. Geoffrey of Vigeois stressed the wide extent of Guy Niger's *principatus*, which was based on the castles of Hautefort, Lastours, and Terrasson; and he noted that Guy fortified Pompadour against the viscount of Ségur (that is, the viscount of Limoges) and died in battle at Limoges.[54] Whatever the exact nature of the *casus belli*, it is note-worthy that Geoffrey of Vigeois's account of what he called the 'very greatest hostility' recalls very closely themes found earlier in Adhemar of Chabannes's chronicle. The sparks of war, as both observers saw matters, were rivalries amongst princes expressed through disputes over marriages, alliances, and castles. It is significant, for example, that Guy Niger developed Pompadour as a fortified base and sustained his hostility towards the viscounts of Limoges by seeking the help of another senior prince, the count of Périgueux. Geoffrey of Vigeois implies that the war zone extended south to the River Vézère and as far as Sarlat on the borders of Périgord and Quercy.[55] The burning of the monastery of Uzerche in 1031 × 1036 may be connected with the disorder caused by the war; there is evidence that local hostilities caused considerable damage to the property of Uzerche and other religious communities around this time.[56] In short, there are faint traces here of a crisis of authority and endemic princely warfare reaching a peak at exactly the time that there was a revival of the Peace movement in the Limousin from *c.*1025 to *c.*1033.[57]

The political and military history of Aquitaine in this period is apt to be distorted by the fact that our principal source is Adhemar of Chabannes, who was a scion of a noble family related to the viscounts of Aubusson from the eastern Limousin (and most proud of his pedigree).[58] Adhemar also admired William V of Aquitaine to the point of hero-worship, which meant that he tended to exaggerate the duke's personal accomplishments, his political effectiveness, and the range of his military

---

[54] 'Chronica' (chs. 3, 6), 281. Guy died in or shortly after 1036: G. Tenant de la Tour, *L'Homme et la terre de Charlemagne à Saint Louis: essai sur les origines et les caractères d'une féodalité* (Paris, 1943), 326.

[55] 'Chronica' (ch. 6), 281.

[56] 'Historia monasterii Usercensis', 337.

[57] For Peace activity in the Limousin in this period, see Goetz, 'Kirchenschutz', 200 and map at 201. See generally, Hoffmann, *Gottesfriede*, 33–40.

[58] Adhemar, *Chronique* (ch. 45), 167–8; id., 'Epistola de apostolatu Sancti Martialis', *PL* 141. 94; L. Levillain, 'Adémar de Chabannes, généalogiste', *Bulletin de la Société des Antiquaires de l'Ouest*[3], 10 (1934), 251–5 (not all of whose conclusions can now be accepted); D. F. Callahan, 'Adémar de Chabannes et la paix de Dieu', *AM* 89 (1977), 22–3.

activity.[59] Adhemar was perfectly aware that violence might flow from personal and small-scale disputes: he reported, for example, the murder of Jordan of Chabanais by a *miles* whom he had humiliated in combat.[60] His chronicle, however, dealt mainly with the high politics of a small group of closely related princely kindreds. In a passage praising William V, he suggested that the most important cause of the violence which threatened the duke was the restlessness of 'the Aquitanian leading men [*primores*]' or 'the several leading men who resisted his authority [*imperium*]'.[61] Aimo of Fleury's perspective was remarkably similar when he set the war between Boso I of La Marche and Gerald of Limoges against the background of unrest among 'the leading men of the Aquitanian kingdom'.[62] Likewise, Adhemar reported that Abbot Geoffrey of St-Martial allied with Boso II of La Marche to protect the relics of St Valeria from 'certain princes'; the relics were later returned to their original site under the auspices of William V.[63] It is open to debate how successful men such as William V or William of Angoulême were in controlling their regions, attracting powerful men into their allegiance, and supervising the construction and garrisoning of castles. Adhemar stated that William V tended not to kill or mutilate his political opponents—he characteristically put this down to the duke's *pietas*—which suggests that William lacked the brutal resolve of, say, Count Fulk Nerra of Anjou.[64] But at least, as Adhemar's chronicle amply testifies, William was trying to exert effective political control over his duchy.[65] The celebrated letter to him from Fulbert of Chartres concerning the duties owed to a lord by his *fidelis* is one example of how William attempted to define and enforce his authority by all practical means.[66] While making every allowance for the sources' noble bias, it is evident that the Peace of God must be set against the background of conflicts involving the greatest princes of Aquitaine. There

---

[59] Adhemar, *Chronique* (chs. 41, 54, 56), 163–5, 176–7, 182; D. F. Callahan, 'William the Great and the Monasteries of Aquitaine', *Studia Monastica*, 19 (1977), 322–3. For a revisionist attempt to deflate William's reputation, see B. S. Bachrach, 'Toward a Reappraisal of William the Great, Duke of Aquitaine (995–1030)', *Journal of Medieval History*, 5 (1979), 11–21.

[60] Adhemar, *Chronique* (ch. 42), 166.

[61] Ibid. (ch. 41), 165; Lair, 183 [*H*].

[62] *Miracles de saint Benoît* (II. 16), 118.

[63] Adhemar, *Chronique* (ch. 43), 166; Lair, 185–6 [*C*].

[64] Lair, 183–4 [*H*].

[65] For treatments of the two Williams' power, see Garaud, *Châtelains*, 29 ff.; Debord, *Société laïque*, 105–7, 108–15, 265–6.

[66] Fulbert of Chartres, *The Letters and Poems*, ed. and trans. F. Behrends (Oxford, 1976), no. 51, pp. 90–2.

is no need to posit an anarchic free-for-all in which all arms-bearers, irrespective of social and political status, participated as equals. Consequently, there is a prima-facie case for arguing that the Peace of God in south-western France is unlikely to have been based on an appeal to the ethics of all fighting men.

The Role and Status of the Lay Leaders Involved in the Peace

The question of whether the Peace movement created an ethos for all arms-bearers may also be addressed by studying the types of laymen who involved themselves in the Peace councils. The best documented layman in this respect is Duke William V, whose Peace activity reached its apogee at a council held at Poitiers in 1010 × 1014.[67] The surviving text of the council's canons states that Duke William summoned the gathering, which was attended by Archbishop Seguin of Bordeaux, four bishops, and twelve abbots. Ecclesiastical business was transacted: the council's stated purpose was 'the restoration of the church', and canons 2 and 3 dealt respectively with payment for the sacraments and clerical marriage.[68] The council also addressed itself, however, to the problem of the renewal of 'peace and justice', an Augustinian phrase which became common jargon for the aims and methods of the Peace of God.[69] An important feature of the Poitevin provisions was the close interaction of ecclesiastical and secular sanctions. The duke and other *principes* undertook (*firmaverunt*) to keep the Peace both by giving hostages—exactly to whom and on what terms are unclear—and by recognizing that transgressors would be excommunicated. A procedure was established, or more probably confirmed, whereby property disputes which had arisen in the previous five years would be heard by the local *princeps* or his subordinate judge (*judex*). The basic unit of jurisdiction which governed the resolution of disputes was the *pagus*, a long-established element of public administration at the local level. If the prince were unable to enforce judgement, the princes and bishops present at the council might be summoned to act together 'to the destruction and confusion' of the

---

[67] The date of the Council of Poitiers has been the subject of debate: see Hoffmann, *Gottesfriede*, 31–2.

[68] Mansi, 19. 267, 268.

[69] Cf. ibid. 530; L. Delisle, 'Notice sur les manuscrits originaux d'Adémar de Chabannes', *Notices et extraits des manuscrits de la Bibliothèque Nationale et autres bibliothèques*, 35 (1896), 268.

offender.[70] The clear implication is that ecclesiastical punishments would serve to bolster the potentially violent operation of secular authority, which was conceived as a necessary element of the Peace. A document from the Poitevin abbey of St-Maixent, most probably dating from 1010, demonstrates how the system of *judices* and recourse to the comital court might operate when religious communities sought the protection of local princes.[71]

Because Poitiers was the dukes' major urban residence, and the county of Poitou the centre of their territorial power, it is not surprising that this Peace had a strongly secular character. It seems, however, that the aims and methods of the Poitevin Peace were not untypical within northern and western Aquitaine as a whole, the same interpenetration of princely and ecclesiastical authority being reflected in Peace initiatives in other areas where the dukes were able to exert some power. The role of Duke William IV Fier-à-Bras at the first known Peace council held in this area, at Charroux in La Marche in 989, is problematical. Our principal source is the text of three canons issued (*decrevimus*) in the name of Archbishop Gunbald of Bordeaux, four of his five suffragans, and Bishop Hildegar of Limoges. Only the ecclesiastical punishments of anathema and excommunication are mentioned. But it is quite probable that the council was connected to a gathering of bishops at William's court which is known to have taken place at some point between 988 and 991; and in general terms it is highly unlikely that the bishops assembled at Charroux, who included Gislebert of Poitiers, would have acted without the duke's consent or contrary to the exercise of his authority.[72] The evidence is much clearer for Duke William V's presence at the Peace council which was held at Limoges in 994 (the first such gathering in the city). In one of his accounts of the council, Adhemar of Chabannes recorded that William came to Limoges 'with the subject counts, townsmen, and great men [*optimates*] under his governance'; his statement that the counts of Toulouse and Bordeaux (i.e. William Sancho of Gascony) attended with large noble followings from their *imperium* is most

[70] Mansi, 19. 267–8.

[71] *Chartes et documents pour servir à l'histoire de l'abbaye de Saint-Maixent*, ed. A. Richard (Archives historiques du Poitou, 16 and 18; Poitiers, 1886), i. 74. For the date see *La Chronique de Saint-Maixent 751–1140*, ed. and trans. J. Verdon (Les classiques de l'histoire de France au Moyen Âge, 33; Paris, 1979), 108, and Hoffmann, *Gottesfriede*, 32–3.

[72] Mansi, 19. 89–90; Hoffmann, *Gottesfriede*, 25–7. See also Richard, *Histoire des comtes*, i. 126–7. Lethald of Micy, 'Delatio corporis S. Juniani in synodum Karrofensem', *PL* 137. 823–5 provides further information on the conduct of this council by the bishops and senior clergy.

probably an exaggeration of the truth, but is also a sign of the importance he attached to the presence of the great princes and respect for public authority.[73] Elsewhere Adhemar wrote that it was the 'unanimous wish of all the princes of the kingdom of Aquitaine' to submit to the bishops' decrees.[74] In a sermon referring back to the events of 994, Adhemar stated that a system was established by which property disputes, the fundamental impediment to peace and justice, were to be judged by *legis docti*. It is unclear precisely what Adhemar meant by this term, but it would seem probable that the 'men skilled in the law' performed functions similar to those of the Poitevin *judices* of *c.*1010.[75] In the years before 994 William V's father, William IV, had gradually retreated from active participation in ducal affairs.[76] The council of 994 must therefore be understood in the context of William V's early efforts to stamp his authority upon the duchy of Aquitaine with the help of his natural allies amongst the senior clergy. Later Peace councils were similarly realized by ducal initiative or at least depended upon ducal co-operation. At Charroux in 1028 William V summoned a council of the senior clergy and princes of Aquitaine and ordered (*praecepit*) that peace be established.[77] At Poitiers in 1029 × 1031 Duke William VI presided with Bishops Isembert of Poitiers, Jordan of Limoges, and Arnald of Périgueux.[78]

Such instances of princely involvement in Peace councils accord with what is known of the dukes' authority over the church in northern and western Aquitaine. For example, a speech which Adhemar of Chabannes put into the mouth of Abbot Odolric of St-Martial at the Council of Limoges in November 1031 refers to the examination, at a gathering held at Poitiers seven years before, of a codex which had been sent to William V by King Canute of England. Many of the details which Adhemar supplies are unquestionably spurious, for he claimed that William gave a learned speech to the assembled clergy on the apostolic status of St Martial, mentioned in the codex, such that 'no one was able to contradict

---

[73] 'Translatio beati Martialis de Monte Gaudio', ed. E. Sackur, *Die Cluniacenser in ihrer kirchlichen und allgemeingeschichtlichen Wirksamkeit* (Halle, 1892–4), i. 393.

[74] Delisle, 'Notice sur les manuscrits originaux d'Adémar', 290.

[75] Adhemar of Chabannes, 'Sermones tres', *PL* 141, no. 1, col. 117; cf. *Chartes de Saint-Maixent*, i. 91 at pp. 110–11. See Hoffmann, *Gottesfriede*, 28.

[76] Richard, *Histoire des comtes*, i. 135–8.

[77] Adhemar, *Chronique* (ch. 69), 194. See T. Körner, *Iuramentum und frühe Friedensbewegung* (10.–12. Jahrhundert) (Münchener Universitätsschriften: Juristiche Fakultät, Abhandlungen zur rechtswissenschaftlichen Grundlagenforschung, 26; Berlin, 1977), 87–8.

[78] *Chronique de Saint-Maixent*, 114–16; *Chartes de Saint-Maixent*, i. 91.

his opinion'.[79] But the broad picture of the duke presiding over a major clerical gathering is very plausible. The supposed council may have a basis in fact. The cathedral of St-Pierre, Poitiers was consecrated on 17 October 1025 in the presence of what must have been a large gathering of senior churchmen; it is known that Archbishop Arnulf of Tours was one important figure invited to attend.[80] Alternatively, if in fact Adhemar was correct to place the council fully seven years before 1031, it is possible to link the gathering to William V's efforts to settle the affairs of his duchy before he undertook an expedition to Lombardy in early 1025.[81] Adhemar provides a further example of William's authority over the church when he reports that, shortly after the consecration of Bishop Gerald of Limoges (1014), opposition to his uncanonically swift passage through the major orders (Gerald was a porter when elected) was ventilated by a cleric at Poitiers. The actions of the bishops who had ordained Gerald were vindicated because 'the prevalent will of Duke William' overrode the technical niceties.[82]

William's determination in this last example was consistent with a long tradition of ducal influence over the see of Limoges. That influence had been most potent under Bishop Ebles (944–69), who was the son of Count Ebles Manzer of Poitou (the ducal title of Aquitaine was not claimed by the counts of Poitou during his period of rule) and the brother of William III Tête-d'Étoupe (d. 963).[83] Possibly because Ebles's protégé Benedict was unelectable because of blindness, the choice of his successor may reveal some loss of control by William IV and an accommodation to local interests, Bishop Hildegar (969–90) being Viscount Gerald of Limoges's son.[84] Even in this instance, however, ducal influence was probably at work to secure allies against the expansionist Boso I the Old of La Marche.[85] At all events, by the time of Bishop Hilduin's election in 990 William IV's authority was clearly effective. Hilduin was the brother of his predecessor and of Viscount Guy I of Limoges, but Adhemar of Chabannes reports that he succeeded 'by the hand of Duke William' and was consecrated not by his metropolitan at Bourges but by Archbishop

[79] Mansi, 19. 521–2.
[80] Fulbert of Chartres, *Letters*, no. 110, p. 196; see also no. 107, pp. 190–2.
[81] For William's connections with Lombardy, see Richard, *Histoire des comtes*, i. 181–6.
[82] Lair, 198 [C].
[83] Adhemar, *Chronique* (ch. 25), 146; Becquet, 'Évêques', 104. 82–3.
[84] Geoffrey of Vigeois, 'Chronica' (ch. 41), 300.
[85] Debord, *Société laïque*, 75.

Gunbald of Bordeaux.[86] Similarly, Bishop Gerald (1014–21) was Bishop Hilduin's *nepos*.[87] He was consecrated at Poitiers by Archbishop Gunbald of Bordeaux, and the fact that he held the office of treasurer of St-Hilaire, Poitiers demonstrates that he had close ties with William V.[88] The tight control over the major ecclesiastical offices of Limoges enjoyed by the vicecomital family in alliance with the dukes of Aquitaine was increased when Viscount Guy's brother Geoffrey was abbot of St-Martial (991–8). According to Adhemar of Chabannes, Geoffrey was the third leader alongside Bishop Hilduin and William V who instigated the 994 Peace.[89] Such was Viscount Guy's kindred's influence over the Limousin church that he retained the *abbatia* (the abbot's rights and revenues) of St-Martial when Geoffrey died, before selling it to Bishop Hilduin.[90]

The election of Bishop Jordan of Laron (1022–51) is usually accounted a turning-point in the history of the bishopric of Limoges, for unlike his four immediate predecessors Jordan was not a close kinsman of either the dukes of Aquitaine or the viscounts of Limoges.[91] There is some evidence that the smooth succession of related bishops was disrupted. Adhemar of Chabannes's account of Jordan's accession reveals that the bishopric became the object of factional fighting among the princely families of the Limousin.[92] The fact, moreover, that Duke William V and Count William of Angoulême held a court to settle the dispute at St-Junien, eighteen miles west of Limoges, suggests that Viscount Guy I was just one of the contending parties and that his family's control over the bishopric had loosened. Jordan, who was provost of St-Léonard, Noblat,

---

[86] Adhemar, *Chronique* (ch. 35), 157; Lair, 166–7 [C]. Version A of Adhemar's chronicle states that Bishop Hugh of Angoulême performed the consecration, but the fuller version C, which has Bishop Hugh enthroning Hilduin, is to be preferred. For our purposes, what is significant is that both versions demonstrate how Limoges was being pulled westwards into the orbit of the ecclesiastical province which corresponded closely to the duchy of Aquitaine.

[87] Adhemar, *Chronique* (ch. 49), 172. It is probable that Gerald was the son of a sister of Bishop Hilduin and Viscount Guy. See, *contra*, Becquet, 'Évêques', 105. 89.

[88] Adhemar, *Chronique* (chs. 49, 50), 172, 174.

[89] Ibid. (ch. 35), 158; id. *et al.*, 'Commemoratio abbatum Lemovicensium basilice S. Marcialis, apostoli', ed. H. Duplès-Agier, *Chroniques de Saint-Martial de Limoges* (Paris, 1874), 6.

[90] Adhemar, *Chronique* (ch. 49), 171. For the viscounts' relationship with the Limousin church, see also Sohn, *Abbatiat Ademars*, 22–7.

[91] F. de Fontette, 'Évêques de Limoges et Comtes de Poitou au XIᵉ siècle', in *Études d'histoire du droit canonique dédiées à Gabriel Le Bras* (Paris, 1965), i. 555–6; Becquet, 'Évêques', 104. 69.

[92] Adhemar, *Chronique* (ch. 57), 182.

may have been a compromise candidate acceptable to Guy. Whatever pressures may have led William V to support Jordan's election at St-Junien, however, it is significant that he was able to enforce his will. William led Jordan into Limoges in January 1023 and orchestrated the public ceremonial which consolidated the new bishop's position in the city; he was sufficiently confident of his power to leave the ordination (Jordan was a layman) to his son, the future William VI the Fat, while he was absent on pilgrimage to Rome; and Jordan was consecrated at St-Jean d'Angély in the Saintonge by Bishop Islo of Saintes and the bishops of Périgueux, Angoulême, and Poitiers. The rights of Jordan's metropolitan, the archbishop of Bourges, were at first ignored.[93]

A clear lessening of the powers of the dukes over the bishopric of Limoges is only discernible in mid-century. Between c.1044 and 1051 Duke William VII Aigret and Bishop Jordan negotiated an agreement which established the respective rights of the duke, the bishop, and the canons of the cathedral of St-Étienne in the choice of future bishops.[94] Jordan's longevity had the practical effect of keeping ducal power over the see in abeyance. When the election of Jordan's successor was disputed in 1051–2, Duke William VII was only one of a number of lords, including Viscount Adhemar II of Limoges and Adhemar of Laron, who were involved in reaching a settlement.[95] But this development could not have been anticipated in 1022–3. Jordan of Laron was no nonentity. Although his exact antecedents are not clear, it is certain that he belonged to one of the most important noble kindreds in the Limousin.[96] His older contemporary Roger of Laron was a powerful lord who had been instrumental in re-establishing the monastic community at Uzerche (which a kinsman had founded) in the face of opposition from Bishop Hilduin.[97] Roger is prominent in the witness list of a charter of Viscount Guy I dated 1025.[98] His son Guy was later to be bishop of Limoges (1073–86).[99] The sister of Bishop Itier (1052–73) married Roger's son Adhemar le

[93] Ibid. 182–3; Geoffrey of Vigeois, 'Chronica' (ch. 12), 284; Becquet, 'Évêques', 105. 95.

[94] GC 2, Inst., cols. 172–3 (no. 12). The terminal dates for the agreement are the end of William VII's minority and Jordan's death.

[95] SEL 121[102].

[96] Geoffrey of Vigeois, 'Chronica' (ch. 14), 284 describes Bishop Guy (1073–86) as Jordan's nepos. Guy was the son of Roger of Laron. It is possible that Jordan was the son of one of Roger's brothers, Gerald and Vivian, who are attested c.1000: Uz. 433 (999 × 1000), 574 (996).

[97] 'Historia monasterii Usercensis', 335–6.

[98] Uz. 47.

[99] Ibid. 431, 435; Mart. I. 32.

Comtor of Laron, who by his first marriage to the heiress of Guy Niger of Lastours was the progenitor of the powerful lords of Lastours.[100] Thus the bishop who presided over the second wave of Peace activity in Limoges between the mid-1020s and early 1030s belonged to an emerging network of powerful noble kindreds. He also owed his position in large measure to Duke William V. Given this combination of circumstances, it would seem quite wrong to speak of the Limousin Peace as inimical to the existence and methods of princely authority.[101] On the contrary, the Peace was predicated on close co-operation between the senior clergy and the most powerful laymen of the area.

The Peace and Contemporary Hagiography

The place of Duke William V and other major princes in the Limousin Peace finds some reflection in hagiographical material produced at the abbey of St-Martial, Limoges during the period. This material was associated with attempts to enhance the status of St Martial by placing him historically in Palestine before and after the Resurrection. By the time of Abbot Hugh of St-Martial (1019–25) a campaign had begun to celebrate Martial as nothing less than an Apostle.[102] Despite being absolutely without historical foundation, the cult of St Martial the Apostle survived into remarkably modern times.[103] It is only more recently still that scholars have begun to clarify the stages by which the cult was developed. Briefly, the position as now understood is as follows. By the middle of the ninth century at the latest the community of St-Martial possessed a short Life of their patron (the *Vita Antiquior*), according to which Martial had been a disciple of St Peter at Rome sent by his master to evangelize the Limousin.[104] Possibly after a fire at St-Martial in *c*.952 destroyed the abbey's copy or copies of this Life, but

---

[100] Geoffrey of Vigeois, 'Chronica' (ch. 6), 281.

[101] See R. Bonnaud-Delamare, 'Les Institutions de paix en Aquitaine au XIe siècle', *Recueils de la Société Jean Bodin*, 14 (1961), 415–16.

[102] For St Martial's cult until the 10th C. see D. F. Callahan, 'The Sermons of Adémar of Chabannes and the Cult of St. Martial of Limoges', *RB* 86 (1976), 252–3.

[103] Martial was in fact a mid-3rd-C. Roman missionary sent to preach in the Limousin: see Gregory of Tours, *Libri Historiarum X* (I. 30), *MGH Scriptores rerum Merovingicarum*, rev. edn., I[1]. 22–3.

[104] This Life is printed as 'Ancienne vie anonyme de saint Martial (VIe siècle)', ed. J. C. E. Bourret, *Documents sur les origines chrétiennes du Rouergue: saint Martial* (Rodez, 1887–1902), 4–6.

more probably after the first Limousin Peace council in 994, a much longer Life (the *Vita Prolixior*) was produced which presented Martial as a Palestinian Jew, related to St Peter, who became an active follower of Christ and consequently one of the first confessors of the Christian faith. Finally in about 1028, when it appeared that the cult of St Martial the Apostle might be formally recognized by Bishop Jordan of Limoges, Adhemar of Chabannes made revisions to the *Prolixior* text with the principal object of presenting Martial in historical situations and in language consistent with his supposed apostolic status.[105]

In the *Prolixior* a major place, second in importance only to that of Martial himself, was accorded to a Duke Stephen, ruler of Aquitaine.[106] After Stephen had been converted by Martial, the Life stated, he became a model pious prince, supporting the saint's ministry, revering the church, dispensing alms, and adopting a chaste, quasi-monastic way of life.[107] In one significant passage Stephen was extolled in language redolent of the aims of the Peace programme: he never showed partiality, and was always mindful of the need to relieve 'the poor and pilgrims, widows, and orphans'.[108] He also gave prodigiously to the church, thereby conforming to the belief widely held in the eleventh century that churches should be 'ennobled' by lavish ornamentation.[109] Stephen's importance to the fledgling church in Aquitaine was based on more than his personal piety

---

[105] L. Duchesne, 'Saint Martial de Limoges', *AM* 4 (1892), 302, 329–30; Callahan, 'Sermons of Adémar', 253–4, 258–63; Landes, 'Dynamics', 473–8. The *Confessor* and *Apostle* versions of the Life are known collectively as the Aurelian legend, for they purport to have been written by a disciple of St Martial named Aurelianus. The fire of *c.*952 is probably a red herring in terms of the origins of the *Vita Prolixior*: see, *contra*, Callahan, 'Sermons of Adémar', 253–4. Belief in the fire's importance rests upon Adhemar's statement in his 'Epistola de apostolatu Sancti Martialis', 95, that opponents of Martial's apostolicity in 1029 supported their belief that the *Prolixior* was less than a century old by blaming the fire for its production. Landes, 'Dynamics', 474–8, 481–3 argues persuasively that the *Prolixior* should be dated after 994. If the fire did destroy all traces of the *Antiquior*, it would have been most odd for a major monastic community to exist for 40+ years without a Life of its patron. The gestation of the *Prolixior* therefore seems to have been a protracted and largely obscure process. For our present purposes, however, what is most noteworthy is that the surviving versions of the Aurelian legend betray the powerful influence of Peace ideas.

[106] References to the *Prolixior* are from 'Vita S. Martialis Episcopi Lemovicensis et Galliarum Apostoli', ed. L. Surius, *De Probatis Sanctorum Vitis* (Cologne, 1618), vi. 365–74. This reproduces one recension of the *Apostle* version. *Vita Sanctissimi Martialis Apostoli*, ed. W. de G. Birch (London, 1872) is an edn. of an Anglo-Norman group of texts of the *Apostle* version from the 12th C. which depart marginally from the Surius text. Orderic Vitalis inserted a slightly abridged *Apostle* version into his *Ecclesiastical History: Historiae Ecclesiasticae*, ed. A. Le Prévost and L. Delisle (Paris, 1838–55), i. 360–82.

[107] 'Vita S. Martialis' (chs. 14 [*recte* 12], 16, 22), 368, 370, 372.

[108] Ibid. (ch. 16), 370.

[109] Cf. Callahan, 'Sermons of Adémar', 290–2.

and devotional tastes, however, for he proffered vital material assistance to Martial.[110] He and other prominent lay persons, moreover, were exemplars of one of the most important themes of the Life, the value of giving alms and providing for pilgrims and the poor.[111]

It was not only Stephen's wealth but also his authority which was mobilized for the church's benefit. The duke, it was stated, converted to Christianity when his retainer, who had just carried out his master's orders to execute St Valeria, was first struck down dead and then revived by St Martial. Stephen's lead in requesting baptism was followed by 'all his counts and dukes, the whole army, and all the people'.[112] In a passage which throws important light upon how the Peace programme might be put into practical effect, Stephen required (*praecepit*) his army to stock up well with provisions for a march to Rome 'so that none should beg or steal from anyone'. This was not at all an appeal to his troops' morality or the evocation of an ethical code, but rather an official order—described as a *decretum* and *constitutio*—which was reinforced by the threat of the death penalty for the disobedient.[113] At Rome Stephen led his soldiers in prostrating themselves as penitents before St Peter.[114] Martial used Stephen's powers of command to send messages throughout those regions 'which lay under his authority', ordering that pagan idols were to be destroyed.[115] It was the duke's practice to go on pilgrimage to Limoges four times annually 'with all the peoples subjected to him'.[116] The Life emphasized the wide geographical extent of Stephen's *principatus* and *regendi potestas*, terms which stressed the legitimacy, public quality, and effectiveness of his rule.[117] In short, Stephen was the church's strong

---

[110] 'Vita S. Martialis' (chs. 13, 22), 368, 372.

[111] Ibid. (chs. 8, 11 [*recte* 9], 14 [*recte* 12], 13, 14, 15, 16, 22), 367, 368, 369, 370, 372.

[112] Ibid. (chs. 14 [*recte* 12], 13), 368.

[113] Ibid. (ch. 14), 368. Cf. H. Conrad, 'Gottesfrieden und Heeresverfassung in der Zeit der Kreuzzüge: Ein Beitrag zur Geschichte des Heeresstrafrechts im Mittelalter', *Zeitschrift der Savigny-Stiftung für Rechtsgeschichte: Germanistische Abteilung*, 61 (1941), 71–7.

[114] 'Vita S. Martialis' (ch. 14), 368–9.

[115] Ibid. (ch. 16), 370.

[116] Ibid.

[117] Ibid. (chs. 14 [*recte* 12], 16), 368, 370. The precise boundaries of Stephen's princedom varied in various recensions of the *Prolixior*. In the Surius text the principal borders are the Atlantic, the Loire, and the Rhône. Callahan, 'Sermons of Adémar', 261–2, 261 nn. 1–4, 262 n. 1 demonstrates that the *Apostle* version expanded the duke's area of authority to comprise all Gaul rather than merely the Limousin and the rest of Aquitaine. This was done so as to be consistent with the enlarged claims made in the *Apostle* text for Martial as the *patronus Galliarum*. For a slightly different geographical description, referring to the Loire, see *Vita Sanctissimi Martialis*, 15.

right arm, the 'father of the Christians and most fierce pursuer of the pagans' who created the conditions in which Martial's mission could flourish.[118]

Duke Stephen was presented as the lay supporter of the church *par excellence*,[119] but other powerful laymen also play an important part in the *Prolixior*. For example, when Martial first came to the Limousin, he was received at the castle of Toulx by one of its residents, Arnulf. Arnulf's wealth ('Arnulfo diuite') is emphasized; so too the authority ('princeps . . . castelli') and nobility (kinship with Emperor Nero) of Toulx's lord, Nerva. Mass conversion based around the castle was prompted by Martial's revival of Nerva's son (who had been suffocated by a demon) after the boy's kinsmen had prostrated themselves at the saint's feet. (Martial had previously cured Arnulf's demoniac daughter without the same knock-on effect.)[120] The Life noted the salient qualities of a paralytic cured by the saint in these terms: 'He was . . . from a great breed of men, and very rich in gold, silver, and possessions.'[121] Martial's first supporter in Limoges itself was Susanna, a 'most noble matron', who led her large household to convert when the saint effected the cure of a manic.[122] When his son Hildebert was drowned by demons in the River Vienne, Count Archadius of Poitou (whom the text implies was a loyal lieutenant of Duke Stephen) led his army to Limoges in order to beg Martial for his help. Hildebert's accident had taken place at a large open-air muster of 'all the princes and counts'. Between Martial banishing the river demons who had pulled the boy under and performing a miraculous revival, Duke Stephen led the *vulgus* and all the army in prostrating themselves at the saint's feet.[123] The evangelization of the Bordelais was originally prompted by an approach made to Martial by one Benedicta on behalf of her ailing husband Count Sigebert. In consultation with the leaders of the community Benedicta smashed the idols of Bordeaux; she also cured her husband with Martial's staff, in gratitude for which Count Sigebert led a large crowd to be baptized by the saint.[124] When later Martial came to

---

[118] 'Vita S. Martialis' (ch. 16), 370. See also Callahan, 'Sermons of Adémar', 285 n. 3. Cf. Adhemar's accounts of Kings Hugh Capet and Robert the Pious: Adhemar, *Chronique* (chs. 30, 31), 150–2, 154–5; Lair, 161 [C].

[119] The parallels between Duke Stephen and William V of Aquitaine are developed by Callahan, 'William the Great', 332–6, and Landes, 'Dynamics', 479–81, 486–7.

[120] 'Vita S. Martialis' (chs. 7–8), 366–7.

[121] Ibid. (ch. 11), 367.

[122] Ibid. (ch. 12), 367.

[123] Ibid. (ch. 15), 369.

[124] Ibid. (ch. 17), 370–1.

preach at Mortagne on the northern shore of the Gironde, Sigebert arrived to bring victuals at the head of 'a large force of *milites*'.[125]

There are passages in the *Prolixior* which portray Martial evangelizing and effecting cures before large crowds without the mediation of secular nobles, or which describe Martial's audience in terms which emphasize its social diversity.[126] The Life was above all, of course, a celebration of Martial's sanctity and his mission to whole peoples. None the less, the above examples point to a significant secondary theme of close co-operation between the church and secular authority. This was not strictly speaking the interaction of equals. In fact, the *Apostle* version attempted to assert Martial's master–servant relationship with Duke Stephen in stronger terms than did the *Confessor*.[127] Furthermore, the prostration of important laymen at the feet of the saint is a recurring image in the Life which was clearly understood as a symbol of the relationship between the church and the laity. Even so, the *Prolixior* provides a picture—albeit idealized and one-sided—of the interdependence of ecclesiastical and princely authority from exactly the period when the Limousin Peace movement was at its zenith.

The Classes of Persons to whom the Peace was Directed

The argument that the Peace movement turned upon a dialogue between the church's hierarchy and the very highest levels of lay society is confirmed by an examination of the status of those laymen who were most directly addressed by the Peace legislation. Much the longest narrative source for the Aquitanian Peace is a description of the Council of Limoges in November 1031 which was written by Adhemar of Chabannes.[128] The extant text is incomplete, for details of the council's third, and presumably final, day are missing. What survives, however, is an important source full of social vocabulary. The account needs to be used with circumspection, for a considerable part of it, purporting to be a report

---

[125] Ibid. (ch. 20), 371.

[126] e.g. ibid. (chs. 7, 15, 21, 22) 366 (Martial attracts 'multus . . . populus' to Toulx), 369 (Martial ministers to 'multae turbe Gothorum et Vasconum'), 371 (Martial cures and baptizes a 'multitudo magna languentium, ac diuersis oppressa infirmitatibus'—a phrase very reminiscent of the crowd of sufferers who attended the Council of Limoges in 994), 372 (Martial summons 'omnem populum . . . a minimo usque ad maximum' to his new basilica: the phrase also occurs in ch. 13 [*recte* 11], 367).

[127] Callahan, 'Sermons of Adémar', 262 and n. 2.

[128] Mansi, 19. 507–48.

of a discussion among the senior clergy which vindicated St Martial's apostolic status, is entirely the product of Adhemar's obsessive imagination.[129] Consequently, the remainder of the text, much of which bears upon the Peace, cannot be treated as an accurate report of what was actually said. It is even possible, taking a very rigorist view, to argue that the Council of Limoges never took place in the form in which Adhemar described: the only early corroborative evidence for it comes from other of Adhemar's writings.[130] Yet it would seem excessively cautious to throw out the whole text. It is not making too great a demand of Adhemar's middling intelligence to argue that he calculated that the interests of his pious forgery would be best served if his fiction were conflated with some tolerably hard fact.[131] For our present purposes, too, whether or not the council record is strictly accurate is less important than the fact that it reveals what Adhemar believed might, or should, have been said by the council fathers.

There are some ambiguities in the social vocabulary of Adhemar's account. When Bishop Jordan addressed the laity in the basilica of St-Sauveur towards the end of the first day of the council, for example, he exhorted the *milites* present—a group implicitly distinguished from the *plebs*/*pauperum turba* who filled the church—to restore what they had stolen from churches and the poor. Jordan's appeal was based upon the example of Zacchaeus, a repentant sinner described in the Vulgate as a rich *princeps* who submitted to Christ in a large crowd (*turba*), returned fourfold what he had unjustly taken from the poor (*pauperes*), and was saved.[132] The *milites* were thus encouraged to respect the example of a figure whom they might compare with a senior lord. The appeal was not made to men who exercised public authority, but to those who were at least moderately rich: Bishop Jordan invoked as the twin causes of the *milites'* unjust exactions *calumnia* (a term evoking disputes over landed property) and *potentia* (the ability to use force as opposed to the right to do so).[133] Another, well-known, passage describes how, at the conclusion of the first day's proceedings, the assembled bishops pronounced ex-

---

[129] Ibid. 509–28; cf. 532–3. For Adhemar's forgeries, see the works of L. Saltet cited by Landes, 'Dynamics', 471 n. 15.

[130] See the works cited in Hoffmann, *Gottesfriede*, 36–40.

[131] See R. L. Wolff, 'How the News was brought from Byzantium to Angoulême; or, The Pursuit of a Hare in an Ox Cart', *Byzantine and Modern Greek Studies*, 4 (1978), 177–82, esp. 178 and n. 24.

[132] Luke 19: 2–10.

[133] Mansi, 19. 529.

communication upon 'those *milites* from this diocese of Limoges who refuse, or have refused, to swear peace and justice to their bishop, as he [Bishop Jordan] has required'; a malediction was pronounced on them, their accomplices (*adjutores*) in evil, their horses, and their weapons.[134] The reference to the *milites'* accomplices corroborates the message of the passage citing Zacchaeus: that *milites* might denote men possessed of a social position and resources superior to those of the most simply equipped mounted warrior. The inclusion in the malediction, however, of these *adjutores*—a term which could have covered anyone from grooms and servants to foot-soldiers—demonstrates that the bishops were attempting to place at least some of the responsibility for the Peace on all individuals who might be involved with violence.[135] This is what Bishop Jordan seems to have meant when he had earlier proposed a malediction upon 'those who wish for wars'.[136]

There are other passages in Adhemar's account of the council, however, which suggest that the highest levels of lay society occupied a particular place in Peace ideas. Bishop Jordan's speech which opened the council set out themes which are developed in the text. He noted the need for 'the unity of peace', and added: 'I make complaint about the secular powers [*saeculares potestates*], my parishioners, who do not allow God's church to be at rest, who interfere with the property of the sanctuary, who afflict the poor entrusted to me and the servants of the church.' Patently, Jordan was recasting in rhetorical form the traditional prescriptions of the Peace programme: as he continued, the secular powers 'do not wish to hear about peace from me, their shepherd'.[137]

Exactly who were understood by the term 'secular powers' is revealed by certain of the council's discussions on the second day. A party of monks from Beaulieu in the southern Limousin complained that their abbey had been deprived of a regular abbot because of the influence of local nobles. According to the monks, Beaulieu, once an important and wealthy Benedictine community, had been absorbed into the *potestas* of an unnamed count of Toulouse, who had granted the abbey as a fief to a count of Périgueux.[138] The count of Périgueux had in turn conceded it to

---

[134] Ibid. 530.
[135] Cf. J. Flori, *L'Idéologie du glaive: préhistoire de la chevalerie* (Travaux d'histoire éthico-politique, 43; Geneva, 1983), 140.
[136] Mansi, 19. 529.
[137] Ibid. 509.
[138] Ibid. 536–7. J. K. Beitscher, 'Monastic Reform at Beaulieu 1031–1095', *Viator*, 5 (1974), 201 and n. 6 argues that the enfeoffment took place after 960.

the viscount of Comborn (probably Archambald I Cambaputrida, *fl.* *c*.980). The monks complained that this viscount or a successor had then imposed upon them a lay person (*laicalis persona*); he had done so because Bernard II, bishop of Cahors and abbot of Beaulieu, had been that lay person's uncle.[139] The monks' grievance is expressed in terse Latin, and the chronological sequence it describes is compressed, but it is reasonably clear that what it refers to is the enfeoffment by a viscount of Comborn of a well-connected *fidelis* who had become lay abbot.

An independent source, which throws valuable light upon this poorly documented region, suggests that the transfer of control of Beaulieu through the hands of the various nobles had not been as neatly executed as Adhemar of Chabannes supposed. In his Life of Abbot Abbo of Fleury (d. 1004), Aimo of Fleury reports that Abbo maintained close relations with an Abbot Bernard of Beaulieu. Bernard had first come to Abbo's attention when, during the abbacy of Abbot Richard of Fleury (962–74), his father Hugh had sent him to the monastery to be educated. Abbo became the boy's tutor.[140] It is noteworthy that Bernard was entrusted to the monks in order to be made literate ('litteris imbuendum') but not, it would appear, as an oblate. Why this should have been so was demonstrated some years later when Bernard's father recalled him to be given first the *abbatia* of 'Solemniacensi'[141] and then that of Beaulieu, which, Aimo believed, had come to Hugh by right of conquest.[142] Later, Count William III Taillefer of Toulouse (944 × 961–1037), with the connivance of the archbishop of Bourges, chose Bernard to become bishop of Cahors (as Bernard II).[143]

The antecedents of Hugh and Bernard's family are not wholly clear. Writing in a part of France far removed from the scene of these events, Aimo of Fleury stated vaguely that Hugh was descended from high-ranking Aquitanian *proceres*, which at least suggests that some memory of the nobility and prestige of Bernard's kindred survived at Fleury.[144] A

[139] Mansi, 19. 537. Bernard II was dead by 1031; his successor Deusdedit is recorded at the consecration of the basilica of St-Sauveur, Limoges, in 1028: 'Annales Lemovicenses', *MGH SS* 2. 252. For Bernard, see also B. 154.

[140] Aimo of Fleury, 'Vita Sancti Abbonis' (ch. 10), *PL* 139. 397–8.

[141] Deloche (B., p. cclii) took this to refer to Solignac, near Limoges. The Latin fits this identification, but Solignac seems too northerly to have been within Hugh's orbit. Souillac, 17 miles W. of Beaulieu, may be what Aimo had in mind.

[142] Aimo of Fleury, 'Vita Sancti Abbonis' (ch. 10), 398. A co-abbot Bernard is attested in 984 or 985 and an abbot Bernard in *c*.990: B. 85, 128.

[143] Aimo of Fleury, 'Vita Sancti Abbonis' (ch. 10), 398. Bernard became bishop in *c*.1000: *GC* 1. cols. 125–6.

[144] 'Vita Sancti Abbonis' (ch. 10), 397.

charter of between 994 and 1028 records the donation to Cluny of a church in Quercy by Bishop Bernard II, his brother Robert, and the latter's wife Matfreda, son Peter, and an unnamed daughter; another church was granted by Bernard to Robert as a fief.[145] The family was clearly very powerful locally—because of his name it is possible that Bishop Bernard I of Cahors, apparently attested in 960 (but whose existence is problematic), was a member of the kindred[146]—and had close ties to the comital family of Toulouse, though exactly how is not known. The bishops of Cahors exercised some comital powers in Quercy under the overlordship of the counts of Toulouse,[147] and it is therefore significant that, as Aimo emphasized, Count William sold the bishopric to Bernard II.[148]

By 1028 the title of abbot of Beaulieu was being accorded to a man named Hugh.[149] It is he who must be the lay person mentioned by the monks at the council of Limoges. In a charter of 1076 a man named Hugh of Castelnau (a lordship near Beaulieu) referred to a Bishop Bernard of Cahors as his uncle; this would corroborate the monks' complaint in 1031 if, as seems likely, the uncle is identified with Bernard II.[150] To complicate matters, however, in 1030 a man named Bernard appears as abbot of Beaulieu.[151] It was certainly this man who was brought before the council in 1031 and made to promise that a regular abbot would be chosen by the following Christmas and the monastic regime at Beaulieu allowed to revert to its 'pristine dignity'.[152] It has been argued that Hugh of Castelnau himself was made to submit to the council, but this cannot be correct.[153] The unnamed abbot at the council was described by Adhemar as a *clericus*, albeit one noted for his nobility and wealth; furthermore, he was an *abbas saecularis* rather than an *abbas laicus*. The monks' complaint was that their community was being overrun by 'seculars'; their objection to the abbot was that he was not a monk, not that he was a layman; and the council granted permission for those

---

[145] *GC* 1, Inst., p. 30 (no. 6). Cf. B. 154 (*c*.1000 × 1028).

[146] *GC* 1, col. 125.

[147] J. Dunbabin, *France in the Making 843–1180* (Oxford, 1985), 85–6.

[148] 'Vita Sancti Abbonis' (ch. 10), 398.

[149] B. 122.

[150] C. iv. 3491.

[151] Uz. 441. The suggestion of Deloche (B., pp. cclv–cclvi) that this Abbot Bernard was a regular abbot expelled by Hugh of Castelnau cannot be sustained in the light of Adhemar's account of the discussion at Limoges in 1031.

[152] Mansi, 19. 537.

[153] Beitscher, 'Monastic Reform', 202.

regulars who remained at Beaulieu to remove themselves if the seculars could not be reformed.[154] It does not appear that, as has sometimes been supposed, Hugh of Castelnau had simply taken over Beaulieu for his private benefit. Some sort of institutional religious life had continued at the abbey under two related abbots, one lay, the other a cleric.

The council of 1031 did not tackle the problem of the lay abbacy head on, but only through its symptomatic effect on the observance at Beaulieu. The power of the lords of Castelnau was and would continue to be well entrenched. When Hugh of Castelnau (almost certainly the same man as that recorded shortly before the Council of Limoges)[155] granted Beaulieu to Cluny in April 1076, he stated that the abbey had descended to him by inheritance.[156] As late as 1095–6 Pope Urban II had to confront Hugh's control of the monastery.[157] The Abbot Bernard of 1030 is almost certainly to be identified with the Bernard 'pontifex, abbas' who was bishop of Cahors between c.1047 and c.1054.[158] Thus the close family ties which linked the bishopric of Cahors, Beaulieu, and the lords of Castelnau had clearly survived the council of 1031. The Peace of God addressed institutionalized, 'slow-motion' abuse of the church as well as more robust acts of violence: as the monks of Beaulieu complained, they suffered under 'tyranny', and a regular abbot would allow them to live 'in peace'.[159] The Council of Limoges failed to correct the abuses it perceived at Beaulieu, but at least it recognized that a powerful castellan's dynastic interests and the political manœuvrings of some of the most important lay princes of south-western France lay at the heart of the abbey's problems. The forces resistant to the council's attempts to reform Beaulieu were, in effect, the mirror image of the alliance of senior churchmen and princes which made the Peace possible.

A large part of Adhemar's report of the 1031 council dealt with the practical problems of enforcing excommunications against violators of the Peace. The abbot of Uzerche, in the southern Limousin, was accused

---

[154] Mansi, 19. 536–8.

[155] Beitscher, 'Monastic Reform', 204 n. 17 cautiously follows Deloche (B., pp. cclviii–cclxv) in believing that the Hughs recorded in 1028 and 1076 were the same man. A Hugh of Castelnau is also recorded in a charter of 1100 (B. 39). If this is the same man, he must have been at least 85 years old by then. The grant of 1100 was made for the soul of Hugh's father Robert. It is tempting to identify this man with the brother of Bishop Bernard II of Cahors.

[156] C. iv. 3491; cf. ibid. 3490.

[157] Beitscher, 'Monastic Reform', 204.

[158] B. 104; GC 1, col. 127.

[159] Mansi, 19. 537.

of having allowed the burial in his monastery of the excommunicated viscount of Aubusson, who had been killed in the act of looting. According to Adhemar, the viscount had been borne to Uzerche by his *milites*. The abbot was able to argue that he had in fact denied the request of the viscount's men to bury their lord, so that they had been forced to carry the body across the River Vézère and dispose of it themselves.[160] Hearing this dispute being ventilated prompted Bishop Deusdedit of Cahors to tell the story of an excommunicated *eques* from Quercy who had recently died unreconciled. Like the viscount of Aubusson, this man had been a violator of the Peace, a *praedans*. He was buried by *milites* in consecrated ground, but miraculously the corpse was repeatedly ejected from its grave until the *milites* realized that it should be laid to rest elsewhere.[161] These stories have two significant features. First, the *milites* were described in situations which made them appear strictly the subordinates of the dead men.[162] The fact that *eques* was used of the dead Cahorsin most probably denotes his relatively elevated social status.[163] Second, the fate of the two excommunicates was intended to serve as a warning to their 'amicis et parentibus' and 'propinquis', meaning their kinsmen and *fideles* of comparable status. Thus, the bishop of Cahors reported of the immediate impact of the *eques* story: 'Struck by this terror, the princes with authority [*principes militiae*] certainly did not delay in swearing peace to us, as was our intention.'[164]

Bishop Deusdedit's contribution led Abbot Odolric of St-Martial to ask what action might be taken against the 'principes militiae Lemovicensis' who obstructed the operation of the Peace. He suggested to the bishops that, if the princes refused to observe the Peace, the entire Limousin should be placed under a 'public excommunication' in order to force the 'principes, capita populorum' to submit. These leaders were distinguished from an inferior group of men described as 'privatis militibus, sive minoribus principibus' (the second element possibly referring to lesser castellans); if they broke the Peace they would suffer only personal excommunication.[165] Although, of course, excommunication was serious enough on an individual level, the difference between the effects of the two types of spiritual sanction was enormous. The public excommunication, or interdict, amounted to the church taking the offensive against the laity as vigorously as its responsibility for the cure of souls

---

[160] Ibid. 539–41.   [161] Ibid. 541.
[162] Cf. ibid. 548.   [163] Flori, *Idéologie du glaive*, 141.
[164] Mansi 19. 539, 541.   [165] Ibid. 541, 542.

would allow. Burials would be denied to all except a few restricted categories of persons; the divine offices would be conducted behind closed doors; once daily all church bells in the diocese would ring out simultaneously to summon the penitent; altars would be stripped and church ornaments removed so that time would appear like an indefinite extension of Good Friday; similarly, Lenten observances would be respected at all times.[166] Bishop Aemilius of Albi and Abbot Azenar of Massay asked what should happen if a minority of *primores* should submit to the council's decrees. They suggested that such men should be reconciled on an individual basis. Abbot Odolric agreed that even a single major *princeps* should be reconciled whenever possible, but he argued that this was insufficient: the reconciled prince's area of authority and family property ('terram honoremque') should be freed from the interdict. The interdict would then remain in force in those parts of the Limousin 'which are held by rebellious princes'.[167] Throughout this exchange the notion that the senior princes were chiefly responsible for the operation of the Peace was consistently emphasized.

The loss of Adhemar of Chabannes's description of the final sessions of the 1031 council means that it is impossible to know whether or how the ideas expressed on the second day were put into practical effect. It would seem likely that, had Abbot Odolric's idea of a patchwork interdict been framed as a decree, it would have proved impossible to enforce effectively. The manner in which Adhemar presented the debates, however, reveals that he favoured Odolric's approach. He did so because of his firm belief in a two-tier system of responsibility for observing the Peace, according to which the group he described as 'omnes majores potestates Lemovicenses'[168] predominated by reason of its social position, landed wealth, and powers of command. According to this scheme, the *milites* had some limited role of their own, but largely in so far as they were one element among all the classes of persons engaged in violence; the main thrust of the Peace programme was directed over their heads. It would, therefore, be mistaken to imagine that the Peace in northern and western Aquitaine was intended to stimulate a code of conduct applicable equally to all arms-bearers.

---

[166] Ibid. 541–2. See also Adhemar of Chabannes, 'Sermones tres', no. 3, col. 120; Delisle, 'Notice sur les manuscrits originaux d'Adémar', 295–6.

[167] Mansi, 19. 542.

[168] Ibid.

The Time-Scale and Impact of the Aquitanian Peace

Of itself, the hierarchical structure of aristocratic society in the early eleventh century might not have prevented the gradual diffusion of a Peace ethic among all mounted warriors over the following sixty or seventy years, had the Peace had time to mature and react to social changes. After *c.*1035, however, the Peace movement in Aquitaine went into a steep decline. It is true that Adhemar of Chabannes, by far our most important source of information, died in 1034, which means that the relative silence of other sources concerning the Peace may not be as significant as it appears.[169] Even so, there is good circumstantial evidence that the Aquitanian Peace petered out in the mid- and late 1030s. The removal from the scene of Duke William V in 1029 (he retired into a monastery) robbed the Peace movement in northern and western Aquitaine of its most important prop. This factor would not have been immediately apparent to William V's contemporaries—hence, for example, the attendance of William VI at the Council of Limoges in 1031[170]—but it soon became decisive. William VI's long captivity in an Angevin gaol (1033–6), the brief rule of Duke Odo (1038–9), the minority of William VII Aigret (1039–*c.*44), and the uncertainty which beset the ducal succession between 1038 and the consolidation of authority by Guy Geoffrey (William VIII (1058–86)), all undermined the dukes' ability to encourage and exploit the Peace movement.[171] By the time Guy Geoffrey was in a position to do so, the traditions of the Peace in Aquitaine had been allowed to decay.

Some late Peace initiatives shed further light on the significance of princely participation. It is possible that Bishop Jordan of Limoges was present when, in or shortly before 1038, Archbishop Aimo of Bourges instituted a Peace oath to be sworn by all the men of his diocese aged 15 years or more.[172] This oath and the 'Peace militia' it is supposed to have created have excited a great deal of scholarly interest, not least because Andrew of Fleury devoted a long passage in book v of the *Miracles of St Benedict* to the episode.[173] Carl Erdmann relied very heavily upon the

---

[169] Bernard Itier, 'Chronicon', 47.

[170] Hoffmann, *Gottesfriede*, 36.

[171] For the dukes between 1030 and 1058, see Richard, *Histoire des comtes*, i. 220–65.

[172] *Miracles de saint Benoît* (v. 2), 193, states that Archbishop Aimo decided upon the oath 'comprovincialibus adscitis episcopis'.

[173] Ibid. (v. 2–4) 192–8; Bonnaud-Delamare, 'Institutions', 474–86; Hoffmann, *Gottesfriede*, 105–8; Devailly, *Berry*, 144–8 (a salutary revisionist interpretation); Flori,

Bourges oath in arguing for the emergence in the eleventh century of militias which used violence at the church's behest and in furtherance of Peace ideals.[174] There is no evidence, however, that Bishop Jordan, even supposing that he had been exposed to Archbishop Aimo's ideas, attempted to introduce a similar oath into the Limousin. Nor is he likely to have done so. The Bourges oath attempted to mobilize all levels of lay society as active guardians of the Peace. This was quite unlike any programme attempted in the Limousin, where memories of the prince-centred Peace must still have been strong in 1038. Furthermore, the spectacular failure of Aimo's militia, which was crushed in battle by Odo of Déols, is unlikely to have encouraged imitations.[175]

The last known Peace in Poitiers, as recorded by the St-Maixent chronicler, also reveals the value of princely co-operation. Bishop Isembert of Poitiers convoked a synod 'where he established a great peace'.[176] This action should be linked with the chronicle's following statement that Bishop Isembert and Countess Eustacia of Poitou forced local religious communities to contribute towards the ransom of William VI.[177] William had been captured by Geoffrey Martel of Anjou in September 1033, and was released for a large ransom in 1036.[178] What may therefore seem to have been a Peace inspired by independent episcopal initiative was in fact an example, prompted by exceptional circumstances, of the usual co-operation between the comital line of Poitou and the senior clergy.

Idéologie, 143–4, 170; D. W. Rollason, 'The Miracles of St Benedict: A Window on Early Medieval France', in H. Mayr-Harting and R. I. Moore (edd.), *Studies in Medieval History Presented to R. H. C. Davis* (London, 1985), 77–8. See now T. Head, 'Andrew of Fleury and the Peace League of Bourges', *HR* 14 (1987), 513–29.

[174] *Origin*, 63–4.

[175] *Miracles de saint Benoît* (V. 4), 196–8; 'Chronicon Dolensis Coenobii', extr. *RHGF* 11. 387–8. It is interesting that Odo of Déols's reputation for personal piety and respect of the church was considerable. In 1027–8 he accompanied Count William IV of Angoulême and Abbot Richard of St-Cybard, Angoulême, on pilgrimage to Jerusalem, and in 1025 Fulbert of Chartres praised him fulsomely: Adhemar, *Chronique* (ch. 65), 189; Fulbert of Chartres, *Letters*, no. 109, p. 194.

[176] *Chronique de Saint-Maixent*, 116.

[177] Ibid.

[178] Glaber (IV. 9), 212; William of Poitiers, *Histoire de Guillaume le Conquérant* (I. 15), ed. and trans. R. Foreville (Les classiques de l'histoire de France au Moyen Âge, 23; Paris, 1952), 32; L. Halphen, *Le Comté d'Anjou au XIᵉ siècle* (Paris, 1906), 57–8; O. Guillot, *Le Comte d'Anjou et son entourage au XIᵉ siècle* (Paris, 1972), i. 52–3. In the 2nd half of the 12th C. Richard the Poitevin believed that the ransom was 200,000*s*.: 'Chronicon', extr. *RHGF* 11. 285. Although both the St-Maixent chronicler and Richard the Poitevin state that William died soon after his release, he in fact survived until Dec. 1038: Richard, *Histoire des comtes*, i. 233 and n. 2.

After 1036–8 the sources for the Peace in northern and western Aquitaine fall largely silent. This cannot be because the princes of the region had been effectively pacified. Two particularly well-documented cases suggest that violence remained endemic. Sometime before 1047 Jordan III of Chabanais occupied and fortified the buildings of the community of canons at Lesterps, in the northern Limousin. Count Hildebert II of La Marche attacked the monastery and burned it to the ground, causing great loss of life and forcing the surviving canons to take refuge nearby. The Life of St Walter of Lesterps records that soon afterwards Jordan's conscience was ' moved by troublesome dreams, prompting him to prostrate himself before the abbot in abject surrender. On a more practical level, it was probably the case that Hildebert's attack had been so fierce that Lesterps had lost its immediate value as a fortress despite Jordan's efforts to rebuild it.[179] The manner in which Count Hildebert, for his part, came to repent of his actions demonstrates both the survival of traces of Peace ideas and the Peace's limitations. The canons of Lesterps appealed to Pope Benedict IX, who placed all those responsible for the disorder under anathema until they submitted to Bishop Jordan of Limoges and Abbot Walter. It is not so stated directly, but it is very probable that Hildebert had earlier been excommunicated by Bishop Jordan, with no result. Eventually Hildebert was brought to make what was described about fifty years later as a 'placitam emend-ationem', a phrase which connotes a negotiated deal thrashed out in a spirit of compromise rather than the imposition of the church's implacable will or the decisive intervention of a higher secular authority.[180] Another, more widespread, eruption of violence occured in 1061–2 when Duke Guy Geoffrey of Aquitaine and Count Fulk IV of Anjou went to war over the Saintonge.[181] The Poitevins were heavily defeated at the Battle of Chef-Boutonne in March 1061, but in the following year Guy Geoffrey was able to reduce Saintes, held by the Angevins, by a vigorous policy of castle-building and scorched-earth tactics.[182] These examples,

[179] Marbod of Rennes, 'Vita sancti Gualterii abbatis et canonici Stirpensis in dioecesi Galliarum Lemovicensi' (ch. 3), *PL* 171. 1574. For the date, Thomas, 'Comtes', 593.

[180] 'Charte d'Almodis, comtesse de La Marche, en faveur de l'abbaye de l'Esterps (12 novembre 1098)', ed. G. Babinet de Rencogne, *Bulletin de la Société Archéologique et Historique de la Charente*[3], 4 (1864), 411–14; Thomas, 'Comtes', 593–4.

[181] For Angevin interest in the Saintonge, see Debord, *Société laïque*, 107–8, 164–5.

[182] *Chronique de Saint-Maixent*, 134–6; Richard, *Histoire des comtes*, i. 285–6; Halphen, *Le Comté*, 136–7.

which could be multiplied, help to demonstrate that Aquitaine was not significantly pacified after *c.*1035.[183]

The impact of the Peace was thus limited in chronological terms. There are also geographical factors which suggest that the Peace's significance was not as great as has often been supposed. Gascony serves as a useful control by which to judge the importance of the Peace in areas such as the Limousin. The Bordelais was doubtless subjected to some Peace influence because its archbishop was metropolitan of the province which corresponded most closely to the effective limits of the Aquitanian dukes' authority. Archbishops of Bordeaux were present, for example, at the Peace councils of Charroux (989), Limoges (994), Poitiers (1010 × 1014), and Limoges (1028).[184] Lack of evidence means that it is unclear whether and how the Peace decrees were enacted in the diocese of Bordeaux itself. The guiding influence of the dukes of Aquitaine to the north was complicated in Bordeaux by the fact that the city was controlled by the dukes of Gascony. The election in 1028 of Archbishop Godfrey, from northern France, is evidence of Bordeaux's unusual position between two spheres of influence, for the choice of prelate was made at a meeting between Dukes William V and Sancho William held at Blaye, on the border of their respective territories.[185]

If the Bordelais was an exceptional middle ground, the experience of southern Gascony seems to have been quite different from that of northern Aquitaine. Even after entering a caveat about the relative paucity of sources for this region, there is no reason to believe that the Peace ever took firm root there while the Aquitanian Peace movement was at its height. It is most unlikely that this was so because Gascony was exceptionally free from violence, or because the dukes' power was so effective that the Peace was unnecessary. On the other hand, the ties of kinship between princely and episcopal authority in Gascony were similar to those which existed in Aquitaine, where they served to encourage and bolster Peace initiatives. The simple fact is that, before Archbishop Austinde of Auch introduced comprehensive administrative reforms in the 1050s and 1060s, there were few solid foundations of episcopal organ-

---

[183] For other violent episodes see Geoffrey of Vigeois, 'Chronica' (chs. 19, 29), 289, 296; *Chronique de Saint-Maixent*, 150. See also E.-R. Labande, 'Situation de l'Aquitaine en 1066', *Bulletin de la Société des Antiquaires de l'Ouest*[4], 8 (1966), 349–53.

[184] Adhemar of Chabannes, 'Sermones tres', nos. 1, 2, cols. 117, 119; Mansi, 19. 89, 90, 267; Geoffrey of Vigeois, 'Chronica' (ch. 10), 283.

[185] Adhemar, *Chronique* (ch. 69), 194.

ization in southern Gascony upon which the Peace could be erected.[186] By the time such foundations existed the Peace in south-western France had run its course.

Significantly, the first direct evidence for the extension of the Peace into southern Gascony appears soon after the Council of Clermont. In 1097 Count Bernard III of Bigorre, emerging from his minority, established the customs regulating the governance of his county at a large gathering of senior clergy and nobility.[187] The customs betray the possible influence of northern Peace ideas. Clerics, monks, women and their companions, peasants, and pilgrims were to enjoy permanent protection; so too fishermen, a peculiarly prominent element of the Gascon rural scene. Furthermore, restrictions were placed on armies' foraging.[188] In a system recalling the Poitevin Peace provisions of 1010 × 1014, but allowing for the relative absence in Gascony of Carolingian administrative survivals, a breach of the Peace was to be justiciable in the first instance by the offender's lord; only if the lord refused to do justice could the count of Bigorre intervene.[189]

The most important feature of the customs is their emphasis upon the operation of secular authority. Count Bernard spoke of his *regimen* over the 'land committed to him' and of his own duty to 'defend and refresh the poor' (the 'poor' probably being understood to include pilgrims). The count's powers were limited by an oath sworn on taking office not to exceed the customs, and by a promise to involve nobles in the comital court of justice, but it is noteworthy how much his authority still stood to be enhanced in such matters as the licensing of castle-building and the levying of military service from his subjects.[190] It is therefore not surprising that the Peace was treated as part of the count's powers. The Peace, for example, was to be sworn by the 'terrae principibus' (the phrase 'procerum terrae' was used of the group which gave the customs its active consent) who became liable to a fine of 65s. to the count if they broke the oath; rape and abduction were punishable by the same sum.[191] In short, the customs were an example of princely lawmaking with the consent of

---

[186] See A. Breuils, *Saint Austinde archévêque d'Auch (1000–1068) et la Gascogne au XIᵉ siècle* (Auch, 1895), 35–44, 192 ff., 294 ff.; R.-A. Senac, 'Essai de géographie et d'histoire de l'évêché de Gascogne (977–1059)', *Bulletin philologique et historique* (1983), 11–25.

[187] P. de Marca, *Histoire de Béarn* (Paris, 1640), 814, 813, 814 (*sic*). For discussion, see R. Mussot-Goulard, *Les Princes de Gascogne* (Marsolan, 1982), 214–15.

[188] Marca, *Histoire de Béarn* (chs. 9, 24, 29, 30), 813, 814(ii).

[189] Ibid. (ch. 25), 813.

[190] Ibid. (preamble and chs. 1, 3, 4, 6, 15, 16), 814(i), 813.

[191] Ibid. (chs. 21, 22, 42), 813, 814(ii).

the local ecclesiastical and secular élites.[192] Typically of medieval legis-lators, Bernard claimed that he was simply re-establishing the usages of his grandfather Count Bernard II (d. *c.*1075), but this does not nec-essarily mean that the possible Peace influences evident in the customs were anything but very recent.[193] Rather than attempt to project the Peace elements within the legislation of 1097 back into the obscure political conditions of Gascony earlier in the eleventh century, it is more useful to look for more immediate factors behind Bernard III's actions. In this context it is surely significant that Bernard's half-brother, immediate neighbour, and protector, Viscount Gaston IV of Béarn, had recently taken the cross.[194]

Further possible evidence for the introduction of the Peace into Gascony appears in a charter of 1104 which was recorded in the lost cartulary of Lescar. Its witness and dating clauses recall a gathering attended by Gaston of Béarn, Count Bernard III of Armagnac, and men from their respective territories, who all swore on oath to observe a perpetual 'peace and truce'. Bishop Sancho of Lescar was present, but the scene of the assembly, Diusse, was an apparently unimportant church in a rural area on the marches of Béarn and Armagnac. The oath-taking ceremony probably represented the formal conclusion of a recent local war; in this connection it is significant that there is solid evidence that Gaston was very active militarily in the early 1100s. Whatever borrowings this sworn peace may have owed to the formal Peace and Truce—its terms are not recorded—it was again the influence of Gaston, four years after his return from crusade, which may have been decisive. Like the customs of Bernard of Bigorre, the 1104 agreement suggests that Peace ideas made a late and only partial entry into southern Gascony, after the needs of the crusade (as the following section of this chapter demonstrates) had extended the geographical range of Peace institutions.[195]

The Peace and the First Crusade

The foregoing has argued that the Peace in Aquitaine was of limited duration and effectiveness, geographically uneven in its impact, and

[192] Cf. *Les Fors Anciens de Béarn*, ed. and trans. P. Ourliac and M. Gilles (Paris, 1990), 111–13.

[193] Marca, *Histoire de Béarn* (preamble), 814(i).

[194] For the relationship between Bernard and Gaston, J. de Jaurgain, *La Vasconie* (Pau, 1898–1902), ii. 379.

[195] Marca, *Histoire de Béarn*, 397 (no. 12).

dominated by a group of senior nobles who were not representative of all arms-bearers. Taken together, these considerations make it very difficult to imagine circumstances in which any coherent knightly ethos could have been central to the ideology and administration of the Peace programme. It remains to test this conclusion against the evidence linking the Peace to the preaching and prosecution of the First Crusade.

It is incontestable that on the level of practical arrangements the Peace and the crusade were associated from the beginning. The Council of Clermont marked the first attempt by the papacy to extend the Peace throughout Latin Christendom, and many of the surviving collections of the council's canons record details of Peace legislation.[196] At Clermont permanent protection was offered to monks, clergy, women and their companions, peasants, and merchants.[197] All animals except horses used in war were also safeguarded.[198] The practice whereby churches and wayside crosses were regarded as places of sanctuary was confirmed.[199] Overlaying the Peace canons, the council probably ordered that a Truce be observed at certain periods on both a weekly and yearly cycle; the evidence suggests that this requirement was adapted slightly when enacted at the local level.[200] The broad relevance of the Peace to the crusade is clear, for domestic stability in the West was bound to aid the recruitment and organization of the expedition. The council fathers at Clermont shrewdly recognized that the complete pacification of France was impossible, and in some situations even undesirable. According to Bishop Lambert of Arras's note of the Truce, free rein was allowed to warriors fighting amongst themselves between Tuesday and Thursday. In a related text from Arras, the suspension of liability was defined more precisely to cover those acting in self-defence or against violators of the permanent Peace. The extension of the Peace to 'all men who do not wage war' recognized that some fighting would occur; similarly, the protection of all animals except horses used in *werra* implied that, within the limits set by the Peace, feuds would continue to be prosecuted.[201] Nevertheless, it seems to have been calculated that the crusade's chances of success depended upon the achievement of maximal order in the West. Thus a

---

[196] See *Councils of Urban II*, 143.

[197] Ibid. 73, 73–4, 124.

[198] Ibid. 73, 74.

[199] Ibid. 81, 96. But cf. OV vi. 250, which suggests that respect was rarely shown to this custom. See also *Miracles de saint Benoît* (VIII. 17), 300–2.

[200] *Councils of Urban II*, 73, 74, 94, 124. The synchronization with the liturgical calendar seems to have varied from place to place: ibid. 94, 116; *RHGF* 14. 391 (Council of Tours 1095 or 1096, canon 1); OV v. 20.

[201] *Councils of Urban II*, 73–4.

version of the canons of Clermont which was probably based on papal records states that a three-year Peace ('treuga continua') was extended to peasants and merchants 'because many regions of Gaul were labouring under shortages of supplies', a reference to an agricultural depression in France which only began to ease in 1096.[202]

There also existed a connection between Urban II's Peace programme and the temporal privileges granted to the crusaders. It is not always clear from the sources, however, precisely how the two were linked.[203] The Arras versions of the Peace canons of Clermont, for example, omit pilgrims from the list of those permanently protected, although there were many precedents for them to be included.[204] Possibly pilgrims were subsumed under the heading of clergy for Peace purposes. Later versions of Urban's enactments at Clermont were more expansive, however, directly linking the Peace to the protection of crusaders' persons, property, and families.[205] Some of the other canons of Clermont were also clearly formulated with the crusade in mind. For example, the decree against laymen usurping another's inheritance, though not linked to the crusade or the Peace in the Arras manuscript traditions, was clearly relevant, for punishment was denial of the sacrament of penance, which effectively made taking the cross impossible. Decrees against reconciling sinners for some grave crimes while other sins went unconfessed had the same practical result; so too the requirement that anyone guilty of molesting a bishop could not bear arms.[206] The Peace was also bound up with the planned length of the crusade campaign. Guibert of Nogent stated that among the practical details expounded by Urban II immediately after his crusade sermon at Clermont was the pronouncement of anathema upon anyone threatening the kinsfolk of crusaders over the

[202] Ibid. 124; Riley-Smith, *First Crusade*, 34.

[203] Villey, *Croisade*, 151–2; Brundage, *Medieval Canon Law*, 10–14, 161, 165; J. Fried, *Der päpstliche Schutz für Laienfürsten: Die politische Geschichte des päpstlichen Schutzprivilegs für Laien (11.–13. Jahrhundert)* (Abhandlungen der Heidelberger Akademie der Wissenschaften, Phil.-hist. Kl. 1980, 1; Heidelberg, 1980), 105–10; Becker, *Papst Urban II.*, ii. 278.

[204] *Councils of Urban II*, 73, 73–4. But see ibid. 124.

[205] Ibid. 108 (Florentine MS of early 12th C. of uncertain value: see ibid. 110), 124 (probably post-1123 but possibly derived from papal records: ibid. 119–21). Cf. the manner in which Paschal II linked the crusade and Peace: *Chronique de Saint-Pierre-le-Vif de Sens, dite de Clarius*, ed. and trans. R.-H. Bautier and M. Gilles (Sources d'histoire médiévale, 3; Paris, 1979), 146; *Chronique de Saint-Maixent*, 178; U.-R. Blumenthal, *The Early Councils of Pope Paschal II 1100–1110* (Pontifical Institute of Mediaeval Studies, Studies and Texts, 43; Toronto, 1978), 86, 91, 93.

[206] *Councils of Urban II*, 78–9, 81, 108, 115.

following three years.[207] That this was the estimate for the duration of the expedition which was actually made in 1095–6 is confirmed by one of the canons of a council held at Tours, probably in early 1096, which required that a *pax* run for three years from the following Pentecost.[208]

Yet, if there were sound practical reasons for applying the Peace to the organizational needs of the crusade, it does not automatically follow that there was a more deep-rooted ideological linkage. The absence of a fundamental theoretical connection between crusading and the Peace is illustrated by a dispute which erupted in northern France in connection with Bohemond of Taranto's crusade of 1106–8, and which throws into clear relief some of the problems connected with crusaders' temporal privileges. The evidence for the dispute is a series of letters written by Bishop Ivo of Chartres (1090–1115), probably between late 1106 and 1107.[209] Hugh II of Le Puiset, viscount of Chartres, had granted to Ivo, lord of Courville, a fief consisting of property held by two of Hugh's vassals. The land was a rear-fief held from Adela, countess of Blois. Subsequently Count Rotrou of Perche bought part of the property (*fundus*); it is not recorded from whom, but, since Bishop Ivo later came to recognize Rotrou's claim that he held his purchase as an allod, it seems that the vendor was Adela herself.[210] In an act which was almost certainly meant to provoke his new neighbours, Rotrou began to build a small fortification (*munitio*) on the site.[211] Hugh of Le Puiset had meanwhile taken the cross, though at what precise stage in the property's tenurial history is unclear.[212] Hugh and Ivo of Courville then appealed to Bishop Ivo to do justice against Rotrou.[213]

When the case was heard before the bishop, Rotrou claimed in his defence that he had simply exercised his proper rights (*tuitio*) over his acquisition. Unable to resort to a judicial duel, Bishop Ivo ordered the dispute to be referred to Countess Adela. When it came to be discussed in

---

[207] 'Gesta Dei per Francos' (II. 5), 140.

[208] *RHGF* 14. 392 (canon 13). Cf. *Councils of Urban II*, 124; Hugh of Flavigny, 'Chronicon', *MGH SS* 8. 475.

[209] Ivo of Chartres, 'Epistolae', *PL* 162, nos. 168–70, 173, cols. 170–4, 176–7; Hoffmann, *Gottesfriede*, 200–2. For the dating of the letters, see the suggestions of R. Sprandel, *Ivo von Chartres und seine Stellung in der Kirchengeschichte* (Pariser Historische Studien, 1; Stuttgart, 1962), 192.

[210] Ivo of Chartres, 'Epistolae', no. 168, col. 171.

[211] Ibid. nos. 168, 169, cols. 171, 172.

[212] Ibid. no. 173, col. 176, states only that Rotrou began to build the castle 'after he [Hugh] had taken the cross'.

[213] Ibid. no. 168, col. 171.

her *curia* Hugh of Le Puiset backed down.[214] Later a private war broke out between Rotrou and Ivo of Courville, during which Ivo was taken prisoner. *En route* to join Bohemond's crusade army in southern Italy Hugh appealed to Pope Paschal II to 'defend the property of the *Hierosolymitanus* according to his decrees', whereupon the pope referred the case to Archbishop Daimbert of Sens, Ivo of Chartres, and two other local bishops with instructions to do justice.[215] As the bishop in whose diocese the disputed property was located, Ivo of Chartres summoned Rotrou and men acting for Hugh. Stalemate resulted when Hugh's agents complained that Rotrou had only been asked to suspend construction-work on the castle, and not to negotiate Ivo of Courville's ransom, before the case was resolved. For his part, Rotrou appealed to the earlier decision of Adela's court, and introduced a significant new argument into the debate by claiming that, because Ivo of Courville was also his own vassal, Hugh of Le Puiset's crusade privileges did not affect his own feudal rights against Ivo (who was still his prisoner). Bishop Ivo submitted the problem to a panel of clerics, who spent a long time deciding that they were unable to resolve the issue. In obvious desperation Bishop Ivo referred the case back to the pope.[216] At this point in the story the sources end.

The convoluted course of this dispute reveals that there were a number of obstacles to the smooth operation of crusaders' temporal privileges. First, there was confusion concerning the relationship between Hugh of Le Puiset's individual rights to protection as a pilgrim-crusader and the broader question of Rotrou's breach of the Truce when he erected his castle. Bishop Ivo wrote that Hugh and Ivo of Courville had, in the first instance, asked the church 'to render them that justice which was owed to *Hierosolymitani* and to the Peace', claiming that Rotrou had built the castle 'unjustly and contrary to the Peace'.[217] Such confusion only increased when Rotrou took Ivo of Courville prisoner. Secondly, and more fundamentally, there was great uncertainty over where effective jurisdiction lay and the correct procedures to be followed. In particular, Bishop Ivo seemed most sensitive to the rights of the comital *curia*,

---

[214] Ibid. Rotrou subsequently claimed that Adela had found in his favour: ibid. no. 173, col. 176.

[215] Ibid. nos. 168, 169, 173, cols. 171, 172, 176. For Hugh on Bohemond's crusade, see 'Narratio Floriacensis de captis Antiochia et Hierosolyma et obsesso Dyrrachio' (ch. 13), *RHC Occ.* 5. 361.

[216] Ivo of Chartres, 'Epistolae', no. 173, cols. 176–7.

[217] Ibid. no. 168, col. 171. For the suspension of castle building, see Hoffmann, *Gottesfriede*, 78, 96, 201. See also G. Fournier, *Le Château dans la France médiévale: essai de sociologie monumentale* (Paris, 1978), 282–4.

and his problems were compounded by his misgivings about excommunicating Rotrou, as Paschal II required, before the various claims were clarified.[218] Thirdly, the case ran up against Rotrou's right to use force (*guerra*) against a rebellious vassal to the point of ravaging lands and taking prisoners. It is significant that Bishop Ivo did not question the general legitimacy of Rotrou's acts of violence towards Ivo of Courville.[219] Finally, the indecision of the panel of clerics exposed the root of the problem. They dithered, 'claiming that the institution of committing the properties of *milites* going to Jerusalem to the church's care was new', and adding that it was unclear whether the protection extended simply to a crusader's *proprietates* (fiefs held personally or allods) or also to the fiefs of a crusader's vassals when they engaged in violence on their own account and in order to protect their own property interests.[220] In short, the relationship between the crusader's temporal privileges and the operation of the Truce was confused and compromised by entrenched ideas of property rights, feudal obligations, and—as Rotrou's conduct throughout the dispute reveals—personal honour. Bishop Ivo's parting shot to Paschal II was an expression of despair, explaining that the protagonists were so powerful locally that they could not be brought to heel.[221] It is noteworthy that, more than a decade after the Council of Clermont, a case should prove so contentious when its principal actors were the pope, a bishop who had one of the finest legal minds of his generation, a former first crusader (whom we shall see resuming his crusade interest in Spain), a *crucesignatus* whose kindred already had strong crusading traditions,[222] and the widow of one of the most important princes on the 1096 and 1101 expeditions (Stephen of Blois). The dispute reveals that the place of the Peace in Urban's original crusading plans could not have been fully worked out in 1095–6. It would have been remarkable, therefore, if the ideological linkage between the Peace and the crusade had been any clearer at that time.

The absence of any coherent theoretical link between the Peace and the crusade is further demonstrated by the fact that Pope Urban's Peace

[218] Such was Ivo's determination to be scrupulously fair that he had to reassure Daimbert of Sens that he was not showing favour to Rotrou: 'Epistolae', no. 170, cols. 173–4. Ivo believed firmly that hasty excommunication prejudiced his pastoral responsibilities, which helps to explain his obvious caution throughout the dispute: ibid. col. 173.

[219] Ibid. no. 173, col. 176.

[220] Ibid. col. 177.

[221] Ibid.

[222] For the presence of Hugh's kinsman Evrard of Le Puiset on the First Crusade, Riley-Smith, *First Crusade*, 77.

programme was not dictated solely by the needs of the expedition. As well as being the occasion at which the crusade was formally launched, the Council of Clermont enacted a large number of canons dealing with the reorganization of the institutional church. Long before 1095–6 Peace programmes had been associated by the church hierarchy with broader efforts of reform.[223] According to Fulcher of Chartres, for example, Urban spoke at Clermont of a renewal of the Truce 'established long since by the holy fathers'.[224] There also existed a specific and important precedent for a pope coming to France in order to pursue a wide programme of reform. At the Council of Reims in 1049 Pope Leo IX had promulgated a Peace of God, albeit one which applied only to the diocese of Reims itself.[225] As archdeacon of Reims between 1055 × 1060 and 1067 × 1070, the future Urban II must have become aware of this.[226] Some years before the crusade plan could have been conceived, moreover, Urban had attempted to popularize the Truce in southern Italy.[227] Fulcher attributed to Urban two sermons at Clermont. The earlier (that is, not the crusade oration) was addressed to the senior clergy, listing the qualities expected of them as diligent reformers: Urban's encouragement that they be *pacifici* falls within a series of other adjectives which bear no obvious relation to the crusade.[228] Furthermore, the measures taken at Clermont against violators of bishops' and other churchmen's persons and property should be understood principally in the light of the council's canons against simony, lay appointment to ecclesiastical office, and lay control over church property during vacancies.[229] Fulcher summed up Urban's achievements on his tour of France as the restoration of peace, the vindication of the church's rights, and the launching of the crusade—in that order.[230]

It is also noteworthy that the survival of the Peace canons of Clermont in certain manuscript traditions was linked to very localized conditions

---

[223] R. I. Moore, 'Family, Community and Cult on the Eve of the Gregorian Reform', *TRHS*[5], 30 (1980), 52–3.

[224] *Historia Hierosolymitana* (I. 2), 129.

[225] Hoffmann, *Gottesfriede*, 185, 217–18 and n. 5.

[226] For Urban's career at Reims, see Becker, *Papst Urban II.*, i. 33–41.

[227] Hoffmann, *Gottesfriede*, 220–1; Becker, *Papst Urban II.*, ii. 277 and n. 9.

[228] Fulcher of Chartres, *Historia Hierosolymitana* (I. 2), 126.

[229] *Councils of Urban II*, 81, 82, 94–5, 109, 111, 113, 114, 115, 121, 122; Fulcher of Chartres, *Historia Hierosolymitana* (I. 2), 126–9; William of Malmesbury, *De Gestis Regum Anglorum Libri Quinque* (ch. 345), ed. W. Stubbs (RS 90; London, 1887–9), ii. 391–3. Cf. Mansi, 20. 935–6 (Council of Nîmes, 1096, canons 4–8).

[230] *Historia Hierosolymitana* (I. 4), 143.

rather than to any compelling generic connection between the Peace and the crusade. The earliest known record of the canons was made by Bishop Lambert of Arras (1093–1115). In the years immediately before Clermont he and other bishops from north-eastern France and Flanders had begun a series of Peace initiatives; and in the first years of the twelfth century Lambert was very actively involved in Peace legislation, significantly with the close co-operation of the former first crusader Count Robert II of Flanders.[231] There were therefore very personal and immediate reasons why a record of the Clermont Peace should survive at Arras.[232] Similarly, an Anglo-Norman series of conciliar texts referring to the Peace was inspired by the Council of Rouen (1096), which Orderic Vitalis records as having applied the Clermont legislation to the unusually troubled conditions in which Normandy then found itself.[233]

At Clermont Pope Urban decreed that the Peace be extended throughout Latin Christendom, but such a measure was only practicable if it were applied through local councils. The decrees survive of one such council, most probably held at Tours.[234] The record of this gathering is undated, but from its contents it is evident that the council was conducted by arrangement between Archbishop Ralph II of Tours and Count Fulk IV Le Réchin of Anjou, who was then master of the Touraine. From the text's preface it appears that clergy from the area who had been present at Clermont brought back news of the pope's Peace decrees, which were then recast to suit local conditions. (Urban may subsequently have confirmed the local provisions when he visited Tours in March 1096.)[235] The crusade is not expressly mentioned in the Tours decrees, but pilgrims were granted permanent protection, and crusaders must have been understood to fall in this category.[236]

---

[231] R. Bonnaud-Delamare, 'La Paix en Flandre pendant la Première Croisade', *Revue du Nord*, 39 (1957), 147–52; Hoffmann, *Gottesfriede*, 150–1, 191–2.

[232] *Councils of Urban II*, 73–4. See ibid. 46, 56.

[233] OV v. 18–24.

[234] *RHGF* 14. 391–2. The text does not specify where the decrees were enacted. Somerville (*Councils of Urban II*, 103) argues for a local synod convened in or near Count Fulk of Anjou's lands. Given the fact that the document refers to the Touraine, Tours seems the likely provenance: ibid. 131. The surviving MS probably originated at Tours: L. Delisle, 'Notice sur les manuscrits disparus de la Bibliothèque de Tours pendant la première moitié du XIXe siècle', *Notices et extraits des manuscrits de la Bibliothèque Nationale et autres bibliothèques*, 31 (1884), 227–9 (the text is edited ibid. 314–16). Cf. Hoffmann, *Gottesfriede*, 126–8; R. Bonnaud-Delamare, 'La Paix de Touraine pendant la Première Croisade', *Revue d'histoire ecclésiastique*, 70 (1975), 749–57 (the text is edited at 756–7).

[235] *RHGF* 14. 391.

[236] Ibid. (canon 7).

The most striking feature of the Touraine Peace is the profound influence upon it of the needs of secular authority. Responsibility for the *pax Domini* lay in the first instance with the bishop, who might excommunicate offenders and was expected to mediate between kindreds engaged in feuds. Violators of the Peace were justiciable in the bishop's *curia*.[237] The count, however, had an important role. If murder were committed in a comital castle, he shared the fine with the bishop; the archbishop of Tours had the right to mediate between the count and his senior vassals, but if he found himself unable to enforce judgement the count could use force against a vassal without violating the Peace himself; the count might be summoned by the archbishop to pursue any violator of the Peace; the count's vassals and provosts were to offer themselves as hostages at Tours twice each year; castle-building was to be suspended.[238] The count's authority was thus directly enhanced by these provisions. The enactments bear comparison with pre-crusade Peaces in Normandy and Flanders in which ecclesiastical and secular jurisdictions co-operated to their mutual advantage.[239] The Tours Peace cannot be seen as the triumph of church-sponsored pacifism generated in the wake of the pope's journey through France. Rather it represents, in an area of France with no indigenous Peace tradition, a process of trimming and accommodation to local conditions. (This is particularly significant in light of the fact that Urban chose Tours as the site of his first major council after Clermont.) Orderic Vitalis saw the Rouen Peace of 1096 in the same light, for he noted that it proved quite ineffective precisely because Duke Robert of Normandy was unable to give it proper support.[240] Against such a background of accommodation to local interests and overlapping jurisdictions, it is difficult to argue that the practical dispositions made for the Peace at Clermont were intended to revive any supposed Peace ethic by which to stimulate enthusiasm for the crusade.

There is one important piece of evidence, however, which might seem to suggest that ideas of peace were present in the original formulation of the crusade message in 1095–6. This is the so-called *Encyclical of Pope Sergius IV*, which is now accepted as an apocryphon produced at the Cluniac abbey of Moissac (which Urban II visited in May 1096) in order

---

[237] *RHGF* 14. 391 (canons 2, 3, 5).

[238] Ibid. 391–2 (canons 3, 5, 8–10).

[239] Hoffmann, *Gottesfriede*, 128; Cowdrey, 'Peace', 59–61.

[240] OV v. 24. The obvious contrast is with the Lillebonne Peace of 1080, which was effective because the weight of Duke William II's authority was put behind it: Cowdrey, 'Peace', 60.

to generate interest in the crusade.[241] This curious document is in the form of an appeal, purportedly addressed by Pope Sergius IV (1009–12) to all the clergy and faithful, to liberate the Holy Land. The *Encyclical* is clearly valuable as a summation of many of the ideas being voiced in 1095–6, for it refers to themes which can also be extracted from the narrative and charter evidence for crusaders' beliefs. The appeal, for example, evoked how Christ died for the sake of sinners; how the Holy Land, and especially the Holy Sepulchre, were relics because of Christ's presence on Earth; and how pilgrims, who formed a brotherhood, abandoned their homelands as an expression of love and took up their crosses in acts of penitence.[242] The crusade's purpose was also expounded in familiar terms. The Holy Sepulchre had been destroyed; it was the task of Christians to leave their families, kill the *gens Agarena*, and restore the Sepulchre; the war was divinely inspired, and was considered an act of vengeance, both of the Lord acting through the faithful and by the faithful themselves.[243] The spiritual benefits on offer comprised eternal life, and the listeners/readers were urged to turn their minds to the Day of Judgement.[244]

This rather muddled combination of cause, intention, and reward occupies most of the document and forms the core of its message. There is then a concluding passage which is more technical. It recalled the Petrine Commission (which established the pope's capacity to summon the faithful to arms), justified the expedition as an example of faith through works, and evoked the image of the cross as the symbol of how Christ ended the Devil's reign on Earth by offering men the chance of redemption. Then the pope ordered on God's, the saints', and his own authority that all men should observe peace, because otherwise it was impossible to serve God. This wished-for peace among Christians had deep roots as an extension and continuation of the perfect state of peace which Christ had enjoined upon the Apostles. In this way peace furthered (*proficit*) salvation and confounded the Devil.[245] There is in these statements a faint suggestion of a vocational link between the Peace programme and

---

[241] A. Gieysztor, 'The Genesis of the Crusades: The Encyclical of Sergius IV (1009–1012)', *Medievalia et Humanistica*, 5 (1948), 3–23, esp. 8 ff.; 6 (1950), 3–34, esp. 19–32; the text is edited in 6. 33–4. See also A. Müssigbrod, *Die Abtei Moissac 1050–1150: Zu einem Zentrum cluniacensischen Mönchtums in Südwestfrankreich* (Münstersche Mittelalter-Schriften, 58; Munich, 1988), 150–1.

[242] Gieysztor, 'Genesis', 6. 33. Cf. Riley-Smith, *First Crusade*, 20–5.

[243] Gieysztor, 'Genesis', 6. 33–4.

[244] Ibid.

[245] Ibid. 34.

crusading. It must be emphasized, however, that this idea was only slightly developed compared to the vivid exhortations of the earlier part of the text. There is little in the concluding passage to suggest that its references to peace were anything more than a grandiloquent statement of the practical and immediate value of domestic stability for ensuring the expedition's success.

Some of the commentators writing about the First Crusade attempted to project the practical connection between the Peace and the crusade on to an ideological level. In this way the Peace became one of the elements believed to have contributed to the crusade vocation. It is important to note, however, that the narrative accounts of the crusade were exercises in justification after the event and should therefore not be given exaggerated weight. Fulcher of Chartres made the link between the Peace and crusading most clearly. Giving the reasons for Urban II's decision to journey to France, he sandwiched between descriptions of the French church and the Turkish occupation of Christian lands in the East a vivid passage on the turbulence of French society. The pope, he believed, acted because he saw that constant fighting between the princes made peace unattainable; agriculture was disrupted by warfare; men were captured and then held to ransom or left to suffer; churches, monasteries, and villages were destroyed by arson.[246] Fulcher described two speeches made by Urban at Clermont, and peace was the bridging theme between them.[247] It is possible, moreover, that before the second speech, the crusade oration, Urban required his audience to give vocal assent to the Peace canons enacted during the council.[248] Fulcher's second speech opened with a reference to the audience's promise to swear to keep the peace and respect the rights of the church. This was linked causally (*quoniam*) to 'a certain other business of God and you', namely the crusade, which was possible because the audience had submitted itself to divine correction.[249] After describing the travails of eastern Christians under Turkish occupation, promising the crusade's spiritual rewards, and appealing to his audience's sense of honour, Urban urged what amounted to conversion: those who had previously directed their violent energies against their fellow Christians and even their own kinsfolk should now

---

[246] *Historia Hierosolymitana* (I. 1), 120–1. Cf. William of Tyre, *Chronicon* (I. 14), ed. R. B. C. Huygens (Corpus Christianorum, Continuatio Mediaeualis, 63/63A; Turnhout, 1986), i. 130.

[247] *Historia Hierosolymitana* (I. 2–3), 129–32.

[248] *Councils of Urban II*, 103–4.

[249] *Historia Hierosolymitana* (I. 3), 132.

legitimately set out against the barbarians; those who had once been looters and pillagers could become *Christi milites*.[250] Finally, Fulcher stated, Urban closed the council with instructions that the Peace be sworn 'throughout all the provinces'.[251]

The other accounts of Urban's speech which were written before *c*.1110 did not integrate the crusade as firmly as did Fulcher into the context of the Council of Clermont's Peace decrees. Guibert of Nogent simply contrasted the new crusade opportunity with traditional warfare.[252] Baldric of Bourgueil was more expansive. Playing on the contrast between *militia* and *malitia*, he made Urban appeal to the arms-bearing classes by listing a series of faults to which they were prone and which were redolent of the language of Peace legislation: oppression of orphans/wards (*pupilli*) and widows, murder, sacrilege, brigandage, and general endemic violence. Because it involved fighting 'exterior' nations, war against Saracens could be good—provided it was conducted with the proper intention, love. Implicitly, therefore, correct intention was absent from wars in the West. As in Fulcher's version, the core of Urban's message was conversion to the status of *Christi milites*.[253] Robert the Monk's use of Peace ideas, however, was much more restrained. Robert was the First Crusade chronicler with the most developed sense of pride in Frankish military renown. He thus made Urban appeal to the Franks' self-esteem as warriors and dynasts, with particular reference to the military achievements of the Carolingian kings.[254] Consequently, Urban had to appear more indulgent towards his lay audience's violent habits, which, Robert believed, were caused by morally neutral factors such as overpopulation and competition for supplies. Even so, Urban called on wars and feuds to end, and this led directly into a summons to go on the *iter*.[255] A generation later Orderic Vitalis was largely content to summarize Baldric, his principal source, at this point, but he linked the swearing of peace by the magnates and knights of the West to the taking of the cross.[256] William of Malmesbury, in contrast, had Urban declare that the crusade was a remedy for all forms

---

[250] Ibid. 136. Cf. William of Tyre, *Chronicon* (I. 15), i. 134.

[251] *Historia Hierosolymitana* (I. 4), 140. Cf. William of Tyre, *Chronicon* (I. 15), i. 135.

[252] 'Gesta Dei per Francos' (I. 1), 124.

[253] 'Historia Jerosolimitana' (I. 4), 14–15.

[254] 'Historia Iherosolimitana' (I. 1), 727, 728. See also ibid. (prologue) 723. Cf. William of Malmesbury's emphasis in his version of Urban's speech—patently coloured by hindsight—that Frankish military tactics and mental toughness were superior to those of the Turks: *De Gestis Regum* (ch. 347), ii. 395–6.

[255] 'Historia Iherosolimitana' (I. 1), 728.

[256] OV v. 14.

of sinful conduct, for example adultery as well as those actions addressed by the Peace programme.[257] Although the majority of these writers saw some link between the Peace and the crusade, only Fulcher, and to a lesser degree Baldric followed by Orderic, committed themselves to projecting this connection on to the nature of the crusade vocation. Guibert and particularly Robert were much more reserved. In the absence of near-unanimity among the crusade chroniclers closest to the events they described (some of them had been present at Clermont), it would be hazardous to conclude that the theoretical linkage between the Peace and the crusade was prominent among the ideas voiced in 1095–6.

The crusade chroniclers' comments deserve some respect; but they were all written with the benefit of hindsight about the crusade's success, and particularly those writers who were monks were embarking on a course which has been labelled 'theological refinement', an attempt to explain the crusade in the context of providential history.[258] This process was made possible by conflating ideas from three sources: the original crusade appeal; the beliefs of the participants as they developed during the crusade; and the predisposition of religious to project their own experiences and values on to the outside world. Thus Baldric and Fulcher, for example, evoked the term *miles Christi*, which Urban II may not have used at Clermont and which lost its overtly monastic connotation anyway when rendered in the vernacular. It would seem that the chroniclers' references to peace were similarly the result of combining ideas from 1095–6, from the crusade campaigns themselves, and from monasticism, monks within their communities being accounted exemplars of fraternal accord and the crusaders regarded as equivalent to temporarily professed religious. In the hyperbole encouraged by the seemingly miraculous outcome of the crusade, even the most unremarkable of Urban II's enactments assumed great significance. Fulcher of Chartres was a secular cleric, not a monk, but his version of Pope Urban's speech at Clermont inspires little confidence as a verbatim report (even though he was an eye-witness), and his words cannot be used to reconstruct a supposedly non-monastic version of the crusade vocation. In sum, the evidence for the vocational connection between the Peace and the crusade reveals that Peace ideas were, at most, peripheral. Once we subtract from our calculations the localized reasons why much of the Peace programme of Clermont was recorded, the fact that domestic stability was a desideratum

[257] *De Gestis Regum* (ch. 347), ii. 393–4.
[258] Riley-Smith, *First Crusade*, 135–52.

of the crusade's planners, and the distorting influences to which the accounts of Urban's crusade appeal were subjected, little remains on which to build a case for arguing that the crusade appeal, and the enthusiasm it generated, would have been impossible without the influence of Peace ideals.

Simply because the Peace movement is a relatively well-documented episode in the relationship between the eleventh-century church and aristocratic society, it does not follow that it must have been an important factor in the response of laymen to the First Crusade. Stability in the West was a desirable precondition of the crusade, and it was natural that the French Urban II should channel that desire through the, to him, familiar institutions of the Peace/Truce. After the crusade's success, Urban's practical measures were worked up to form part of the crusade ideology by the monks and seculars whose predecessors had been the Peace's champions. But those who argue that the Peace directly influenced nobles and knights who took the cross in 1095–6 cannot have it both ways. If the Peace had succeeded in restraining arms-bearers' violence, why should some closely contemporary writers have insisted that the crusade was a remedy for endemic disorder in France? If the Peace had failed in this task (as the evidence from Aquitaine suggests it did), why should it be assumed that some folk-memory of its prescriptions survived —in Aquitaine over at least two generations—to prey upon nobles' and knights' consciences? And how could this have been so in areas such as Gascony which had no Peace tradition but which supplied first crusaders? Themes which the Peace pursued, such as disrespect for the church and, to a lesser extent, the spiritual dangers of bloodshed, did worry men at the time of the First Crusade; but their roots lay much deeper than the Peace.

The emphasis which many treatments of the origins of the crusade have accorded the Peace appears symptomatic of a tendency to concentrate on the theoretical potential of ecclesiastics' pronouncements rather than the actual experiences of the lay rank and file. The significance for the early crusade movement of the Spanish *Reconquista*, discussed in the following chapter, has been distorted by a similar process.

The French Military Contribution in Spain
1064–c.1130 and its Bearing upon the Popular
Ideology of the First Crusade

THE participants on the First Crusade came to believe that they were fighting for and with God in a divinely inspired cause.[1] It follows that a discussion of the popular ideology of the first crusaders must address the question of whether laymen had been exposed to, and become familiar with, holy war influences before the First Crusade. There are intriguing possibilities. Were, for example, the motivations of Norman first crusaders from southern Italy uniquely influenced by their experience of wars of expansion against Muslims in Sicily? It is, however, beyond the scope of this study to examine incidences of warfare against non-believers across the Latin world. Rather, it is a useful exercise to concentrate on what was, from a south-western French perspective, the nearest and most accessible potential theatre of holy war: Spain.

Many historians studying the antecedents of the First Crusade have believed in the formative influence of the eleventh-century Spanish *Reconquista*. The peninsular wars against the Moors, it is argued, created opportunities for ecclesiastics to expound theories of meritorious violence, and familiarized 'French' (that is, non-Spanish Romance-speaking) chivalry with ideas of warfare against infidels in the service of the church.[2] Certain accounts of Pope Urban II's crusade plans stated that he

---

[1] J. S. C. Riley-Smith, *The First Crusade and the Idea of Crusading* (London, 1986), 91–2, 99–107, 111–19.

[2] M. Villey, *La Croisade: essai sur la formation d'une théorie juridique* (L'Église et l'État au Moyen Âge, 6; Paris, 1942), 63–73; P. Rousset, *Les Origines et les caractères de la Première Croisade* (Neuchâtel, 1945), 31–5, 49–51; C. Erdmann, *The Origin of the Idea of Crusade*, trans. M. W. Baldwin and W. Goffart (Princeton, NJ, 1977), 136–40, 155–6. Erdmann (ibid. 289) believed that, 'The Spanish war was where the knighthood of France had manifested its crusading sentiments.' See now also I. S. Robinson, *The Papacy 1073–1198: Continuity and Innovation* (Cambridge, 1990), 323–4.

drew parallels between the Muslim threats in Europe and the Levant;[3] and the most recent study of Urban's policies has argued that he took a keen interest in the progress of the *Reconquista*, regarded Spain as one theatre in a Mediterranean-wide effort of Christian expansion, and developed ideas of linking spiritual rewards to service of the peninsular church against the Muslim menace.[4] From the peninsular perspective, until recent times a dominant tradition in Spanish (and particularly Castilian) historiography maintained that the *Reconquista* acquired a religious, quasi-crusading character long before the First Crusade, and that, by extension, the Spanish experience of wars against Muslims directly influenced crusade ideas. For example, the foremost Spanish medievalist in the first half of the twentieth century, Ramon Menéndez Pidal, argued that the First Crusade was inspired by Castilian resistance to the Almoravid invasions of Spain from 1086, and that the foundation of the kingdom of Jerusalem was a conscious repetition of the creation by Rodrigo Díaz of Vivar (the Cid) of a Christian frontier lordship in Valencia in 1094.[5] It was axiomatic to the nationalistic school represented by Menéndez Pidal that the *Reconquista* and the supposedly religious impulses which inspired it were predominantly the result of peninsular conditions.[6] Yet, if one pursues the argument that Spain was an exemplar to crusaders from France, it would have to follow that there were contacts between the two areas which drew 'proto-crusade' ideas northwards over the Pyrenees.

Links between Spain and France were certainly many and varied in the eleventh century because of, for example, the development of the pilgrimage routes to Santiago de Compostela, urban colonization, and movements of ecclesiastical personnel.[7] South-western France in particular was bound to be influenced by such exchanges, being the corridor into all of Spain west of Catalonia. By concentrating, however, on the strictly

---

[3] Guibert of Nogent, 'Gesta Dei per Francos' (II. 1), *RHC Occ.* 4. 135; see also ibid. (I. 4) 130; William of Malmesbury, *De Gestis Regum Anglorum Libri Quinque* (ch. 347), ed. W. Stubbs (RS 90; London, 1887–9), ii. 395.

[4] A. Becker, *Papst Urban II. (1088–1099)* (MGH Schriften, 19; Stuttgart, 1964–88), i. 227–30, 251–3; ii. 337–9, 341–2, 346–9, 351–3.

[5] *La España del Cid*, 5th edn. (Madrid, 1956), ii. 578.

[6] See P. Linehan, 'Religion, Nationalism and National Identity in Medieval Spain and Portugal', in S. Mews (ed.), *Religion and National Identity* (Studies in Church History, 18; Oxford, 1982), 161–99, esp. 166–7. Menéndez Pidal's nationalism should not be construed pejoratively: see R. A. Fletcher, *The Quest for El Cid* (London, 1989), 204–5.

[7] The standard treatments are P. Boissonnade, *Du nouveau sur la Chanson de Roland* (Paris, 1923), esp. 3–68, and M. Defourneaux, *Les Français en Espagne aux XIe et XIIe siècles* (Paris, 1949).

military contribution in Spain of Frenchmen, particularly Aquitanians and Gascons, it becomes apparent that the peninsula's significance as a cradle of crusading ideology has hitherto been exaggerated.

It is worth noting immediately that the evidence for French involvement in the eleventh-century *Reconquista* is very fragmentary. No French chronicler whose work has survived addressed the subject at any length. The two best informed narrative traditions are preserved in the *Chronicle of Saint-Maixent* and a text which most probably originated at Fleury as an expansion of the writings of that abbey's celebrated historian Hugh of Ste-Marie.[8] These two narratives were quite well informed about important peninsular events such as major battles and the deaths of kings, but they usually provide nothing more than a series of very brief references. In their surviving form both post-date the First Crusade. Spanish chronicles also provide very little information about the French in eleventh-century Spain. The quantity and quality of the narrative evidence is not an entirely satisfactory guide—the later eleventh and early twelfth centuries were a low point in the production of historical works in south-western France, for example—but it is still noteworthy that most French and Spanish observers seem not to have been interested in, or able to obtain much news about, each others' lands. Such apparent mutual indifference is a salutary preliminary warning not to exaggerate the scope and ideological content of Franco-Spanish military contacts. It remains to see whether the warning is justified.

Frenchmen Fighting in Spain before the First Crusade

The obvious point at which to begin a discussion of the French contribution to the eleventh-century *Reconquista* is the campaign in 1064 in which a force of Normans, Aquitanians, Burgundians, and other Frenchmen helped an army of Catalans and, probably, Aragonese to besiege and capture the Muslim town of Barbastro, a strategically im-

---

[8] See J. Verdon, 'Une source de la reconquête chrétienne en Espagne: la Chronique de Saint-Maixent', in P. Gallais and Y.-J. Riou (edd.), *Mélanges offerts à René Crozet* (Poitiers, 1966), i. 273–82. The Fleury narrative is edited as 'Historiae Francicae Fragmentum', extr. *RHGF* 10. 210–12; 11. 160–2; 12. 1–8. See the comments of Waitz in his introd. to Hugh of Fleury, 'Opera historica', *MGH SS* 9. 343, and of A. Vidier, *L'Historiographie à Saint-Benoît-sur-Loire et les Miracles de saint Benoît* (Paris, 1965), 82–4. The Fleury Anonymous states that he himself visited the south-west in 1108: 'Historiae Francicae Fragmentum', 12. 7.

portant site to the north of the Ebro valley which was one of the principal defences of the *taifa* kingdom of Saragossa. It is worth noting in passing that this was not the first instance of French military activity in Spain. In the years around 1020, for example, the Norman lord Roger of Tosny had acted as a trouble-shooter in Catalonia during a period of political instability.[9] The Barbastro campaign, however, was the first co-ordinated expedition, involving Frenchmen from a wide area, for which evidence survives. As such it has attracted a good deal of attention from historians.[10] The events of 1064 pose many problems which have often been discussed in the past and do not need to be fully rehearsed here.[11] (In fact, it will be argued that, unless important new evidence comes to light, the campaign must remain an enigma.) Instead, it is worth concentrating upon those particular aspects of the Barbastro expedition which have invited comparisons with the crusades.

In this connection the most important piece of evidence is the surviving fragment of a letter which was sent by Pope Alexander II (1061–73) to the *clerus Vulturnensis*. Alexander's letter states:

With fatherly love we exhort those who are intending to journey to Spain [*qui in Ispaniam proficisci destinarunt*] that they take the greatest care to achieve those aims which, with divine admonishment, they have decided to accomplish. May each of them confess, according to the quality of his sins, to his bishop or spiritual father, and, lest the Devil be able to accuse them of impenitence, may a measure of penance [*modus penitentie*] be enjoined upon them. We, however, by the authority of the holy Apostles Peter and Paul, lift the penance from them [*penitentiam eis levamus*] and grant a remission of sins [*remissionem peccatorum facimus*]; our prayers go with them.[12]

The pope thus offered spiritual rewards to persons about to go to Spain. As James Brundage has observed, what was on offer was a remission of penance, not an indulgence as this came to be understood in later

---

[9] Adhemar, *Chronique* (ch. 55), 178–9; *Chronique de Saint-Pierre-le-Vif de Sens, dite de Clarius*, ed. and trans. R.-H. Bautier and M. Gilles (Sources d'histoire médiévale, 3; Paris, 1979), 112; William of Jumièges, *Gesta Normannorum Ducum*, ed. J. Marx (Rouen, 1914), 157 (interpolation by Orderic Vitalis); OV ii. 68. The story recounted in *Liber Miraculorum Sancte Fidis* (III. 1), ed. A. Bouillet (Paris, 1897), 128–30 may be connected with Roger's travels to Spain.

[10] Boissonnade, *Du nouveau*, 23–8; Defourneaux, *Français*, 131–5; Erdmann, *Origin*, 136–40.

[11] A. Ferreiro, 'The Siege of Barbastro 1064–1065: A Reassessment', *Journal of Medieval History*, 9 (1983), 129–44, reviews previous scholarship.

[12] BL Add. MS 8873, fo. 48^{r-v}; *Epistolae Pontificum Romanorum Ineditae*, ed. S. Loewenfeld (Leipzig, 1885), no. 82, p. 43.

theology.[13] When it is considered, however, that the First Crusade 'indulgence' was predicated on the idea that participation in the expedition was a satisfactory penance,[14] there would appear to be a strong prima-facie case for arguing that Pope Alexander's statement directly anticipated crusade ideas.

This conclusion presents a number of difficulties, however. For one thing, the letter is not dated. Ewald, the first scholar to study the surviving manuscript in detail, believed that the fragment, taken from a lost register of Alexander II, formed part of the collection belonging to 1063.[15] Ewald's argument required that some of the 1063 correspondence had been mislocated between the letters from the years 1065 and 1066–7, and contrary to the otherwise fairly consistent chronological arrangement of the collection.[16] Although this hypothesis is not implausible, it cannot be conclusively proven, and the uncertainty surrounding the letter's date must therefore serve as a warning not to read too much into Alexander's words. Moreover, even if 1063 is accepted as the date (so that the letter could have been connected with the preliminaries of the Barbastro campaign) further problems remain to be resolved.

First, the letter makes no mention of warfare, and there is nothing in it to preclude some other form of journey to Spain, perhaps a pilgrimage.[17] It has been argued that the phrase 'qui in Ispaniam proficisci destinarunt' excludes the possibility that Alexander was referring to a group pilgrimage to Compostela, because a more specific term such as 'ad sanctum Iacobum' would have been more appropriate.[18] This argument is not wholly convincing, for the terminology of contemporary charters reveals

---

[13] *Medieval Canon Law and the Crusader* (Madison, Wis., 1969), 24–5.

[14] Riley-Smith, *First Crusade*, 27–9.

[15] The collection of Alexander's letters is headed 'Ex Registro Alexandri Papae': BL Add. MS 8873, fos. 3ʳ, 38ᵛ. The MS mostly contains collections of letters by a number of popes up to 1088–9. It would seem to represent a rare survival of *excerpta*, extracts made with the intention of providing materials for compilations of canon law. The collection probably reached its surviving form *c.*1090, but the MS, in a single hand and neatly produced with few textual corrections, appears to be a copy of the original, probably produced in Italy in the early 12th C. See P. Fournier and G. Le Bras, *Histoire des collections canoniques en Occident depuis les Fausses Décrétales jusqu'au Décret de Gratien* (Paris, 1931–2), ii. 155–63, esp. 155–7, 162–3.

[16] P. Ewald, 'Die Papstbriefe der Brittischen Sammlung', *Neues Archiv der Gesellschaft für ältere deutsche Geschichtskunde*, 5 (1880), 344–9.

[17] The MS copy bears no marginalia or additions absent from Loewenfeld's edition. It is listed simply as 'de proficiscentibus in ispaniam': BL Add. MS 8873, fo. 4ʳ.

[18] A. Noth, *Heiliger Krieg und Heiliger Kampf in Islam und Christentum: Beiträge zur Vorgeschichte und Geschichte der Kreuzzüge* (Bonner Historische Forschungen, 28; Bonn, 1966), 119.

that 'Spain' was sometimes used as shorthand for the shrine of St James.[19] It must be admitted, however, that pilgrimage may not square entirely satisfactorily with the text as it survives. The fact that the letter was written at all suggests that the pope was responding to a scheme which struck him or his correspondents as unusual. The practice of going on pilgrimage to Compostela, however, was becoming well established by the time of Alexander's pontificate. It is not altogether easy, therefore, to imagine circumstances in which such a generous spiritual reward would have had to have been spelt out so carefully to intending pilgrims. Nevertheless, it is at least feasible to argue that a group pilgrimage to Spain, of the sort led to Jerusalem during Alexander's pontificate by Bishop Gunther of Bamberg,[20] but on a smaller scale, is as likely to lie behind the letter as a military expedition motivated by ideas of meritorious violence.

The possibility that pilgrimage was involved is lent some support by the identity of the *clerus Vulturnensis*. Pursuing the argument that Alexander's letter was a proto-crusade encyclical evidencing papal preaching of the Barbastro campaign, Erdmann maintained that *Vulturnensis* was the garbled name of a French diocese.[21] Castel Volturno in Campania has also been suggested, however, and this is much more plausible.[22] The same manuscript which contains the supposed indulgence also preserves two fragments of letters addressed by Alexander II to respectively the 'Vulturnensis aecclesiae clericis' and the 'clero Vulturnensi'.[23] The former, which occurs early in the collection within that part of it which belongs to 1062, concerns a man named Theoderic (a predominantly Italian name, it is worth noting) who had killed his son. It states that a seven-year penance had been imposed, and describes in detail how the penitent should behave during this time.[24] The second fragment immediately follows the 'indulgence' letter in the manuscript. It addresses the question of the appropriate penance for a man who had killed a priest in self-defence.[25]

Thus, in two letters almost certainly addressed to the same people as

---

[19] e.g. SEL 60[56] (1052 × 1060).
[20] See E. Joranson, 'The Great German Pilgrimage of 1064–1065', in L. J. Paetow (ed.), *The Crusades and Other Historical Essays Presented to Dana C. Munro by his Former Students* (New York, 1928), 3–43.
[21] *Origin*, 138 n. 71.
[22] Ferreiro, 'Siege of Barbastro', 132.
[23] BL Add. MS 8873, fos. 40ᵛ–41ʳ, 48ᵛ.
[24] Ed. in Alexander II, 'Epistolae et diplomata', *PL* 146, no. 117, col. 1405.
[25] Ibid., no. 116, cols. 1404–5.

the recipients of the 'indulgence' (and at least one of them was most probably written in the same period) Pope Alexander had tackled queries put to him about the application of penitential discipline to serious and rather unusual offences. Neither letter, as it survives, makes the merest hint of a war, Spain, or expiatory violence.[26] It is possible, therefore, to interpret the 'indulgence' as Alexander's response to yet another query put to him by the same confused correspondents: in this instance, about the expiatory value of pilgrimage to Compostela (which, from Campania, was potentially as arduous as a journey to, say, Jerusalem). Pursuing this argument, Alexander's statement can be construed as an authorative declaration, based on his Petrine powers, that the penances granted to unknown persons by their confessors should be of a sufficient severity, and no more ('modus penitentie'), to make them readily commutable to a long pilgrimage. Consequently, the phrase 'remissionem peccatorum facimus' can be interpreted, not as a declaration that journeying to Spain merited a full remission of all sins (which would have anticipated the later crusade indulgence), but as a guarantee that the spiritual benefits which could have been achieved through the performance of the particular penances originally imposed would also be secured by the alternative action (pilgrimage). As the 'indulgence' letter stands, therefore, it would seem more realistic to treat it as a straightforward exercise in pastoral advice rather than as a proto-crusade bull.

A further problem with the supposed indulgence stems from the fact that Pope Alexander was responding to an initiative of others, whose decision to go to Spain had already been made. The pope approved of their scheme, for he described it as divinely inspired; but, if warfare were involved, it is difficult to imagine circumstances in which the pope would have so generously favoured a military campaign which he had not himself instigated. Alexander did bestow his approval upon expeditions which had not been begun by papal initiative, such as the Norman invasion of England in 1066;[27] but he did not do so by promising anything resembling a commutation of penance. In fact the Normans, who may have fought under a papal banner at Hastings, were encouraged by a papal legate during Alexander's pontificate to do penance for the deaths and injuries they had caused.[28]

[26] The rubrics in the MS are 'de penitentia parricidii' and 'de interfectore presbiteri': BL Add. MS 8873, fos. 3ᵛ, 4ʳ.

[27] Erdmann, *Origin*, 154–5, 188–9.

[28] H. E. J. Cowdrey, 'Bishop Ermenfrid of Sion and the Penitential Ordinance following the Battle of Hastings', *JEH* 20 (1969), 233–5, 241–2.

It might be argued that there was an obvious and substantial difference between the spiritual value of fighting Spanish Muslims and Anglo-Saxon Christians (albeit perjurors and schismatics by reason of their association with respectively Harold Godwinsson and Archbishop Stigand of Canterbury). Certain other of Pope Alexander's pronouncements reveal that he did direct his thoughts to the particular problems of violence against non-Christians. In one letter, to Archbishop Wifred of Narbonne, Alexander observed that all laws, those of the church and of secular authority, forbade bloodshed unless it served as a punishment for crime or, as might be the case with Saracens, there was a need to counter hostile aggression.[29] In a letter to the bishops of Spain Alexander congratulated his correspondents for protecting local Jews from persecution at the hands of those 'who were proceeding into Spain against the Saracens': to kill Jews, he declared, was wrong because they had been saved by God's mercy to live in eternal penitence and subjugation, whereas the Saracens were able actively to persecute Christians and expel them from their lands.[30] In another letter the pope congratulated Viscount Berenguer of Narbonne for protecting Jews who lived under his authority.[31]

Two themes of these letters invite comment. First, Alexander's concern for the safety of Jews at the hands of non-Spaniards intending to fight Muslims may seem to tie in with the well-documented eagerness of some first crusaders to persecute those Jews whom they encountered on the march.[32] Consequently, it might be argued that the ideas of those 'proceeding into Spain'—perhaps on the Barbastro campaign, but it is impossible to be certain—were fuelled by the same sort of religious zeal, leading to an unwillingness to distinguish between Jews and Muslims as enemies of the faith, which informed the anti-Semitism of many first crusaders. It must be emphasized, however, that the origins of the anti-Jewish sentiments revealed by Alexander's letters are obscure. Foreigners coming across Spanish Jews were at least as likely to abuse them in order to rob them and because of inveterate prejudice as out of any misdirected anti-Muslim zealotry. Alexander's letters make no suggestion that the motives of those who molested the Jews were laudable ideals turned sour.

---

[29] *Epistolae Pontificum*, no. 83, p. 43.
[30] Alexander II, 'Epistolae et diplomata', no. 101, cols. 1386–7.
[31] Ibid., no. 102, col. 1387. The letters to Berenguer and Wifred appear consecutively in BL Add. MS 8873, fo. 48ᵛ. They would therefore appear to refer to one incidence or period of anti-Jewish feeling.
[32] Riley-Smith, *First Crusade*, 34–5, 50–7.

Rather, the pope condemned them point-blank as greed and stupidity.[33] The attractive link between this episode and the First Crusade must therefore be treated, in all prudence, as a chimera.

The second significant feature of these letters lies in the fact that, by distinguishing between the Christian postures towards Jews and Muslims in order to protect the former group, Alexander recognized that violence against Muslims might be licit in certain situations. It is noteworthy that Alexander made this point by clearly appealing to the language of just war. He told the bishops of Spain that the 'cause' of the Jews was unquestionably different from that of the Saracens; when the Muslims actively persecuted Christians they might be repelled 'justly'.[34] The pope's observations, however, need to be placed in context, for, as he told Berenguer of Narbonne, 'God is not pleased by the spilling of blood, nor does he rejoice in the perdition of the evil [*mali*].'[35] The *mali*, so termed without qualification, must have been meant to include those Muslims who did not immediately merit resistance by posing a threat to Christians. Alexander's statements were not, therefore, a declaration that wars against Muslims were *ipso facto* just;[36] and his scrupulousness about the cause of any given war against Muslims seems incompatible with the idea that the 'indulgence' to the *clerus Vulturnensis*, which if connected to the Barbastro campaign must be assigned to early in his pontificate, reveals papal sponsorship of a divinely inspired holy war which attracted to it spiritual rewards. It is also worth noting that none of these three letters, though they seem connected, can be dated more precisely than to the first six years of Alexander's pontificate (the period covered by the collection of correspondence in the British Library). This confirms the conclusion that to link them to the behaviour and motivations of the Barbastro campaigners is a very hazardous exercise.

Narrative sources provide only a few clues to the motivations of the Frenchmen at Barbastro. The testimony of the Italian chronicler Amatus of Montecassino is worthy of attention, not least because he wrote before the First Crusade. According to Amatus, the capture of Barbastro was the work of God granting victory to his people, a statement which might

[33] 'Epistolae et diplomata', no. 101, col. 1386.

[34] Ibid., col. 1387. Cf. Pope Gregory VII's later comments to King Alfonso VI of Leon-Castile concerning Christians' relations with peninsular Jews: *Das Register* (IX. 2), ed. E. Caspar, 2nd edn. (MGH Epistolae Selectae, 4–5; Berlin, 1955), ii. 571.

[35] 'Epistolae et diplomata', no. 102, col. 1387.

[36] See, *contra*, P. David, *Études historiques sur la Galice et le Portugal du VIe au XIIe siècle* (Collection portugaise, 7; Lisbon, 1947), 372.

seem to suggest that the campaign was conceived as a meritorious war between faiths.[37] It is worth noting, however, that Amatus appears not to have been well informed: the relative importance at Barbastro of his hero Robert Crispin, whom he made the leader of the Christian forces, has been questioned by recent research;[38] and his account of the campaign comes at the end of an enthusiastic panegyric of Normandy and Norman military achievements throughout Europe, which raises the possibility that he exaggerated the significance of the capture of Barbastro by working it into his triumphalist scheme.[39] Furthermore, Amatus's belief that the leaders of the expedition were inspired to extend the Christian faith by driving back 'the hateful folly' of the Muslims suggests that he deliberately projected into Spain his experience of the recent Norman expansion into Sicily.[40] The only narrative evidence that similar views were voiced in France is the fairly neutral comment of the *Chronicle of Saint-Maixent* (which in its surviving form post-dates the First Crusade) that Duke Guy Geoffrey of Aquitaine, who was one of the leaders of the army, took Barbastro 'for the Christian name'.[41]

One statement by Amatus which is corroborated by French narratives is that the capture of Barbastro provided great opportunities for personal enrichment. Like Amatus, the Fleury Anonymous believed that Barbastro was a notably rich city which yielded large amounts of movable wealth.[42] Implying that the plunder available was considerable, both the Anonymous and the St-Maixent chronicler also recorded that the massacre of its inhabitants was great, a view confirmed by non-Christian sources.[43] It is superficially tempting to link this bloodshed with the great massacre of Jews and Muslims which took place when Jerusalem fell to the first

---

[37] *Storia de' Normanni* (I. 5), ed. V. de Bartholomaeis (Rome, 1935), 14.

[38] Ibid. (I. 5–6), 13–15; Ferreiro, 'Siege of Barbastro', 137–8.

[39] *Storia de' Normanni* (I. 1–4), 9–13. See H. E. J. Cowdrey, *The Age of Abbot Desiderius: Montecassino, the Papacy, and the Normans in the Eleventh and Early Twelfth Centuries* (Oxford, 1983), pp. xx, 25–6.

[40] *Storia de' Normanni* (I. 5), 13.

[41] *La Chronique de Saint-Maixent 751–1140*, ed. and trans. J. Verdon (Les classiques de l'histoire de France au Moyen Âge, 33; Paris, 1979), 136. Cf. *Cartulaire de l'abbaye royale de Notre-Dame de Saintes*, ed. T. Grasilier (Cartulaires inédits de la Saintonge, 2; Niort, 1871), 12, 229; *Cartulaire de l'abbaye de Saint-Cyprien de Poitiers*, ed. L. Rédet (Archives historiques du Poitou, 3; Poitiers, 1874), 569.

[42] Amatus of Montecassino, *Storia de' Normanni* (I. 6), 14; 'Historiae Francicae Fragmentum', 11. 162.

[43] 'Historiae Francicae Fragmentum', 11. 162; *Chronique de Saint-Maixent*, 136; R. Dozy, *Recherches sur l'histoire et la littérature de l'Espagne pendant le Moyen Âge*, 3rd edn. (Paris, 1881), ii. 341–2, 344; J. Bosch Vilá, 'Al-Bakri: dos fragmentos sobre Barbastro en el "Bayan Al-Mugrib"', *EEMCA* 3 (1947–8), 255.

crusaders in July 1099; but it is more realistic to see the killings in 1064 as the result of panic and competition for human and material booty rather than as an expression of religious ardour and settlement policy, the operative factors in 1099. The only other motive for the Barbastro campaigners suggested by the French sources lies in Hugh of Fleury's cryptic observation that the French nobles left for Spain since King Philip I had not yet reached adulthood, which is more likely to mean that they were exploiting Philip's minority than that an invitation to lead an expedition had been extended to the young king or to his father Henry I (d. 1060).[44] Hugh is the only French chronicler to link the campaign to political conditions in France; and the Aquitanians, Burgundians, and Normans who formed the bulk of the French forces came from areas where royal authority was largely ineffective. The comment of the *Chronicle of St-Pierre-le-Vif* that another expedition to the peninsula, in 1087, was begun on Philip's order points to the potential importance of Capetian influence over French involvement in Spain; but Hugh of Fleury's meaning concerning the events of 1064 is not clarified by this.[45] For our purposes, what is significant is that, when trying to place the Barbastro campaign in its broad context, Hugh made no mention of ecclesiastical involvement nor of religious ideas among the participants.

It is important not to underestimate the tactical importance of the French at Barbastro. Although it is impossible to measure their numbers relative to those of the local Christian forces, it is reasonable to suppose that they made a significant contribution to the capture of the city, and that their departure after the siege undermined the ability of local troops under Count Ermengol III of Urgel (who died in the subsequent fighting) to prevent its recapture by the Muslims in 1065.[46] Non-Spanish narratives agree, moreover, that the French contingent was large (though it is important to allow for exaggeration). On the other hand, the notion that the Barbastro campaign was a proto-crusade inspired by nascent crusading ideals is not based on secure evidence. It is worth repeating that there can be no conclusive answer to the problems associated with the events of 1064. Rather, it must be emphasized that discussion of the expedition has been coloured by two factors: the hindsight of the First Crusade and the central importance to it, as motivating forces, of spiritual rewards and religious fervour; and the suspicion that, because

---

[44] 'Liber qui modernorum regum Francorum continet actus' (ch. 11), *MGH SS* 9. 389.

[45] *Chronique de Saint-Pierre-le-Vif*, 136.

[46] See Antonio Ubieto Arteta, *Historia de Aragón: la formación territorial* (Saragossa, 1981), 63–6.

superficial similarities exist between the crusade movement and the eleventh-century *Reconquista*, Spain ought to have been a breeding-ground for crusade ideals. In fact the operation of hindsight seems also to have influenced medieval commentators writing after 1100. By the twelfth century, north of the Pyrenees, the Barbastro campaign was accorded a significance out of all proportion to its strategic achievements. The Fleury Anonymous, for example, numbered subsequent French expeditions to Spain from it.[47] It also formed the very loose factual basis for a late twelfth-century *chanson*, part of the William of Orange cycle, which was most probably based on earlier, and distorted, oral traditions.[48] *Mutatis mutandis*, medieval observers fell into an interpretative trap similar to the one which has caught out some of their modern counterparts.

The most obvious way to make sense of the Barbastro campaign is to see it in the context of other French expeditions to Spain in the later eleventh century. If we place the maximal interpretation upon the evidence in order to argue that the 1064 expedition was a proto-crusade for which the church offered spiritual rewards analogous to a crusade indulgence and in which religious fervour was prominent, it becomes very difficult to reconcile with the scope and purposes of French military activity in the peninsula over the following thirty years. This activity was infrequent, predominantly motivated by a desire for material gain, and effectively controlled by peninsular rulers far more than by those external forces, such as the papacy, which might conceivably have introduced a consistent religious element into French participation in the *Reconquista*.

Two of the earliest letters of Gregory VII's pontificate (1073–85) refer to an expedition to Spain under Count Ebles of Roucy.[49] This was planned as a war of conquest, for in one of the letters Gregory informed unnamed nobles about to leave for Spain that a formal agreement had been reached between Ebles and the papacy to the effect that all land taken from the Muslims was to be held as a fief of the Holy See.[50] Plans were obviously in hand for expeditions independent of that under Ebles;

---

[47] 'Historiae Francicae Fragmentum', 12. 1, 2.

[48] *Le Siège de Barbastre: chanson de geste du XIIe siècle*, ed. J. L. Perrier (Les classiques français du Moyen Âge, 54; Paris, 1926); E. Muratori, 'L'assedio di Barbastro, prima crociata di Spagna, e la canzone di gesta omonima: occasioni della storia e scarto retorico', *Francofonia*, 8 (1985), 23–35. See also Geffrei Gaimar, *L'Estoire des Engleis*, vv. 6077–8, ed. A. Bell (Anglo-Norman Texts, 14–16; Oxford, 1960), 193; T. A. Archer, 'Giffard of Barbastre', *EHR* 18 (1903), 303–5.

[49] Gregory VII, *Register* (I. 6–7), i. 8–12. Both are dated 30 Apr. 1073, eight days after Gregory's election.

[50] Ibid. (I. 7) 11–12.

and Gregory hinted at what he thought was their motivation when he warned that retention of conquered land was an injury to St Peter, and expressed the fear to Alexander II's former legates that freelance nobles would act in ignorance of the pact with the count.[51] Gregory's principal purpose was to reform the errors of Christians living in Muslim territory (that is, to suppress the native Mozarabic liturgy in favour of the Roman rite and to bring the peninsular church more directly under papal control) through the efforts of clergy directed by Hugh Candidus and Abbot Hugh of Cluny. This aim and the limited spiritual advantages offered— the promise that those furthering papal interests would be protected by St Peter and earn 'the deserved rewards of fidelity'—may have restricted interest in the expedition. Gregory probably miscalculated by failing to distinguish between desire for land and desire for movable wealth as possible motives for French nobles and knights.[52] He most probably miscalculated, too, by neglecting to consult adequately the wishes of the peninsular rulers.[53] There is no firm evidence that Ebles's expedition ever reached Spain.[54] It seems probable that by April 1073 the plans for the campaign were already in abeyance; and the only French or Spanish writer to mention that Ebles raised an army was Suger, writing about seventy years later, who used this 'fact' in a bitter attack on Ebles's violent ambition.[55]

Even if Ebles did reach Spain, the hostility of the peninsular kings towards unsolicited or strategically inconvenient French aid may explain the silence of the Spanish sources. On occasion the antipathy might be considerable. For example, the only Spanish reference to an attempt by William IX of Aquitaine to enter Spain, probably in 1088, is the dating clause of an Aragonese charter: 'In this year the count of Poitou arrived in Spain, and the glorious King Sancho [Sancho Ramírez of Aragon-Navarre] made him return to his homeland.'[56] In contrast, an expedition

---

[51] Ibid. (I. 6, 7) 10, 12.

[52] H. E. J. Cowdrey, *The Cluniacs and the Gregorian Reform* (Oxford, 1970), 221–2.

[53] David, *Études historiques*, 380.

[54] Antonio Ubieto Arteta, *Historia de Aragón*, 81–2 suggests that a testament recording the death of a Bernardo Ramón in September 1073 returning from Rome 'cum militibus aliis Hispaniam ingressus a sarracenis . . . ibi interfectus' refers to this expedition. The date is late, and the text's meaning is ambiguous.

[55] *Vie de Louis VI le Gros* (ch. 5), ed. and trans. H. Waquet (Les classiques de l'histoire de France au Moyen Âge, 11; Paris, 1929), 24–6.

[56] *Documentos correspondientes al reinado de Sancho Ramírez*, ed. J. Salarrullana de Dios and E. Ibarra y Rodríguez (Colección de documentos para el estudio de la historia de Aragón, 3 and 9; Saragossa, 1907–13), ii. 49 at p. 134. The charter is dated 1080, in which event the Spanish venture would have been the work of Duke Guy Geoffrey (William VIII). Ubieto

led in *c.* 1076 by Duke Hugh of Burgundy seems to have been more successful because it was welcome to Sancho Ramírez, and the pattern set at Barbastro of the looting of movables, taking of captives, and prompt return home to France was allowed to be repeated.[57]

At first glance a French expedition to Spain in 1087, requested by King Alfonso VI of Leon-Castile (1065–1109) after his heavy defeat by the Almoravids at Sagrajas (Zalaca) in October 1086, seems significant for the prehistory of the First Crusade.[58] Failing to engage the Muslim invaders, who had withdrawn to north Africa, the French campaigned inconclusively, possibly linking up with Aragonese forces near Tudela, and soon returned home, having apparently achieved nothing of permanence.[59] The potential significance of this expedition lies not in its achievements, however, nor indeed in the supposed introduction of religious zealotry into Spain by the Almoravids, but in the identity of its leaders. A number of future first crusaders were involved: Viscount William the Carpenter of Melun; Hugh VI of Lusignan; Duke Odo of Burgundy; and, probably, Raymond of St-Gilles, count of Toulouse.[60]

Two charters preserved in the records of the Poitevin abbey of Nouaillé refer to this expedition. One states that Hugh of Lusignan made grants to Nouaillé and Notre-Dame de Lusignan when 'about to go to Spain against the Saracens'.[61] The charter bears some resemblance to crusading documents. The grant was made at a large and formal gathering at Lusignan attended by the young Duke William IX of Aquitaine and

---

Arteta, *Historia de Aragón*, 97, however, argues for 1088. The statement in *Historia Pontificum et Comitum Engolismensium* (ch. 33), ed. J. Boussard (Paris, 1957), 27 that Bishop Adhemar of Angoulême (1076–1101) 'cum duce Aquitanorum in Hispaniam contra Saracenos militavit' may refer to this episode, but it is equally possible that, as a canon of Lesterps, Adhemar was present at Barbastro in 1064.

[57] 'Historiae Francicae Fragmentum', 12. 1. This was the Anonymous's *secunda expeditio* after Barbastro. Defourneaux, *Français*, 142 interprets this incorrectly to mean that Hugh campaigned in Spain twice.

[58] According to the 'Chronicon Lusitanum', *ES* 14. 418, Alfonso fought at Sagrajas alongside a large force of Frenchmen. This is singularly improbable: see B. F. Reilly, *The Kingdom of León-Castilla under King Alfonso VI 1065–1109* (Princeton, NJ, 1988), 180–1, 191.

[59] *Chronique de Saint-Maixent*, 148; 'Historiae Francicae Fragmentum', 12. 2; *Chronique de Saint-Pierre-le-Vif*, 136.

[60] William of Melun is probably the 'Guillelmus . . . Normannus' identified as the expedition's leader by the *Chronique de Saint-Maixent*, 148. For Raymond of St-Gilles, see J. H. and L. L. Hill, *Raymond IV Count of Toulouse* (Syracuse, NY, 1962), 19–20 and n. 48. For Odo, see 'Chronicon Trenorciense', extr. *RHGF* 11. 112–13. For Hugh, see the following para.

[61] *Chartes de l'abbaye de Nouaillé de 678 à 1200*, ed. P. de Monsabert (Archives historiques du Poitou, 49; Poitiers, 1936), 157.

Bishop Peter of Poitiers; Hugh secured the consent of his wife and sons to his alienations; he made provision for two vassals in the event of his death; and he asked that Bishop Peter punish with excommunication anyone who attacked the transferred property (but not, it is worth noting, all his lands). The second charter deals with a much humbler layman, Peter Abrutit, but is more informative.[62] One of its clauses suggests that the Almoravid invasion of Spain in 1086 was considered, at least in Poitou, as an attack on Christians generally: 'In which year the Saracens invaded Spain against us.' Peter stated, moreover, that he was setting out 'with other Christians'. Doubtless in order to fund his journey, and anticipating later crusade practice, Peter granted allods to Nouaillé in return for 150*s*., and secured the consent of his mother and two brothers. A third document, from the Limousin, also suggests that the Almoravid menace made some impression north of the Pyrenees. It records that in May 1087 Catard Vicarius went to Spain, possibly to fight, and raised a loan by pledge of 200*s*. from the canons of St-Étienne, Limoges. The canons obviously respected Catard's intentions, for they had already advanced 1,190*s*. on the pledged property (though they also refused to countenance further loans).[63] These charters reveal that news of the Almoravid invasion of Spain struck a chord of Christian fellow-feeling in France, attracted the attention of important laymen and the church, and moved some religious communities to support Frenchmen volunteering to fight. What they do not reveal, however, is that the 1087 expedition was preached by the church or even encouraged as a meritorious war against the enemies of the faith. Furthermore, it is noteworthy that, compared to the number of surviving charters associated with the First Crusade, the numbers involved here are tiny.[64]

The expedition of 1087 most probably forms the background to the

---

[62] Ibid. 158.

[63] SEL 114[95]; cf. ibid. 88[81]. The only suggestion in the charter that Catard intended to fight is the statement in the dating clause 'Christianis pergentibus contra Sarracenos', but this must have been linked to Catard's plans. By May, however, the expedition had already broken up: Reilly, *Alfonso VI*, 191.

[64] Boissonnade (*Du nouveau*, 34 n. 3), following the editor's suggestion, supposed that a reference in a document from the Poitevin abbey of Talmond to the deaths of members of a local family and certain companions 'martyrio suffocatorum' recalled fatalities on the 1087 expedition. Because the document refers to Abbot Evrard (*c*.1079–*c*.1091) it is probable that the deaths occured in the 1080s and certainly before the First Crusade, but no mention is made of Spain. Further, the relevant document is a compendium of entries extending into the 12th C. It would therefore be rash to speculate about the circumstances behind this most curious document: 'Cartulaire de l'abbaye de Talmond', ed. L. de la Boutetière, *Mémoires de la Société des Antiquaires de l'Ouest*, 36 (1873), 16.

only specific reference to the Spanish theatre found in the early narrative accounts of the conduct of the First Crusade (as opposed to words put into Urban II's mouth at Clermont): the widely reported story of Bohemond of Taranto's reproach of William the Carpenter of Melun, who had tried to desert the Christian forces at Antioch in 1098. The wording of the *Gesta Francorum*'s version of Bohemond's harangue may suggest that during the First Crusade (this part of the *Gesta* was very probably written towards the end of 1098) the 1087 expedition came to be remembered in quasi-crusading terms: 'you wished to betray these knights and this army of Christ, just as you betrayed the others in Spain.'[65] Even if this sort of assimilation had taken place, however, it is misleading evidence for the actual motivations and aims of the 1087 campaigners. Guibert of Nogent, who was the crusade chronicler best informed about William of Melun and able to add details absent from the *Gesta*, described the army of 1087 as a 'Frankish expedition' fighting in self-defence against pagans from Africa; but this did not prompt him to make a direct comparison with the First Crusade beyond stating that William had betrayed the 'Lord's people' in Spain.[66] The thrust of these two accounts was William's violence towards noble codes of conduct: where he had first showed himself cowardly was largely an incidental detail. Significantly, Baldric of Bourgueil, Robert the Monk, Ralph of Caen, and Albert of Aachen all used the story of William's cowardice in their accounts of the crusade but omitted any reference to Spain.[67]

In terms of its date, the identity of its leaders, and its impact on a few French charters, the 1087 expedition is the phase in the eleventh-century *Reconquista* which might appear to have the most resonances in the First Crusade. It is significant, therefore, that even in this instance the parallels between the crusade and the peninsular wars can only be drawn very approximately and were, in fact, missed by a number of closely contemporary observers. Like the Barbastro campaign, the events of 1087

---

[65] *Gesta Francorum et aliorum Hierosolimitanorum*, ed. and trans. R. Hill (London, 1962), 33–4. Cf. the similar formulation in Peter Tudebode, *Historia de Hierosolymitano Itinere*, ed. J. H. and L. L. Hill (Documents relatifs à l'histoire des croisades, 12; Paris, 1977), 68–9.

[66] 'Gesta Dei per Francos' (IV. 7, 9), 174, 175.

[67] Baldric of Bourgueil, 'Historia Jerosolimitana' (II. 12), *RHC Occ.* 4. 43–4 (but see ibid. 45 n. 1 recording a variant reading); Robert the Monk, 'Historia Iherosolimitana' (IV. 12), *RHC Occ.* 3. 781–2; Ralph of Caen, 'Gesta Tancredi in expeditione Hierosolymitana' (ch. 60), *RHC Occ.* 3. 650–1; Albert of Aachen, 'Historia Hierosolymitana' (IV. 37), *RHC Occ.* 4. 414–15. Fulcher of Chartres and Raymond of Aguilers did not mention the incident.

cannot be treated as evidence of French arms-bearing society preparing itself for the crusade message.

Marriage Alliances and War in Spain

The family networks which lay behind most French participation in the eleventh-century *Reconquista* also point to the conclusion that Spain was not the breeding-ground of ideological impulses, built around the attraction of holy war, which were then absorbed into the motivations of French first crusaders. The frequency of French campaigns into Spain, and the identity of their leaders, were closely connected with marriage ties between peninsular rulers and noble kindreds from north of the Pyrenees.[68] It is not clear in every case whether a marriage alliance preceded, and therefore provided the context for, a French noble's presence in Spain, or whether it was a reward for military aid which had already been given. Problems of dating are increased by the possibility of long intervals between betrothal and nuptials, so that references to French wives in Spanish charters can usually provide only one approximate terminal date for the creation of an alliance. For example, Ebles of Roucy's sister Felicia married King Sancho Ramírez of Aragon, a match which was to prove extremely important for Franco-Spanish contacts in subsequent generations. The meagre charter evidence for this marriage is confused, however, by the likelihood that Sancho Ramírez had earlier married a daughter of Count Ermengol III of Urgel called Felicia or Isabella.[69] A Queen Felicia appears in Aragonese documents from 1072 and 1080.[70] There is no evidence that Ebles or a close kinsman fought at Barbastro.[71] It has been suggested that the marriage was negotiated by Pope Alexander II, with a view to future military co-operation, when Sancho Ramírez visited Rome in 1068; but the future King Pedro I, who was not Felicia of Roucy's son, was not born until *c.* 1069, and 1071 or 1072 is the more

---

[68] Boissonnade, *Du nouveau*, 13–16; Defourneaux, *Francais*, 140–1.

[69] Herman of Laon, 'De miraculis B. Mariae Laudunensis libris tribus' (I. 2), extr. *RHGF* 12. 267; *Colección diplomática de Pedro I de Aragón y Navarra*, ed. Antonio Ubieto Arteta (Escuela de Estudios Medievales, Textos, 19; Saragossa, 1951), pp. 25–7, 28; M. González Miranda, 'La condesa doña Sancha y el Monasterio de Santa Cruz de la Serós', *EEMCA* 6 (1956), 186–7.

[70] *Documentos de Sancho Ramírez*, i. 6, 51; ii. 50. The document of 1094 is suspect. See 'Necrologio del monasterio de San Victorian', *ES* 48. 278.

[71] The suggestion of Antonio Ubieto Arteta in *Colección de Pedro I*, pp. 30–1.

likely date for the creation of the Roucy alliance.[72] Ebles's ties by marriage to Robert Guiscard and his resulting contact with Rome would help to explain Sancho Ramírez's choice of bride from a family which was descended from a daughter of King Robert the Pious of France but which was of the second rank and based in the distant north. The precise dynamics of the Roucys' interest in Spain, however, are obscured by the problems of dating.

Other contacts are also problematical. The marriage of Sancho Ramírez's heir Pedro to Agnes, the daughter of Duke Guy Geoffrey of Aquitaine and Aldeardis of Burgundy, was celebrated in January 1086.[73] At that time Agnes was about 13 years old, Pedro about 16, which means that the betrothal could have been planned years before.[74] Agnes was certainly still queen in 1094, and probably so in May 1097; Pedro remarried in August 1097.[75] In this instance it is quite possible that the marriage alliance lay behind war contacts. Hugh VI of Lusignan, who campaigned in Spain in 1087, was closely associated with the young Duke William IX of Aquitaine's first years of rule (William succeeded his father in 1086), and, as was noted earlier, William himself may have attempted some role in peninsular affairs in 1088.[76]

There are further problems with the synchronisms between French military aid and marriage ties with the kings of Leon and Castile. None of King Alfonso VI's five wives were Spanish, and the first two in particular created alliances with great military potential.[77] Alfonso married Guy Geoffrey of Aquitaine's daughter Agnes (an elder half-sister of Pedro of Aragon's wife) in about 1073.[78] She last appears in Spain in

---

[72] David, *Études historiques*, 376; Antonio Ubieto Arteta, *Historia de Aragón*, 81.

[73] *Colección de Pedro I*, p. 31 and no. 1.

[74] Agnes's elder brother, the future William IX, was born in October 1071: *Chronique de Saint-Maixent*, 140.

[75] *Colección de Pedro I*, 14, 34, 35. Agnes may then have married Count Helias of Maine: OV v. 306 and n. 5.

[76] A. Richard, *Histoire des comtes de Poitou 778–1204* (Paris, 1903), i. 387–8, 389–90, 391–2.

[77] Pelayo of Oviedo, *Crónica*, ed. B. Sánchez Alonso (Textos latinos de la edad media española, 3; Madrid, 1924), 86; *Crónicas Anónimas de Sahagún* (ch. 9), ed. Antonio Ubieto Arteta (Textos Medievales, 75; Saragossa, 1987), 16–17; *Primera Crónica General de España* (ch. 847), ed. R. Menéndez Pidal, 3rd edn. (Fuentes Cronísticas de la Historia de España, 1; Madrid, 1977), ii. 520–1.

[78] *Chronique de Saint-Maixent*, 132, 138–40; Martin of Montierneuf, 'De constructione Monasterii novi Pictavis' (ch. 9), ed. F. Villard, *Recueil des documents relatifs à l'abbaye de Montierneuf de Poitiers (1076–1319)* (Archives historiques du Poitou, 59; Poitiers, 1973), p. 425. For the date, see David, *Études historiques*, 387, corrected by Reilly, *Alfonso VI*, 79–80.

May 1077 and seems to have been repudiated shortly thereafter, having probably borne no children.[79] By May 1079 Alfonso had married Constance, the aunt of Duke Hugh of Burgundy (who had campaigned in Spain within the previous three years)[80] and the widow of Count Hugh II of Chalon, who had died in Spain in 1078, probably *en route* to Compostela;[81] Hugh of Chalon's father, Count Theobald, had also died in Spain, possibly as he returned home from the Barbastro campaign.[82] Alfonso's remarriage did not entail a complete rupture with the Poitevins, for in about 1068 Guy Geoffrey (who was related through his mother Agnes to the counts of Burgundy) had married Aldeardis, the daughter of Duke Robert I of Burgundy and thus Constance's sister.[83] Constance's marriage was the basis of a series of important alliances in the 1080s and 1090s; her nephew Henry and cousin Raymond of Amous both married daughters of Alfonso VI.[84] She died in 1093.[85]

The identity of the leadership of the 1087 expedition provides particularly clear evidence of the importance which marriage ties assumed in stimulating French interest in the peninsula. Duke Odo of Burgundy was Constance's nephew.[86] Raymond of St-Gilles and Hugh of Lusignan were half-brothers whose mother (the much-travelled Almodis of La Marche) had also married Count Ramon Berenguer I of Barcelona.[87] It is likely that shortly before 1087 Raymond of St-Gilles's niece Philippa, the daughter of Count William IV of Toulouse (and future wife of William IX of Aquitaine), had married Sancho Ramírez of Aragon-Navarre.[88] Later alliances cemented the kindred networks linking Spain to France. Raymond of St-Gilles married Elvira, natural daughter of Alfonso VI, in

---

[79] C. iv. 3508.

[80] 'Chronicon Trenorciense', 112; B. F. Reilly, *The Kingdom of León-Castilla under Queen Urraca 1109–1126* (Princeton, NJ, 1982), 11.

[81] *Cartulaire du prieuré de Paray-le-Monial*, ed. C. U. J. Chevalier (Collection de cartulaires dauphinois, 8²; Montbéliard, 1891), 11; David, *Études historiques*, 384. C. iv. 3530 (1078) was possibly connected with Hugh's Spanish journey.

[82] *Cartulaire du prieuré de Paray-le-Monial*, 10.

[83] *Chronique de Saint-Maixent*, 110, 138; C. B. Bouchard, *Sword, Miter, and Cloister: Nobility and the Church in Burgundy 980–1198* (Ithaca, NY, 1987), 258.

[84] Reilly, *Urraca*, 13–14.

[85] Ibid. 11.

[86] David, *Études historiques*, 385 n. 2; Bouchard, *Sword, Miter*, 256.

[87] William of Malmesbury, *De Gestis Regum* (chs. 382, 388), ii. 447, 455–6; *Chronique de Saint-Maixent*, 132; *Gesta Comitum Barcinonensium* (ch. 4), ed. L. Barrau Dihigo and J. Massó Torrents (Cròniques Catalanes, 2; Barcelona, 1925), 7.

[88] *Cartulaire de l'abbaye de Saint-Victor de Marseille*, ed. M. Guérard (Collection des cartulaires de France, 8–9; Paris, 1857), ii. 686; Hill, *Raymond IV*, 16.

or shortly before 1094, and in June 1095 Raymond's son Bertrand married Odo of Burgundy's daughter.[89]

Marriage alliances cannot have been concluded solely in connection with the requirements of military planning. The fact, for example, that Constance of Burgundy was the niece of Abbot Hugh of Cluny suggests that dynastic ties might be formed against a complicated background of ecclesiastical and diplomatic contacts. The range of the marriage ties is relevant to the specific problem of co-operation in war, however, in that it suggests that the extent of French military involvement in Spain before the First Crusade was placed within strict limits. The fact that most of the French nobles known to have fought in Spain were already, or soon became, closely related both to each other and to peninsular rulers reveals that, by the final quarter of the eleventh century, active interest in the *Reconquista* was largely confined, north of the Pyrenees, to a small, easily identified group. Moreover, the fact that the Spanish end of the marriage market was dominated by kings and their immediate kindred suggests that the alliances were seldom the result of random encounters between French visitors and their hosts, but rather, from the early 1070s and possibly because of the experience of Barbastro, the subject of close royal control. As far as the peninsular rulers were concerned, foreign marriages created opportunities to summon and, no less importantly, to limit military aid from France. Their policy made sense if they wished simply to create the potential for occasional, moderately sized expeditions disciplined by the control of relatives and compatible with the pursuit of their own peninsular policies. In such circumstances it would have been remarkable if Spain could have become a suitable arena for significant numbers of Frenchmen to express any supposed feelings of nascent crusading fervour.

Gascony: A Special Case?

Awareness of Spanish conditions is unlikely to have been uniform throughout France. Geographical proximity made it likely that Gascons would be particularly closely involved in peninsular affairs in the eleventh century (although it should also be borne in mind that contemporaries had few illusions about the Pyrenees, or 'Alps' as they were sometimes

---

[89] Pelayo of Oviedo, *Crónica*, 86; *HGL* 3, pp. 470–1, 474; ibid. 5, no. 311.

known, as a formidable barrier).[90] Some French historians have worked from the assumption that, from a French perspective, Gascony was a special case, a sort of ultramontane extension of Spain sharing much of its identity and many of its aims.[91] On close examination, however, the evidence for contacts between the Gascon and peninsular aristocracies does not in fact provide such a coherent picture.

When reliable evidence becomes available, in the latter half of the tenth century, there are signs of ties between the Gascon comital (ducal) line and the Pamplonese (Navarrese) kings. Before succeeding his brother as count of Gascony, William Sancho was married to Urraca of Navarre, daughter of King Sancho Abarca.[92] It is a possible inference from a confused comment by Ralph Glaber that William Sancho assisted King Sancho in wars against the forces of the Umayyad caliphate under Al-Mansur (d. 1002).[93] Similarly, there is evidence for the occasional attendance at the Pamplonese court of Sancho William of Gascony before he succeeded his childless brother, Bernard William, as duke in 1010.[94] After an apparent hiatus of about thirty years (the surviving evidence is meagre) contacts between Sancho William and Navarre seem to have become close once more between the mid-1020s and the duke's death in 1032. Sancho William, moreover, may have accompanied King Sancho el Mayor (1004–35) on the latter's only recorded journey north of the Pyrenees, to attend the ceremonial invention of the head of St John at Angély in the Saintonge in c. 1014.[95] Contacts of this sort would seem to lie behind Adhemar of Chabannes's belief that King Sancho fought one particularly successful campaign, probably in the mid-1020s, with Gascon help.[96]

[90] See *Colección diplomática de la catedral de Huesca*, ed. A. Duran Gudiol (Fuentes para la Historia del Pirineo, 5–6; Saragossa, 1965–9), i. 17, 39, 50; *Documentación medieval de Leire (siglos IX a XII)*, ed. A. J. Martín Duque (Pamplona, 1983), 114; *Historia Compostellana* (II. 20), ed. E. Falque Rey (Corpus Christianorum, Continuatio Mediaevalis, 70; Turnhout, 1988), 260; 'Altera S. Adelelmi Vita' (ch. 12), *ES* 27. 451.

[91] J. de Jaurgain's 2-vol. work *La Vasconie* (Pau, 1898–1902) is in essence an extended essay arguing this point. See also Boissonnade, *Du nouveau*, 13, 21–2; Defourneaux, *Français*, 15, 127–9.

[92] R. Mussot-Goulard, *Les Princes de Gascogne* (Marsolan, 1982), 163.

[93] Glaber (II. 9), 82. But see ibid., n. 2.

[94] *Cartulario de San Millán de la Cogolla*, ed. L. Serrano (Madrid, 1930), 66 (992), 67 (996).

[95] *Cartulario de San Juan de la Peña*, ed. Antonio Ubieto Arteta (Textos Medievales, 6 and 9; Valencia, 1962–3), i, p. 121 n. 12, nos. 43, 44, 47, 48, 51 (ibid., nos. 36 (1014) and 37 (1016) are suspect); Adhemar, *Chronique* (ch. 56), 180. See also *Colección de Huesca*, i. 14, the text of which is clearly corrupt, and *Documentos de Sancho Ramírez*, i. 43.

[96] Adhemar, *Chronique* (ch. 70), 194–5.

Friendly contacts did not necessarily mean subjugation. There is no evidence that Sancho William became Sancho's *fidelis*, at least for lands north of the Pyrenees, which means that it is difficult to discover any basis for Sancho's insertion of Gascony into the *regnante* clauses of his acts shortly after Sancho William's death in September 1032.[97] The protocols of the documents make it clear that Sancho claimed royal authority over Gascony but do not expatiate on why or how this was done.[98] Sancho's claims were most probably an opportunist attempt to exploit the failure of the Mitarrid ducal line, to which he was related. The *regnante* formula featuring Gascony was maintained for less than a year, and was not revived before Sancho's death in 1035. There is no evidence that it was adopted by any of his successors, and later narratives do not mention Gascony at all in their accounts of how Sancho divided his domains among his sons, although this was one of the cardinal events which shaped the political fortunes of eleventh-century Spain.[99] Although a very late source (fourteenth century), the *Chronicle of San Juan de la Peña* makes a significant distinction in a passage lauding Sancho's territorial power. Navarre, Aragon, Cantabria, Castile, and Leon, it states, came to the king by right of his predecessor Sancho Abarca, or through his wife, whereas Sancho's claim to Gascony—placed under his *principatus* as opposed to his direct *dominatus*—arose by reason of his probity and power (*virtus*).[100] Such deliberate vagueness looks like an attempt to express Sancho's dimly remembered claim to authority over Gascony while avoiding impossible constitutional precision. Thus the suggestion which has sometimes been advanced that Sancho pursued long-standing Navarrese claims to Gascon 'sovereignty' is both inaccurate and anachronistic.[101] Contacts between Gascony and Navarre in the first part of the eleventh century should rather be seen as similar to those which

---

[97] Mussot-Goulard, *Princes*, 170 and n. 50. *Cartulario de San Juan de la Peña*, i. 46, purporting to date from 1025 and including Gascony in Sancho's *regnante* protocol, is corrupt; cf. ibid. 58 (1033).

[98] *Cartulaire de l'abbaye de Conques en Rouergue*, ed. G. Desjardins (Paris, 1879), 578 (the date is corrupt); *Cartulario de San Juan de la Peña*, i. 46 (*recte* 1033), 58, 59; Mansi, 19. 412; *Documentación de Leire*, 23. Cf. J. Pérez de Urbel, *Sancho el Mayor de Navarra* (Madrid, 1950), 97–8, 100 n. 15. *Cartulario de San Juan de la Peña*, i. 59 (May 1033) employs the verb *imperare* in Sancho's protocol because Sancho had recently taken control of Leon, not because of any claim to Gascony: cf. ibid. i. 60, 62.

[99] *Historia Silense*, ed. J. Pérez de Urbel and A. Gonzalez Ruiz-Zorrilla (Escuela de Estudios Medievales, Textos, 30; Madrid, 1959), 179; *Crónica de San Juan de la Peña* (ch. 14), ed. T. Ximenez de Embun (Biblioteca de Escritores Aragoneses, Sección Histórico-Doctrinal, 1; Saragossa, 1876), 42.

[100] *Crónica de San Juan de la Peña* (ch. 14), 38; cf. *Historia Silense*, 178–9.

[101] Pérez de Urbel, *Sancho el Mayor*, 92–8, 99–100.

might exist between two neighbouring French principalities, expressed at the princely level by kinship bonds, occasional co-operation in war, and exploitation of the other's political weakness as opportunities arose.

Evidence for Spanish contacts with Gascony becomes thinner for forty years after Sancho William's death, the period in which the Poitevin dukes of Aquitaine very gradually realized their claims to Gascony, and the Gascon nobility adjusted to new influences from the north. There were two significant episodes, however. In 1036 Sancho el Mayor's son King Ramiro I of Aragon married Gilbergis, the daughter of Count Bernard I of Bigorre.[102] The upper Gallego valley, an important pass due north from Aragon, formed part of Gilbergis's dower, which suggests that Ramiro, in seeking the alliance, had acted quickly to improve his communications after the fragmentation of Sancho el Mayor's domains.[103] The marriage was the root of the bonds of kinship between Ramiro's son by Gilbergis, Sancho Ramírez (1063–94), and the viscounts of Béarn and counts of Bigorre. It is noteworthy, for example, that Gilbergis's brother, Count Bernard II of Bigorre, died fighting the Moors in or shortly before 1077.[104]

The second episode of note was a church council at Jaca summoned by King Ramiro in 1063 in order to consecrate the cathedral there and to establish the boundaries of the new see.[105] Austinde of Auch was the only archbishop present, and in the absence of papal legates (who had not yet penetrated Spain west of Catalonia) he most probably presided over proceedings. There is no evidence, however, that, as has sometimes been asserted, he did so on the basis of any metropolitan rights he could claim south of the Pyrenees.[106] Rather, the reconstituted diocese bordered on Austinde's province, and he was a noted reformer who had over the previous decade vigorously pursued in Gascony exactly the sort of structural renewal which King Ramiro wished for the Aragonese church. The same applied to the two other Gascon bishops present, Stephen of Oloron and Heraclius of Bigorre (Tarbes), the brother of Count Bernard II

---

[102] In Spain Gilbergis was known as Ermesindis: *Cartulario de San Juan de la Peña*, ii. 159 at p. 199.

[103] Ibid. ii. 69; *Crónica de San Juan de la Peña* (ch. 14), 45.

[104] *Cartulario de San Juan de la Peña*, ii. 163; Jaurgain, *Vasconie*, ii. 373.

[105] *Colección de Huesca*, i. 27. See generally F. Balaguer, 'Los límites del obispado de Aragón y el Concilio de Jaca en 1063', *EEMCA* 4 (1951), 89–95.

[106] A. Breuils, *Saint Austinde archevêque d'Auch (1000–1068) et la Gascogne au XIe siècle* (Auch, 1895), 270; J. M. Lacarra, 'À propos de la colonisation "franca" en Navarre et en Aragon', *AM* 65 (1953), 333.

and Gilbergis.[107] This instance of ecclesiastical contact across the Pyrenees is therefore analogous to the nature of contemporary secular ties: not 'constitutional' linkage but neighbourly co-operation reinforced by kinship bonds.

It is not surprising that when in the 1070s evidence re-emerges for close ties between major Gascon nobles and peninsular rulers, it involves kinsmen of the second and third generation after 1036. Once Sancho Ramírez had added northern and eastern Navarre to his dominions in 1076–7 after the bizarre death of King Sancho IV el de Peñalen (his brother and sister pushed him over a cliff),[108] and Centulle IV of Béarn had taken control of Bigorre in right of his wife Beatrice in c. 1077, their kinship bonds cemented contacts across most of the western Pyrenees. It is doubtless for this reason that Centulle appears as lord of Ara and Peña in 1080.[109] It must not be assumed, however, that relations were invariably warm. The text has survived of an oath, dated between c. 1077 and 1086, which was sworn by Sancho Ramírez to Centulle.[110] It pre-supposed that Centulle had already become Sancho Ramírez's *fidelis*, but, equally, the very existence of the oath suggests that the relationship between the two men might sometimes be subject to strain. Furthermore, Centulle reserved the fealty which he claimed he had sworn to Duke Guy Geoffrey of Aquitaine and his son, a sign that the political loyalties and search for self-interest of the Gascon nobility could extend northwards as well as to the south. The oath may have been the result of a foray by Sancho Ramírez in 1081 or 1082 into Lavedan and the valley of Cauterets to exploit an outbreak of disorder in that area and possibly to revive property rights inherited from Gilbergis of Bigorre.[111] At all events, relations between Centulle and Sancho Ramírez seem to have improved by the mid-1080s, when the pattern for future relations between their kindreds was established. Centulle's son, the future Gaston IV of Béarn, was married to Talesa, daughter of Ramiro I's natural son Count Sancho Ramírez, in about 1085.[112] (Count Sancho Ramírez provides a close

[107] Jaurgain, *Vasconie*, ii. 371.

[108] Reilly, *Alfonso VI*, 87, 89.

[109] *El Cartulario de Roda*, ed. J. F. Yela Utrilla (Estudios históricos, 1; Lérida, 1932), p. 39; Agustín Ubieto Arteta, *Los 'tenentes' en Aragón y Navarra en los siglos XI y XII* (Valencia, 1973), 202.

[110] 'Cartulaire de Bigorre', Bibliothèque Municipale, Bordeaux, MS 745, fo. 16^{r-v}.

[111] *Cartulaire de l'abbaye des bénédictins de Saint-Savin en Lavedan (945–1175)*, ed. C. Durier (Cartulaire des Hautes-Pyrénées, 1; Paris, 1880), 3 at p. 9.

[112] Jaurgain, *Vasconie*, ii. 219, 546. For Count Sancho Ramírez's parentage, see

family connection with Gaston's participation on the First Crusade, for he went on pilgrimage to Jerusalem in 1092.)[113] Centulle died in Spain, most probably murdered by a rival claimant to the lands which he had been granted by Sancho Ramírez in the valley of la Teña.[114]

The ties between the Navarro-Aragonese court and the princes of Bigorre and Béarn thus reproduced on a smaller scale the contacts between northern French noble dynasties and the peninsular monarchs. Gascony differed from northern and the rest of western France, however, in so far as it was potentially possible for lesser nobles to involve themselves in Spain without prohibitive expense or long absences from home. The evidence reveals a series of isolated contacts with no under-lying pattern other than geographical proximity. The importance of the passes at the western edge of the Pyrenees, for example, meant that the kings of Navarre would cultivate links with lords from the area around Bayonne. The Navarrese Lope Sánchez, who appears as *mayordomo* (chief dignitary of the royal household) in Sancho el Mayor's court between 1011 and 1025 and was descended from a cadet line of the Pamplonese royal dynasty, may be the man of that name who established a lordship in Labourd (Bayonne) around the 1020s and from whose brother, Fortún Sánchez, the later eleventh-century vicecomital line of Labourd was descended. This Fortún Sánchez may be the man of that name who acted as guardian of the young King García IV of Navarre (1035–54); a man with the same name was also Navarrese *mayordomo* in 1067–70.[115] The Lope Garces who appears at the court of King Sancho IV of Navarre in 1057 as 'dominator' of the northern Pyrenean region of Baïgorry may possibly be identified with a Navarrese lord who appears as Sancho's *alférez* (standard-bearer) in 1060. Lope Garces of Baïgorry was succeeded by his brother Eneco Garces, who was the progenitor of the twelfth-century viscounts of that area.[116] Given the recurrence of a small number

Documentos de Sancho Ramírez, i. 23; *Documentación de Leire*, 78; *Crónica de San Juan de la Peña* (ch. 16), 45.

[113] *Documentos de Sancho Ramírez*, ii. 76.

[114] Marca, *Histoire de Béarn*, 327. Centulle was alive in July 1089 and dead by 1091: *Documentos para el estudio de la reconquista y repoblación del valle del Ebro*, ed. J. M. Lacarra (Textos Medievales, 62–3; Saragossa, 1982–5), i. 6; *Cartulaire de Saint-Victor*, ii. 818.

[115] Jaurgain, *Vasconie*, i. 207–8, 223; ii. 233–7; Pérez de Urbel, *Sancho el Mayor*, 98–9; *Colección diplomática de Irache*, ed. J. M. Lacarra and A. J. Martín Duque (Saragossa, 1965), i. 2, 8, 39, 40, 42, 44, 47; *Documentación de Leire*, 83; *Colección diplomática medieval de la Rioja (923–1225)*, ed. I. Rodríguez de Lama (Logroño, 1976–9), ii. 7.

[116] *Documentación de Leire*, 53; *Colección de Irache*, i. 17–18; Jaurgain, *Vasconie*, i. 222; ii. 269–70. See also Agustín Ubieto Arteta, *Los 'tenentes'*, 244–5.

of patronyms among the Iberian aristocracy, identification of individuals cannot always be certain, but it is not unlikely that areas such as Baïgorry and Labourd—remote, mountainous, and Basque-speaking—saw some 'over-spill' of noble interest from Navarre. Thus an elaborate peace agreement concluded between Raymond I, viscount of Soule, and Centulle IV, dated 1077 × c.1085, made allowance for obligations of service to the Navarrese king.[117] Significantly, however, the clauses of the agreement gave the same weight to Raymond's actual or potential obligation of suit of court towards Guy Geoffrey of Aquitaine as that owed Sancho Ramírez. As in the case of Centulle IV noted before, Raymond I realized that his interests might direct his attention to the north as well as to the south.

Soule, Baïgorry, and Labourd lay in the extreme south-west of Gascony on the edges of the Pyrenees. Some measure of contact with Spain was therefore natural. In 1072, for example, William Arnald of Soule, possibly a relation of the local vicecomital family, sold land in Besolla to Eneco Sánchez, grandfather of Lope Lépez of Liédana, for 25*s.*, 100 cows, and a horse worth 50*s.* The nature of the price offered and the presence as witnesses of men 'from beyond the mountain' demonstrate that in the high valleys of the western Pyrenees contacts on a modest scale were possible.[118] Similarly, a grant to the abbey of Leire in 1125 of rights in the valley of Roncal was witnessed by men from that area, from Salazar, and from 'the valley of Soule'.[119] A Gaston of Soule witnessed Alfonso I's testament, drawn up towards the end of the siege of Bayonne in October 1131.[120] At such a local level distinctions between French and Spanish, or even between Gascon and Navarrese, would have been quite meaningless.

Nevertheless, conditions which obtained in the mountainous, Basque fringe of Gascony should not be applied to the whole duchy. Between the level of the major princes and the highly localized dealings of mountain-dwellers, there is no evidence that there was a significant marriage or property market across the Pyrenees before the twelfth century. The impression gained by some historians, that interest in Spanish affairs came as second nature to the Gascon nobility, is therefore only true in

---

[117] Marca, *Histoire de Béarn*, 294. Jaurgain, *Vasconie*, ii. 460 places the agreement *c.*1078. Raymond died *c.*1085: ibid. ii. 462.

[118] *Documentación de Leire*, 97, 257. William Arnald's relationship with the viscounts of Soule (his names recur in the family) is suggested by Jaurgain, *Vasconie*, ii. 462, but he misdates the transaction noted above to *c.*1100.

[119] *Documentación de Leire*, 292.

[120] Ibid. 299.

part. Moreover, the evidence does not exist to support the idea that eleventh-century Gascony was necessarily immersed in the ideology of the *Reconquista*, or that it channelled proto-crusading ideals into France.

Frenchmen in Spain after 1100 and the Impact of Crusade Ideas

Thus far attention has been concentrated on Franco-Spanish contacts before 1095. It is also a valuable exercise to examine the extent to which French perceptions of Spain changed after this time in the light of crusade experiences. The speed with which French, particularly Gascon, attention turned to Spain after the First Crusade may be treated as an index of the significance of Spanish connections before 1095.

A number of factors suggest that the potential existed for a canalization of popular crusading enthusiasm into the peninsula long before the first Spanish crusades were preached. The Spanish episcopate was represented at the Council of Clermont and would have been able to relay news of Pope Urban's appeal.[121] Some writers included Spain in their lists of those parts of Christendom which provided participants on the First Crusade.[122] King Pedro I of Aragon-Navarre seems to have taken the cross, probably in 1100 and possibly with the intention of travelling east with his former brother-in-law William IX of Aquitaine. Quite probably dissuaded from leaving for the East by Pope Paschal II, he campaigned against the *taifa* of Saragossa in 1101 bearing a banner of Christ.[123] Claims that this expedition was a crusade preached by papal legates are, however, unwarranted: Pedro's use of the banner may simply have been a reflection of his existing status as vassal of the Holy See, which status he had renewed as recently as 1099.[124] Furthermore, the possibility that

[121] R. Somerville, 'The Council of Clermont (1095), and Latin Christian Society', *Archivum Historiae Pontificiae*, 12 (1974), 71–2, 73.

[122] Ekkehard of Aura, 'Hierosolymita' (ch. 6), *RHC Occ.* 5. 16 (Galicia); Sigebert of Gembloux, 'Chronica', *MGH SS* 6. 367; 'Notitiae duae Lemovicenses de praedicatione crucis in Aquitania' (no. 1, ch. 2), *RHC Occ.* 5. 350–1.

[123] Charter edn. in *Colección de Pedro I*, p. 113 n. 6; F. Fita, 'El concilio nacional de Palencia en el año 1100 y el de Gerona en 1101', *BRAH* 24 (1894), 232. Reilly, *Alfonso VI*, 304 rejects the possibility that Pedro took the cross, but his argument, based on the chronology of events in 1101, does not allow for the fact that Pedro could have taken the cross the year before, thus allowing time for the pope to intervene.

[124] *Colección de Pedro I*, pp. 114–15; J. M. Lacarra, *Vida de Alfonso el Batallador* (Saragossa, 1971), 20; Antonio Ubieto Arteta, *Historia de Aragón*, 131–2. For the king's

Pedro may have continued to wear his cross during his domestic war is itself no more indicative of an official redirection of crusading than the fact that Count Helias of Maine, a thwarted first crusader, ostentatiously bore the cross motif when he fought against William II of England in northern France.[125] Even so, Pedro's actions at least reveal that interest in the First Crusade penetrated Spain.

A few papal pronouncements suggest that tentative comparisons between the Spanish and eastern theatres were made soon after 1095, but also that the two fronts were not yet treated as equally important. Sometime between 1096 and 1099 Urban II wrote to four Catalan counts insisting that their decision to leave for the East was counter-productive, in that it exposed their homeland to Muslim attack. Urban wished to reassure the Catalans that their domestic war in defence of Tarragona had some spiritual value, but he stopped short of equating it with the crusade; those who died, but not all participants, were granted an 'indulgence of sins' and the prospect of eternal life.[126] In May 1098 Pope Urban II wrote to Bishop Peter of Huesca (which had fallen to the Christians eighteen months earlier) equating the recent *Reconquista* with territorial gains made by the advancing First Crusade, and linking the characteristics of both fronts with the idea of liberation from infidel oppression.[127] Urban was not thereby investing the Aragonese reconquest with the status of a crusade, however. Rather, he seems to have intended Bishop Peter to use his arguments to discourage Aragonese from leaving for the East. Similarly, Pope Paschal II's efforts in the earlier part of his pontificate to dissuade intending Spanish crusaders from journeying to Jerusalem demonstrate that initially the spread of crusading enthusiasm south of the Pyrenees was considered more a problem than an opportunity for the crusade to be extended to new regions.[128]

Evidence survives of a modest but not insignificant number of early peninsular crusaders. The Leonese Pedro Gutiérrez granted property to the abbey of Sahagún before leaving for Jerusalem in 1100; Count Fernando Díaz, from Asturias, returned from the East in the same

vassalic status, see P. Kehr, 'Cómo y cuándo se hizo Aragón feudatario de la Santa Sede', *EEMCA* 1 (1945), 304–5.

[125] J. de Moret, *Anales del reino de Navarra* (Tolosa, 1890–2), iii. 148–9; OV v. 228–32.

[126] *Papsturkunden in Spanien*, i. *Katalanien*, ed. P. Kehr (Abhandlungen der Gesellschaft der Wissenschaften zu Göttingen, Phil.-hist. Kl., NS 18²; Berlin, 1926), no. 23 (ed.'s date incorrect).

[127] Urban II, 'Epistolae et privilegia', *PL* 151, no. 237, col. 504.

[128] *Historia Compostellana* (I. 9, 38–9), 25, 77–8.

year.[129] From Aragon and Navarre there survive charter references to a few possible early crusaders (some names supplied by later writers have no secure basis, and even in the well-documented cases it is sometimes impossible to decide whether a crusade vow or straightforward pilgrimage to Jerusalem was involved).[130] Aznar Garcés of Mendinueta, east of Aragon, departed for Jerusalem in or shortly after 1094.[131] In 1097 Fortún Enecones, lord of Grez, made over property to Leire when he was about to journey (*pergere*) to Jerusalem. Fortún was very probably a crusader, for the charter recorded that his brother Sancho had recently left for Jerusalem, and provision was made for their deaths 'in ista uia'.[132] A second, more problematic case is that of Fortún Sánchez. In 1100 he sold property to the canons of Huesca for the large sum of 1,000*s*. after being unable to shift the property in what may have been a depressed market, and having resorted to a begging tour of Christian neighbours, Jewish money-lenders, and, curiously, Muslims. His wife intended to accompany him to Jerusalem.[133] Aznar Jiménez of Aoiz returned from Jerusalem in about February 1102.[134] As far as one may judge from the moderate number of documentary references, the Spanish response to the First Crusade does not seem to have been considerable. The progress of local conflicts with Muslims must have been an important distraction, for the period between the final siege of Huesca (1096) and the capture of Barbastro (1100) witnessed a peak of activity in the Aragonese reconquest, and in 1097 the Almoravids began a series of major offensives against Toledo and Valencia which occupied the attention of Alfonso VI and the forces of Leon-Castile. But if the number of identifiable Spanish crusaders is not large (differences in diplomatic conventions may hide crusaders who could have been explicitly described as such in French charters), it is no less significant that the evidence for crusading enthusiasm which does survive comes from across the peninsula. It follows that the experiences of a French crusader returning to the West in, say, 1100 would not have met with incomprehension or distrust from Spaniards.

[129] Fita, 'Concilio nacional', 229–31, 232. See also Reilly, *Alfonso VI*, 305. Fernando Díaz was still in Spain in Apr. 1097: *El monasterio de San Pelayo de Oviedo: historia y fuentes*, ed. F. J. Fernández Conde, I. Torrente Fernández, and G. de la Noval Menéndez (Oviedo, 1978–81), i. 5.

[130] See Antonio Ubieto Arteta, 'La participación navarro-aragonesa en la primera Cruzada', *Príncipe de Viana*, 8 (1947), 361–8, 369–72, 373. For an apparent instance of normal pilgrimage, see *Documentación de Leire*, 212 (1105).

[131] *Documentación de Leire*, 146; cf. ibid. 142.

[132] Ibid. 161.

[133] *Colección de Huesca*, i. 78.

[134] *Documentación de Leire*, 192.

It is therefore significant that returning French first crusaders transferred their enthusiasm into the Spanish theatre only very gradually. This can be illustrated by an examination of the careers of three men, two of whom certainly and one very probably went on the First Crusade.[135] The best documented case is that of Viscount Gaston IV of Béarn. His activities in the decade after his return from the East in 1100 reflected two preoccupations. First, he resumed his pre-crusade policy of extending his overlordship over his weaker neighbours in southern Gascony. In 1104 he concluded a formal peace with Count Bernard III of Armagnac, which suggests that he had recently been at war with his most powerful local rival.[136] The establishment by Gaston in 1107 of a fortified centre at Mongiscard is a sign of his territorial ambitions at the expense of the viscounts of Dax, with whom he may have been waging a feud since before the First Crusade.[137] To judge from an entry in the cartulary of St-Vincent, Lucq, Gaston also resumed the Béarnais offensive against Soule which he had inherited from his father and himself vigorously prosecuted before 1096—although the entry also suggests that Gaston returned home with greater misgivings about seizing St-Vincent's property in order to pursue his war effort.[138] The pressure on Soule was potentially significant, for it meant that Gaston's interest in the Pyrenees was bound to increase.[139] For the present, however, his military activity was very localized.

Gaston's second concern was his patronage of two institutions which served as the vehicles for his religious enthusiasm as it had been shaped by his experience of the crusade. In 1101 he assisted Bishop Sancho in introducing canons regular into the cathedral of Lescar. The notice recording this act links it closely to Gaston's recent return from Jerusalem 'with great honour', and states that he was aided by his wife Talesa.[140]

---

[135] The three are Gaston of Béarn, Rotrou of Perche, and Centulle of Bigorre. The first two are securely attested on the First Crusade: e.g. *Gesta Francorum*, 92, 95; *Saint-Denis de Nogent-le-Rotrou 1031–1789: histoire et cartulaire*, ed. Vicomte de Souancé and C. Métais (Archives du diocèse de Chartres, 1; Vannes, 1899), 10, 81. Centulle, who was probably in his mid-teens in 1095–6, is probably to be identified with the 'Centorio de Bieria' mentioned as a crusader by Baldric of Bourgueil: 'Historia Jerosolimitana' (I. 8), 17. See also OV v. 30.

[136] Marca, *Histoire de Béarn*, 396, 397.

[137] Ibid. 275, 400–1; Mussot-Goulard, *Princes*, 226.

[138] *Cartulaire de Saint-Vincent-de-Lucq*, ed. L. Barrau Dihigo and R. Poupardin (Pau, 1905), 16.

[139] For a statement of Soule's significance as a channel of communication into Spain, see *Documentación de Leire*, 114.

[140] Marca, *Histoire de Béarn*, 375; see also, ibid. 380–1, 383–4.

(Her father had been an active supporter of the Augustinians.)[141] The most important end served by the reform of the cathedral community, the notice explains, was the creation of an almonry for pilgrims, supervision of which was to be one of the canons' principal duties. Gaston's endowment of the community was impressive, consisting of allods and rights in southern Gascony and more modest revenues drawn from Jaca. The disparity between the Gascon and Aragonese properties given is an indication of Gaston's relative interest in the two areas at that time. In the early years of the twelfth century he and Talesa also made generous grants to the hospital of Ste-Christine, Somport, an important staging-post on one of the two main routes through the Pyrenees taken by pilgrims to Compostela. According to a later charter their generosity was motivated by dynastic fears—their son Centulle had recently died and Gaston wished to merit a new heir. The subsequent birth of the future Centulle V of Béarn was accounted the direct result of the benefaction. In another document Talesa asked that she and her kindred be remembered in the prayers of the community of Ste-Christine and 'in the alms of the poor passing by'.[142]

Thus, in terms of the outlets found for his piety and as a reflection of his trans-Pyrenean marriage connections, Gaston returned from the crusade interested above all in promoting pilgrimage. For a southern Gascon lord such an interest inevitably meant enthusiasm for pilgrimage to Compostela. Gaston's patronage of Ste-Christine in particular was bound to cement ties with the kings of Aragon-Navarre: Pope Paschal II's privilege for the hospital in 1116 singled out the kings of Aragon and the princes of Béarn as its principal benefactors.[143] But if this community of interest resulted in concern for the progress of the Aragonese reconquest, there is no evidence that Gaston was committing resources to it before 1113. In June of that year Gaston features as King Alfonso I's appointee lord in the important frontier town of Barbastro (which had fallen to the

---

[141] F. Balaguer, 'La vizcondesa del Bearn doña Talesa y la rebelión contra Ramiro II en 1136', *EEMCA* 5 (1952), 84.

[142] Ibid. pp. 86–7, and docs. 1, 2 (the date of which, Era 1166 = 1128, may stand), pp. 110–11.

[143] *Papsturkunden in Spanien*, ii. *Navarra und Aragon*, ed. P. Kehr (Abhandlungen der Gesellschaft der Wissenschaften zu Göttingen, Phil.-hist. Kl., NS 22[1]; Berlin, 1928), no. 28; cf. ibid., no. 29; *Colección de Pedro I*, 82, 164; *Documentos del Ebro*, i. 35, 46. Gaston pursued his interest in pilgrimage by supporting the creation of a number of hospitals north of the Pyrenees which were made subject to Ste-Christine: A. Durán Gudiol, *El hospital de Somport entre Aragón y Bearn (siglos XII y XIII)* (Saragossa, 1986), 30–4.

Aragonese in 1100).[144] Gaston soon relinquished this command, and he does not seem to have involved himself with Alfonso in earnest before the summer of 1117, when he and his half-brother Centulle II of Bigorre accompanied the king on a foray up to the walls of Saragossa.[145] Gaston and Centulle were undoubtedly at the siege of Saragossa, or at least its latter stages, in 1118, and made an important contribution to the Christian victory, for they headed the list of witnesses when Alfonso granted the city its *fuero* (charter of regulations) within a month of its fall.[146] Thereafter, as the Aragonese expansion gathered momentum, Gaston's interest in the defence and extension of Christian territory in the Ebro valley increased significantly. He was granted the lordship of Saragossa and settled a number of followers, including Béarnais, in the city.[147] He was to hold Saragossa until his death in 1130. His styling in the dating protocols of many Aragonese charters emphasized that he became one of King Alfonso's most important and loyal lieutenants.[148] Such was his value to Alfonso that he was granted lordships in Huesca from 1123 and Uncastillo, an important stronghold on a route into the Pyrenees, by April 1124.[149] Gaston's final great religious act north of the Pyrenees reveals how his priorities had changed since the late 1110s. A notice recording the foundation of the abbey of Sauvelade in 1127 states that Gaston established the community before journeying to Spain 'in order to subjugate the Saracens'.[150] When he died Spanish Muslims expressed relief at the demise of a famous and feared opponent.[151]

The progress of Gaston's career reveals that he was attracted to the service of Alfonso I in the years immediately before and after the fall of Saragossa by the opportunities which were created by the opening up of the Ebro valley. In terms of his commitment to the extension of the Christian frontier and the resources put at his disposal, Gaston's experience was unusual among French lords but not wholly without parallels. Some of his enthusiasm for the *Reconquista* seems to have been shared by

[144] *Colección de Huesca*, i. 112.

[145] *Documentos del Ebro*, i. 6; Agustín Ubieto Arteta, *Los 'tenentes'*, 129.

[146] *Documentos del Ebro*, i. 57. A late medieval tradition held that Gaston undertook the siege of Saragossa on his own initiative and was only later joined by Alfonso: text of 15th-C. MS in J. M. Lacarra, 'Gastón de Bearn y Zaragoza', *Pirineos*, 8 (1952), 129. This has no relevance to Gaston's actual conduct in 1118.

[147] e.g. *Documentos del Ebro*, i. 59, 109.

[148] Ibid. i. 72, 75, 78, 90, 99, 164; *Colección de Huesca*, i. 126.

[149] *Documentos del Ebro*, i. 89, 98, 187; *Documentación de Leire*, 283.

[150] Marca, *Histoire de Béarn*, 421. Cf. St-Seurin, 40 (*c.*1118 × 1130).

[151] Mussot-Goulard, *Princes*, 228.

his half-brother Centulle II of Bigorre. Centulle was present at the fall of Saragossa. His Spanish career thereafter is not as well documented as Gaston's, but he features as lord in Tarazona in 1121, 1129, and 1130 (although Alfonso I at various times committed at least partial control of the town to locals such as Gazco, lord of Luesia, and Fortún Aznárez, suggesting that Centulle was not as regularly in Spain as Gaston).[152] In 1127 Centulle accompanied Gaston on an important diplomatic mission for Alfonso I to Alfonso VII of Leon-Castile, a sign of the importance which both men had acquired in Aragonese affairs.[153]

One particular event in Centulle's Spanish career demonstrates the influence of crusading ideals in the early twelfth century. In May 1122 Centulle became Alfonso I's *fidelis* for a series of important honours which included the castle and adjacent settlement of Ruesta and half of Tarazona. The text of the *conuenio* recording Centulle's agreement with the king has two important features.[154] First, it clearly anticipated that Centulle would make an important contribution to future conquests, for Alfonso promised him all of Albarracín, deep to the south on the borders of the *taifa* of Valencia, 'when Almighty God will grant it to me', as well as land sufficient to settle two hundred mounted men when it was liberated, and 2,000*s*. per annum. Second, the *conuenio* was concluded at Morlaas in Béarn. The reason why Alfonso should have troubled to cross the Pyrenees to negotiate with a *fidelis*, however potentially important he might be, is suggested by the fact that the grant to Centulle included half of the lordship held of Alfonso by Galindo Sanz, lord of Belchite. Belchite was the base of a military confraternity, the creation of which has been convincingly dated to between February and May 1122, that is just before Alfonso's Gascon detour.[155] Doubts have been expressed whether Galindo Sanz was the original rector, or head, of the confraternity, but he is almost certainly to be identified with the 'Guarinus Sancio' mentioned by Orderic Vitalis at the head of a group identified as the 'brothers of the Palms' which campaigned alongside Alfonso and Gaston in 1124–5.[156] Galindo was an obvious candidate for the position of rector, a companion

---

[152] *Documentos del Ebro*, i. 80, 106, 124, 141–3, 153, 165, 176–7, 191; Agustín Ubieto Arteta, *Los 'tenentes'*, 163.

[153] *Chronica Adefonsi Imperatoris* (ch. 10), ed. L. Sánchez Belda (Escuela de Estudios Medievales, Textos, 14; Madrid, 1950), 13.

[154] *Documentos del Ebro*, i. 82.

[155] Antonio Ubieto Arteta, 'La creación de la Cofradía militar de Belchite', *EEMCA* 5 (1952), 427–34. See generally, Lacarra, *Vida*, 71–4.

[156] P. Rassow, 'La Cofradía de Belchite', *Anuario de historia del derecho español*, 3 (1926), 211–12; OV vi. 400.

of Alfonso at the fall of Saragossa and his lieutenant in the resettlement of Belchite in December 1119.[157] It thus seems most probable that in 1122 Centulle of Bigorre was approached by Alfonso, with Gaston of Béarn's co-operation, to assume responsibilities which would release Galindo Sanz for the new confraternity.

The confraternity of Belchite is recorded in only one document, a confirmation of its privileges granted by King Alfonso VII of Leon-Castile coupled with a statement of the spiritual rewards available to those who joined or supported it. Both records date from the Council of Burgos held in October 1136.[158] The graded series of rewards set out in the document purport to reproduce the contents of an earlier text which had been drawn up at the time of the confraternity's creation. Some statements in the surviving document seem to echo the words of the original. For example, the text refers to Christian victories on the First Crusade, the Balearic expedition of 1114, and at Saragossa in 1118, but not Alfonso I's famous Andalusian campaign of 1125–6. The clause, however, which states that anyone who donated horses or arms to the confraternity would earn the same remission of sins as if he had so endowed the Hospital or Temple cannot refer to conditions in 1122. Consequently, it is not always easy to distinguish between the confraternity's original features and later accretions. Nevertheless, there is no reason to suppose that the confraternity had fundamentally changed in character in the fourteen years before 1136. By that date its rector was Lope Sanz, Galindo Sanz's brother, who had succeeded to Belchite between December 1124 and August 1126, and must have been well placed to provide the confraternity with continuity.[159] Furthermore, the preamble of Alfonso VII's privilege was manifestly influenced by crusading ideals, describing Gentiles' oppression of the church, the captivity of Christians, and the confraternity's duty to serve God and protect the faithful through unstinting war.[160] Given Alfonso I's interest in crusading,[161] it is likely that this expressed ideas already present in 1122. The confraternity's description as an 'army of God and militia of Christ' is also probably

---

[157] *Documentos del Ebro*, i. 57–8.

[158] The texts are edited in Rassow, 'Cofradía de Belchite', 220–6 and discussed ibid. 200–20. See also O. Engels, 'Papsttum, Reconquista und spanisches Landeskonzil im Hochmittelalter', *Annuarium Historiae Conciliorum*, 1 (1969), 265–6.

[159] *Documentos del Ebro*, i. 106, 133. For the pair's relationship, see *Colección de Huesca*, i. 153.

[160] Rassow, 'Cofradía', 220–1.

[161] See the references to Alfonso's devotion to the True Cross in *Crónicas Anónimas* (ch. 29), 52 and *Chronica Adefonsi Imperatoris* (ch. 52), 43.

original.[162] Thus the *conuenio* drawn up at Morlaas in May 1122 furnishes evidence that Gascon ex-crusaders were closely involved in the establishment of the confraternity of Belchite and the consolidation of crusade ideas in Spain.

The link between Gascons, crusade enthusiasm, and Alfonso I is also demonstrated by the creation in the late 1120s of a second military confraternity, at Monreal del Campo.[163] Monreal was populated by Alfonso I as a frontier town in the autumn of 1124. Most probably failure to attract adequate settlement—efforts to repopulate had to be renewed in or shortly before 1128—soon persuaded Alfonso to switch tack with the founding of a confraternity.[164] The creation of the militia of Monreal is recorded in only one document, an encyclical stating that Alfonso had established a 'militia Christi', or 'confraternitas', which would fight under the king's control.[165] The influence of crusade values is manifest, for the stated purpose of the militia, like that of the confraternity of Belchite, was to expel the Muslims from Spain in order to open up the southern overland route to Jerusalem, an idea to which Alfonso had borne witness in the recent Andalusian expedition of 1125–6.[166] Furthermore, the confraternity was granted the same exemptions as the 'milicia confraternitatis Iherosolimitana', which is probably a very early reference to the presence of the Templars in Aragon-Navarre.[167] The spiritual benefits offered to *confratres* and benefactors were equated with the complete remission of sin ('ab omnibus absoluimus peccatis') believed to have been granted to *Iherosolimitani*. The encyclical was sent to Gascony to solicit recruits and alms; the surviving copy was subscribed by Archbishop William of Auch, who recorded that he had become a *confrater* and clarified the spiritual rewards on offer as graded remissions of penance. The encyclical (or at least the version of it sent to Gascony) also singled out Gaston of Béarn as the king's principal adviser behind the creation of the confraternity. Gaston features as lord of Monreal in May and December 1128, and he most probably remained in this position until his

---

[162] Rassow, 'Cofradía', 221.

[163] See Lacarra, *Vida*, 95–6; A. J. Forey, 'The Military Orders and the Spanish Reconquest in the Twelfth and Thirteenth Centuries', *Traditio*, 40 (1984), 197–8.

[164] *Documentos del Ebro*, i. 102–4, 167–8.

[165] Ibid. i. 173.

[166] Cf. *Historia Compostellana* (II. 78), 378–80. For the conduct of this remarkable feat of arms, see Lacarra, *Vida*, 83–8; A. Huici Miranda, *Historia musulmana de Valencia y su región: novedades y rectificaciones* (Valencia, 1969–70), iii. 51–64.

[167] See A. J. Forey, *The Templars in the Corona de Aragón* (London, 1973), 6–9.

death in 1130.[168] Significantly, the militia of Monreal appears to have become defunct shortly after the Gascon connection was broken.[169]

The third French lord to exploit the Aragonese expansion in the 1120s, and the most problematical, was a northerner, Count Rotrou of Perche. Despite featuring in two narrative traditions, Rotrou's Spanish career before 1120 is very obscure. According to Orderic Vitalis, he was invited south by Alfonso I at some unspecified time after 1104 (when Alfonso succeeded Pedro I) in order to bring Frankish auxiliaries, but, betrayed by Aragonese plotters, returned home unrewarded.[170] There is no documentary confirmation of this story, though if it has a basis in fact it may reveal that Rotrou tried prematurely to exploit the accession of Alfonso I (who, unlike Pedro I, was his cousin) before the Christian drive down the Ebro valley gained momentum in the later 1110s. Orderic is a suspect source. His chronology of Spanish affairs was confused, and he had a highly developed admiration of Rotrou and his family which led him to exaggerate Rotrou's importance in Spain after 1120 and, implausibly, to postulate resentment on Alfonso's part of the French achievements in Spain.[171] The story found in later medieval narratives that Rotrou joined the siege of Saragossa after himself seizing Tudela (which in fact fell to the Christians after Saragossa in February 1119) is demonstrably false.[172] The charters reveal that Rotrou was created lord of Tudela by 1121, and that he held this position, sometimes through castellan deputies, until about 1135.[173] It is noteworthy that the beginning of Rotrou's commitment to Spain coincided neatly with the death in the White Ship (November 1120) of his wife Matilda, the natural daughter of King Henry I and guarantee of his prospects in the service of the Anglo-Norman realm.[174] Like Gaston of Béarn, Rotrou was a former first crusader who went on to do very well from involvement in Spain. Yet two decades separated the Spanish phase of his career from the events of

---

[168] *Documentos del Ebro*, i. 158, 163.

[169] Forey, 'Military Orders', 198.

[170] OV vi. 394–6.

[171] Ibid. iv. 160, 304, 330; vi. 398–400, 404.

[172] L. H. Nelson, 'Rotrou of Perche and the Aragonese Reconquest', *Traditio*, 26 (1970), 121–7.

[173] *Documentos del Ebro*, i. 91, 113, 120, 177; *Documentación de Leire*, 283–4, 295; *Jaca: documentos municipales 971–1269*, ed. Antonio Ubieto Arteta (Textos Medievales, 43; Valencia, 1975), 13, 14; Agustín Ubieto Arteta, *Los 'tenentes'*, 165. See also *Saint-Denis de Nogent-le-Rotrou*, 27, 45, 120; *Cartulaire de l'abbaye de la Sainte-Trinité de Tiron*, ed. L. Merlet (Chartres, 1883), i. 118. Rotrou missed the great Andalusian campaign of 1125–6: see OV vi. 404.

[174] OV vi. 40, 304, 398; William of Malmesbury, *De Gestis Regum* (ch. 419), ii. 497.

1096–1101. Patently Rotrou did not immediately come to treat Spain as an obvious extension of the Holy Land.

A number of other French, particularly Gascon, arms-bearers benefited from the opening up of the Ebro valley.[175] Probably present in at least the final stages of the siege of Saragossa were Viscount Peter of Gavarret, Count Bernard I of Comminges, Auger of Miramont, and Arnald of Lavedan.[176] Arnald was granted property in Saragossa by Gaston of Béarn in 1124 and may have taken part with him in the Peña Catella campaign of 1124–5.[177] Auger of Miramont was present alongside Centulle II of Bigorre and Bishop Bertrand of Bazas when Alfonso I issued a privilege for the Bordelais abbey of La Sauve-Majeure at Uncastillo in March 1125.[178] Bernard of Comminges seems to have sought influence further west; in December 1134 he appears as a witness alongside Alfonso-Jordan of St-Gilles in a charter of Alfonso VII of Leon-Castile.[179] Gascons, Bordelais, and Limousins of inferior status also exploited the Aragonese conquests as colonists. For example, Bonet, Raymond, and Walter of Bordeaux appear as settlers in Saragossa in the 1120s; by 1125 a William of Bordeaux was established in Tudela.[180] The Peter of Limoges who appears in Saragossa in 1122 may have been the man of that name who had settled in Jaca by 1110.[181] The large numbers of Gascon toponyms found in Aragonese charters suggest that settlement from Gascony was particularly common.[182] By c.1120 Gascon immigration was not novel, nor had it been confined to Aragon-Navarre: Gascons, for example, head the list given by the Sahagún Anonymous of foreign artisans who settled in Sahagún in the 1080s.[183] But in terms of the numbers involved and the role of men above the class of artisan/burgess, the offensive against Saragossa and Tudela seems to have been an important watershed.[184] The

[175] J. M. Lacarra, 'Los franceses en la reconquista y repoblación del valle del Ebro en tiempos de Alfonso el Batallador', *Cuadernos de Historia*, 2 (1968), 72–4.

[176] *Documentos del Ebro*, i. 57. See also J. M. Lacarra, 'La conquista de Zaragoza por Alfonso I (18 diciembre 1118)', *Al-Andalus*, 12 (1947), 78–83.

[177] *Documentos del Ebro*, i. 109; ii. 321.

[178] Ibid. i. 117.

[179] Ibid. i. 245.

[180] Ibid. i. 71, 73, 100, 118, 161, 210.

[181] Ibid. i. 88; *Documentación de Leire*, 236; *Cartulario de Santa Cruz de la Serós*, ed. Antonio Ubieto Arteta (Textos Medievales, 19; Valencia, 1966), 19 (late 11th or early 12th C.); González Miranda, 'Condesa doña Sancha', app. 5, p. 202. For another Limousin, in Saragossa, see *Documentos del Ebro*, i. 253.

[182] e.g. *Documentos del Ebro*, i. 59, 88, 100, 130, 134; *Colección de Huesca*, i. 132–3.

[183] *Crónicas Anónimas* (ch. 15), 20–1.

[184] See Auch, 74, which may recall preaching of the crusade in 1118 by Archbishop Bernard of Auch.

Chronicle of San Juan de la Peña identified the preliminary campaigns against Saragossa as the decisive turning-point, for it was then that Alfonso I attracted many nobles and knights from Gascony and elsewhere 'beyond the passes'.[185] (Interestingly, the chronicle linked this achievement with Alfonso's confirmation of the privileges of La Sauve-Majeure, one of the most prestigious abbeys in south-western France.)

Between 1117 and 1130 the French military contribution to the Navarro-Aragonese reconquest was dominated by former first crusaders, most of whom must have been comfortably into their middle age at least: Gaston of Béarn; Centulle of Bigorre; William IX of Aquitaine (who campaigned in Spain in 1120);[186] Rotrou of Perche; Raymond of Turenne.[187] Both for these lords, and for those whose presence on the First Crusade is unrecorded or unlikely, the same sort of family ties operated that had been important before 1095. Gaston and Centulle were Alfonso I's cousins. So was Rotrou; although suspect for much of Rotrou's Spanish career, Orderic was certainly correct to stress his kinship with Alfonso as one reason for his involvement in peninsular affairs.[188] The family network which stemmed from the children of Hilduin of Ramerupt (Ebles of Roucy had been a member of the kindred) also came to include the Limousin Raymond of Turenne, almost certainly as the result of contacts made in 1096–1100.[189] Other marriage ties were also important. Viscount Peter of Gavarret married Gaston's daughter Guiscardis sometime after 1103.[190] Viscount Peter of Marsan married the heiress of Centulle II of Bigorre, and by September 1130 had assumed his father-in-law's rights in Tarazona.[191]

As had been the case before the First Crusade, therefore, French noble interest in Spain was stimulated and channelled by family ties. There was also an important new element, however. The growth of French interest in Spain in this period was connected to the extension of the crusade appeal to the peninsula. As we have seen, the papacy had felt its way

---

[185] *Crónica de San Juan de la Peña* (ch. 19), 65–6.

[186] 'Annales Compostellani', *ES* 23. 321; *Documentación de Leire*, 275; *Chronique de Saint-Maixent*, 188–90.

[187] Raymond's presence on William IX's campaign in 1120 may be inferred from *LSM*, p. 22.

[188] *OV* vi. 396. See also Herman of Laon, 'De miraculis B. Mariae' (I. 2), 267.

[189] Herman of Laon, 'De miraculis B. Mariae' (I. 2), 267; Geoffrey of Vigeois, 'Chronica' (ch. 23), 290.

[190] Mussot-Goulard, *Princes*, 217–18.

[191] *Jaca: documentos municipales*, 14; *Documentos del Ebro*, i. 194; Jaurgain, *Vasconie*, ii. 129.

towards a tentative comparison between the eastern crusade and the Spanish reconquest in the first years of the crusading movement. It had not done so, however, because it believed in the strict equivalence of the two theatres, and its early concern was to suppress peninsular enthusiasm for the crusade to the East. Spain became a recognized crusading theatre quite gradually. It is unlikely that the peninsula could have been anything more than of peripheral concern to crusade strategists before the end of Bohemond of Taranto's crusade in the Balkans in 1108. The first direct evidence for the assimilation by the papacy of the eastern and Spanish crusades comes in 1123, when Canon 10 of the First Lateran Council clarified the action which would be taken against those who were known to have fixed crosses to their clothing for either the *iter* to Jerusalem or that to Spain, but then had taken them off; in the same year Pope Calixtus II extended to Spain the same 'remission of sins' which he had offered to those who went out to defend the eastern church.[192] The formative phase for crusading in Spain fell between these two dates. Pope Paschal II granted the status of crusade to a northern Italian and Catalan expedition which attacked the Balearic islands in 1114 and continued on the mainland in 1116.[193] The Saragossa campaign in 1117–18 should be regarded as an extension of this crusading effort: according to the St-Maixent chronicler, the 'via de Hispania' was 'confirmed' at a council held at Toulouse in the spring of 1118.[194] The two theatres were further linked in the person of the papal legate Boso of Sta Anastasia, who took part in the Balearic campaign and was then sent on a legation to Spain in early 1117.[195] It was almost certainly after this Spanish legation, and in connection with business discussed during it, that Boso visited southern France. In a document from Uzerche he is recorded travelling through

---

[192] *Conciliorum Oecumenicorum Decreta*, ed. J. Alberigo *et al.*, 3rd edn. (Bologna, 1973), 192; *La documentación pontificia hasta Inocencio III (965–1216)*, ed. D. Mansilla (Monumenta Hispaniae Vaticana, Sección Registros, 1; Rome, 1955), 62.

[193] J. Goñi Gaztambide, *Historia de la Bula de la Cruzada en España* (Victoriensia, 4; Vitoria, 1958), 68–70.

[194] *Chronique de Saint-Maixent*, 186. The location of this passage within descriptions of events datable to the early part of 1118 makes it most unlikely that the chronicler was erroneously referring to the council held at Toulouse under Pope Calixtus II in July 1119, which he noted separately (ibid. 188).

[195] R. Hüls, *Kardinäle, Klerus und Kirchen Roms 1049–1130* (Bibliothek des Deutschen Historischen Instituts in Rom, 48; Tübingen, 1977), 147; L. Vones, *Die 'Historia Compostellana' und die Kirchenpolitik des nordwestspanischen Raumes 1070–1130* (Kölner historische Abhandlungen, 29; Cologne, 1980), 348–9. Unfortunately, little is known of Boso's activity in Aragon-Navarre: P. Kehr, 'El Papado y los reinos de Navarra y Aragón hasta mediados del siglo XII', *EEMCA* 2 (1946), 151.

Aquitaine 'in order to gather an army for Spain'; as part of this effort he met Bishop Eustorgius of Limoges, and probably other prelates too.[196] A charter from Vigeois, very close to Uzerche, which records that four brothers from a local family went to Spain is probably linked to Boso's visit.[197] We do not know enough about Boso's movements to assess how far he preached the cross in 1117–18, but it is surely very significant that he later acquired a reputation as the man who had been responsible for the liberation of Majorca and Saragossa.[198] It is therefore reasonable to suppose that it was he who convoked the Council of Toulouse in 1118 and supervised the extension of crusading effort into the Ebro valley.

The earliest direct evidence that the Saragossa campaign was a crusade is a letter of Pope Gelasius II dated 10 December 1118 and addressed to 'the army of Christians besieging the city of Saragossa, and to all believers in the Catholic faith'. Recalling that the besiegers had sent him letters painting a dark picture of the siege's progress, the pope declared that anyone who participated in the campaign as penance, and then died, would be absolved of all his sins; those who survived or contributed to the restoration of the church of Saragossa would be granted a remission of penances appropriate to their efforts (it being the responsibility of local bishops to judge what remission had been earned).[199] At first glance it seems that it was only at this late stage that the Saragossa campaign attracted to it crusade-type rewards. If so, the campaign would effectively not have been a crusade at all, for Gelasius's letter could not possibly have reached Spain before Saragossa's capture on 18 December. On the other hand, from the letter's reference to earlier correspondence it would appear that the besiegers had written specifically to explain that, because the siege was proving tougher than expected (the city had been invested in May 1118), they needed to appeal to all Christians for help. Doubtless this had prompted them to suggest that the spiritual rewards available

---

[196] Uz. 1038. The document bears no date. Hüls, *Kardinäle*, 147 and Z. Zafarana, 'Bosone', *Dizionario biografico degli Italiani*, 13 (Rome, 1971), 268 place the visit to the Limousin during Boso's journey to Spain in 1116–17, but their suggested route would have been difficult to cover in the time available. Hüls has not found any record of Boso between Apr. 1117 (Gerona) and Dec. 1118 (Orange). It would appear likely that the Limousin visit, possibly part of a wider preaching effort in southern France, took place in the earlier part of the intervening period.

[197] V. 220.

[198] *La Chronique de Morigny (1095–1152)* (II. 9), ed. L. Mirot (Collection de textes pour servir à l'étude et à l'enseignement de l'histoire, 41; Paris, 1909), 33.

[199] Gelasius II, 'Epistolae et privilegia', *PL* 163, no. 25, col. 508. Cf. cols. 507–10 n. 10.

should both match their efforts and serve to attract more volunteers. It is therefore reasonable to suppose that Gelasius's letter amplified a promise of spiritual rewards—quite probably a partial remission of penance granted to all participants, whether they survived or not—which had been promulgated earlier.

In sum, the mid- and late 1110s were the formative phase in the gradual extension of crusade ideas to mainland Spain. Given the surviving evidence, it is impossible to say precisely on whose initiative this process occurred, but it is most significant that the actions of the papacy and its legates appear to have been decisive. Spain did not simply witness a spontaneous outpouring of native crusade enthusiasm; external forces were also at work. It is in this context that the careers of men such as Gaston of Béarn, Centulle of Bigorre, and Rotrou of Perche assume significance, for they grafted the traditional patterns of trans-Pyrenean kinship ties on to the new impulse of crusading. That they did so from *c.*1117, not immediately after the First Crusade, demonstrates that Spain could not have been a forcing ground for proto-crusade ideals before 1095.

Attendance at Spanish courts would have exposed Frenchmen to ideas of just territorial expansion which was favoured by God, extended the faith, expelled non-believers, and furthered reform of the church.[200] It is difficult to conceive of circumstances, however, given the limited opportunities for fighting in Spain before the First Crusade, and the restrictive influence of marriage ties, in which ideas of this sort could have formed the basis for an ideology of French military involvement south of the Pyrenees. It is hard to avoid the conclusion that eleventh-century French nobles and knights went to Spain principally in the hope of becoming rich. As French monasteries which did well from Spanish rulers' largess would have been able to communicate to their local nobilities, Spain offered land (though migration by the military classes was much more significant after the First Crusade) and it was a source of movable wealth, especially impressive quantities of gold which were seldom available in the north.[201] This sort of attraction helps to explain an interesting story

---

[200] e.g. *Documentos del Ebro*, i. 12 (*c.*1091); *Documentos de Sancho Ramírez*, i. 43 at pp. 143–4 (1090); *Colección de Pedro I*, 117 (1102); see also *Historia Compostellana* (I. 46), 84. See R. A. Fletcher, 'Reconquest and Crusade in Spain *c.*1050–1150', *TRHS*⁵ 37 (1987), 38.
[201] See e.g. C. iv. 3441, 3509, 3562, 3638; *Cartulaire de l'abbaye de Conques*, 576–8; *Cartulaire de l'abbaye de Saint-Sernin de Toulouse (844–1200)*, ed. C. Douais (Paris, 1887), 450, 453. Gold featured in the *census* paid by the peninsular vassal states to Rome: e.g.

told by Orderic Vitalis concerning the reputation of a Norman knight who probably fought in Spain in 1087. In his account of William the Conqueror's death-bed speech, Orderic stated that the king restored to Baldric fitz Nicholas lands which had been forfeited 'because he foolishly left my service, and went to Spain without my permission'. Orderic wrote this forty years later, and the speech is apocryphal, but the judgement passed on Baldric comes close to condensing the state of mind of French warriors in Spain before 1095: 'I do not consider that any young knight can be found who is better than he in the use of arms, but he is extravagant and unstable, and wanders from place to place.'[202] This is not a comment upon a holy warrior casting around for outlets for his proto-crusading religious zeal.

What is absent from the sources for the eleventh-century French campaigns in Spain is any trace of a coherent religious experience. The once popular idea that holy war in Spain was actively preached in France by Cluniac monks has been shown to be untenable.[203] The most striking feature of the supposed Barbastro indulgence, even if the maximal interpretation is placed upon it, is its isolation. Similarly, Gregory VII's ideas of realizing Petrine rights in the peninsula through foreign armies were swiftly abandoned in favour of direct dealings with the Spanish rulers, and service of the papacy did not re-emerge as an inducement for Frenchmen to go south.[204] When Urban II began to assimilate prosecution of the reconquest with devotional practice, his ideas were expressed to Catalans (who, in this context, should be regarded as Spaniards).[205] Frenchmen who went to Spain to fight may well have taken the opportunity to go on pilgrimage to Compostela, but, apart from the uncertain case of Hugh of Chalon in 1078,[206] scant evidence of this practice

Colección de Pedro I, 21. For the wealth of Spanish rulers, see Pelayo, *Crónica*, 80; J. M. Lacarra, 'Aspectos económicos de la sumisión de los reinos de taifas (1010–1102)', in J. Malaquer de Motes (ed.), *Homenaje a Jaime Vicens Vives* (Barcelona, 1965–7), i. 255–77; C. J. Bishko, 'Fernando I and the Origins of the Leonese-Castilian Alliance with Cluny', *Studies in Medieval Spanish Frontier History* (London, 1980), art. 2, pp. 36–41.

[202] OV iv. 100. Cf. Orderic's remark that Alfonso I promised Rotrou of Perche and his followers 'larga stipendia, et secum remorari uolentibus opima praedia': ibid. vi. 396.

[203] H. E. J. Cowdrey, 'Cluny and the First Crusade', *RB* 83 (1973), 287–90; see also ibid. 297–300. P. Segl, *Königtum und Klosterreform in Spanien: Untersuchungen über die Cluniacenserklöster in Kastilien-León vom Beginn des 11. bis zur Mitte des 12. Jahrhunderts* (Kallmünz, 1974), 7–10 reviews earlier scholarship on this matter.

[204] See e.g. Gregory VII, *Register* (1. 63–4; IV. 28), i. 91–4, 343–7; Urban II, 'Epistolae et privilegia', no. 6, cols. 289–90.

[205] *Papsturkunden*, i. *Katalanien*, nos. 22, 27.

[206] David, *Études historiques*, 384.

survives. There is no case for arguing that before the First Crusade Spain provided a *de facto* fusion of ideas of pilgrimage and meritorious violence; and given the geographically remote position of Compostela and the extent of the Christian frontier by the later eleventh century, such a fusion is singularly unlikely. In general terms, if eleventh-century Frenchmen fighting in Spain had believed that they were involved in an overtly religious and meritorious exercise similar to crusading, it is very difficult to explain why this was not picked up far more fully and coherently in papal correspondence, chronicles, and charters.

It is difficult to estimate the numbers of French warriors in Spain before 1096. Narratives tended to say that the set-piece expeditions were large, but the relative silence of the charters suggests that they were prone to exaggeration; certainly, the numbers involved could not have approached the size of the First Crusade forces. It is sometimes argued that the near-silence of the Spanish narratives with respect to French military aid was the result of incipient national antagonism prompted by what was seen as foreign interference in peninsular affairs.[207] This view is misguided, however. When, after 1100 and especially from the mid- or late 1110s, Frenchmen impinged upon peninsular affairs, Spanish observers took note.[208] The relative silence before then should be taken at face value: there was very little to write about.

It has recently been argued that the wars between Christians and Muslims in Spain gradually acquired a crusading character between the latter half of the eleventh century (especially the 1080s) and the 1120s. The agents of change, it is contended, were French adventurers, the reformed papacy, and the religious fanaticism of the Almoravids.[209] This argument has two great merits: it takes into account revisionist scholarship in Spain which has challenged the assumptions of Menéndez Pidal and his generation about the autochthonous religious quality of the *Reconquista*;[210] and it allows for the impact upon Spain of external influences at a time when the peninsula was becoming more receptive to the

---

[207] See e.g. L. de la Calzada, 'La proyección del pensamiento de Gregorio VII en los reinos de Castilla y León', *Studi Gregoriani*, 3 (1948), 49–50.

[208] e.g. *Historia Compostellana* (I. 83), 132–3; *Crónicas Anónimas* (chs. 33, 35, 41–2), 61, 66, 75–6; *Crónica de San Juan de la Peña* (ch. 19), 66, 68, 76; *Chronica Adefonsi Imperatoris* (chs. 51, 56–7), 42, 46–7; 'Annales Compostellani', 321; *Documentos medievales Artajoneses*, ed. J. M. Jimeno Jurío (Pamplona, 1968), 63; *Documentación de Leire*, 275; 'Annales Toledanos, I', *ES* 23. 389.

[209] Fletcher, 'Reconquest and Crusade', 37–8, 42–3.

[210] See ibid. 32–7.

rest of the Latin world. The formulation of these influences, however, needs to be revised. At least for the pre-crusade period, from the French point of view, the Almoravids must largely drop out of the picture. There is only slight evidence that the 1087 expedition against them was informed by reactive Christian zeal on the part of the French; and there was no further significant opportunity for any such zeal to develop through renewed campaigning before the time of the First Crusade. As for the papacy, having had its fingers burned in 1073, it was not in the business of launching wars in Spain on its own account until crusading expanded its terms of reference. Even then it was not until a lull in crusade activity in the East after 1108 that the papacy began in earnest to extend crusading to the peninsula. As far as French adventurers are concerned, to argue that they helped to introduce crusade ideals into Spain from the 1080s or earlier only begs the question of where their putative proto-crusading ideology had its roots. In absolving the native Spaniards from responsibility for inventing the peninsular holy wars, it is of no value to point the finger at the French, whose experience of anti-Muslim warfare before 1097–1101 was minute in comparison.

The chronology of the introduction of crusading into Spain can be narrowed down fairly precisely: a preliminary phase beginning in the later 1090s, during which the church tentatively made comparisons between the eastern theatre and Spain, and Spaniards grasped the broad principle of crusading (at this stage expressible predominantly through enthusiasm for campaigning in the East); and the period between 1114 and 1123 which saw the definitive introduction of crusade terminology (the vow), symbolism (the cross), and rewards (penitential remissions). In this latter stage the agents of change were arranged in a triangular pattern: the papacy; French former crusaders who could give Spaniards the benefit of their experience; and Spaniards themselves, of whom the paradigm was Alfonso I of Aragon-Navarre.

In these circumstances what had transpired in Spain before the First Crusade—apart from the creation of kindred networks—becomes of little direct significance, from a French perspective, for the history of the Spanish crusades. *A fortiori* it is of minimal relevance to the First Crusade itself. Doubtless contacts with Spain helped to introduce south-western Frenchmen to the Muslim world. (Adhemar of Chabannes, for example, tells the story of twenty Muslim captives, seized by Christian sailors from the Narbonnais, who were presented to Abbot Geoffrey of St-Martial, Limoges (1007–18). Geoffrey kept two as servants and palmed the others on to visiting noble pilgrims.)[211] Yet it is clear that, in terms of numbers and motivation, the Spanish theatre could not have been anything more than a very minor factor behind the response of Aquitanians, Gascons, and others to the First Crusade.

[211] Adhemar, *Chronique* (ch. 52), 175.

3

Recruitment to Religious Communities

THE superficially attractive links between crusading and earlier movements or themes—the supposed existence of an ethical code derived from
the Peace of God, and the *Reconquista* in the eleventh century—become
much less compelling if subjected to close scrutiny. If we remove the
Peace of God or Spain from our calculations, however, there is a danger of
leaving the origins of the crusade *in vacuo*. Both topics were fundamentally important to the approach of scholars such as Erdmann, Delaruelle,
and Villey who studied the crusade's origins through educated clerics'
ideas, for they were considered to be vital points of contact which enabled
the intellectuals' ideas to filter down to the laity. Without channels of
this sort the laity might seem to become isolated from the movements
which are supposed to be working towards the events of 1095–1101.

This is only a problem, however, if we persist with the normal terms of
reference. (Pilgrimage, the third traditional prop of the old crusading
origins argument, has to be retained, as shall be seen in Chapter 5.) If we
postulate that the crusade was not a secular exercise dressed up in
religious garb, but a movement built on the potent influence of the
church upon the laity, then it becomes necessary to identify and explore
the contacts between ecclesiastics and the faithful in order to isolate the
roots of crusade enthusiasm. The Erdmann-type approach misses the
wood for the trees. The important contacts were everywhere, played out
thousands of times every year in chapterhouses, abbey churches, and
cathedrals throughout south-western France (and, it is reasonable to
suppose, the rest of the Latin world) as laymen, especially arms-bearers,
interacted with religious bodies.

The relations between laymen and local churches may be placed in
context by examining the ways in which religious communities drew
upon the human resources of their localities. Two fundamental points
emerge: the importance of family ties and traditions as the cement which
bound religious institutions to lay benefactors; and the elasticity of the
various forms of recruitment to religion, which had the practical effect of

being able to accommodate laymen's spiritual concerns throughout their lives and over a number of generations.

Great importance was attached to the common practice of giving property to the church. Nobles and knights did not treat religious communities simply as targets of material benefaction, however. It is only possible to understand the close relationship between the laity and professed religious when it is considered that many laymen put at the church's disposal their most important resource: their own persons or those of close relatives. From the professed religious point of view there was an unquestionable hierarchy of worth which separated the surrender of one's person to God from the transfer of one's property. The system of recruitment to religious communities was predicated on the laity's acceptance of precisely the same perspective. This becomes evident as the various types of entry into religion are examined in turn.

Oblation

In the eleventh century an important source of recruitment to religious houses, especially Benedictine monasteries, was child oblation. According to this practice a boy (or more rarely a girl) would be offered to a community at an age between infancy and adolescence when he or she would be receptive to the vigorous demands of an education in a monastic school.[1] The system was not without its critics. Reformers of the twelfth century attacked oblation as a flaw in traditional monastic practices, less out of any concern for children's welfare than because of a rigorist belief in their own interpretation of voluntary devotion to the *vita apostolica*.[2] It is also the case that even commentators writing from within the black-monk tradition realized that oblation had its faults. In the early twelfth century, for example, Guibert of Nogent and St Anselm observed that ex-oblates might sometimes be complacent, literally holier-than-thou towards adult converts, and handicapped by their lack of experience of the

---

[1] G. de Valous, *Le Monachisme clunisien des origines au XVe siècle* (Archives de la France monastique, 39–40; Ligugé, 1935), i. 40–4. On the oblates' regime, see P. Riché, 'Les Moines bénédictins, maîtres d'école (VIIIe–XIe siècles)', in W. Lourdaux and D. Verhelst (edd.), *Benedictine Culture 750–1050* (Mediaevalia Lovaniensia¹, 11: Louvain, 1983), 100–10.

[2] J. H. Lynch, *Simoniacal Entry into Religious Life from 1000 to 1260* (Columbus, Ohio, 1976), 37–9.

outside world.[3] What these writers criticized, however, was the weakness of individuals and the poor state of some religious houses, not the system of oblation itself. The evidence from the Limousin and Gascony reveals that, throughout the eleventh century and into the twelfth, oblation was the single most important source of recruitment to established religious communities. By placing a young relative in a monastery, many kindreds amply demonstrated that they had no objection whatsoever to the practice. In fact, oblation is a particularly important area of study, for it, more than any other form of entry into the religious life, reveals the operation of dynastic policy and the belief that members of a kindred bore some collective responsibility for the spiritual welfare of the whole family group. The following discussion will concentrate on youthful entries into Benedictine communities because they are the best documented cases. Oblation to communities of secular canons and, to a lesser extent, those of canons regular followed slightly different conventions; but it is also the case that the religious preoccupations which might lead a family to make one of its number a canon were fundamentally the same as those which informed Benedictine oblation.

It was long conventional wisdom that the practice of oblation flourished because it enabled kindreds decorously to dispose of those excess children whose demands upon the patrimony might otherwise diminish a family's landed power and prestige.[4] More recently it has been argued by Constance Bouchard that in Burgundy in the eleventh and twelfth centuries the placing of oblates in religious houses was far from a cheap option for kindreds because the entry gifts which the communities required might be at least as valuable as the property from which an oblate would have been expected to live had he remained in the world. Furthermore, Bouchard has argued, entry gifts were permanently alienated from the patrimony whereas there was always a chance that properties granted to younger lay sons might revert to the stem-patrimony.[5] The evidence

---

[3] Guibert of Nogent, *Autobiographie* (I. 8), ed. and trans. E.-R. Labande (Les classiques de l'histoire de France au Moyen Âge, 34; Paris, 1981), 50; 'Liber Anselmi archiepiscopi de humanis moribus' (ch. 78), ed. R. W. Southern and F. S. Schmitt, *Memorials of St. Anselm* (Auctores Britannici Medii Aevi, 1; London, 1969), 68. But for favourable comments on former oblates, see OV ii. 96, 102.

[4] For recent statements of this view, see A. Vauchez, *La Spiritualité du Moyen Âge occidental VIIIe–XIIe siècles* (Paris, 1975), 37–8; C. H. Lawrence, *Medieval Monasticism: Forms of Religious Life in Western Europe in the Middle Ages* (London, 1984), 63–4; G. Duby, *The Knight, the Lady and the Priest: The Making of Modern Marriage in Medieval France*, trans. B. Bray (Harmondsworth, 1985), 105.

[5] *Sword, Miter, and Cloister: Nobility and the Church in Burgundy 980–1198* (Ithaca, NY, 1987), 59–60.

from the Limousin and Gascony, regions which were not as wealthy as Burgundy, is more problematical. The relative scarcity of charter evidence from these areas means that we do not possess a sufficiently large sample from which to arrive at an accurate figure for the average cost of oblation in any given period. Instead, a few general points can be made.

By concentrating particularly on the Benedictine communities of the Bas-Limousin, which furnish the fullest evidence, there is some indication that oblation became progressively cheaper between the late tenth and early twelfth centuries. In 986 or 987 Gerald of La Valène offered his son as an oblate to Tulle and gave properties located throughout the southern Limousin.[6] In 996 X 1031 Bernard of Chanac granted Tulle three *mansi* (units of agricultural production, of varying size), of which the abbey already had a claim to two, when he offered his son.[7] Viscount Guy I of Limoges gave Uzerche three *mansi* and two *bordariae* (lesser but similar units of production, of varying size) concentrated in one parish when his son Fulcher became a monk there in about 1020.[8] Around the third quarter of the eleventh century an entry gift at Vigeois comprised property which had earlier been valued at 191*s*.[9] In contrast, from the second quarter of the century one *mansus* became established as a typical entry gift.[10] By the end of the century it had become more common for communities to accept fractions of a *mansus*, or censual or jurisdictional rights attached to a property.[11]

It is important, however, not to read too much into this apparent cheapening of oblation in the eleventh century, for the settlement of an entry gift might be reached through a process of bargaining which introduced a number of variables. In the first place, oblates' kinsmen

[6] T. 46.

[7] Ibid. 31.

[8] Uz. 354.

[9] V. 12.

[10] SEL 72 (*c*.1060); V. 14 (1082 X 1091), 139 (1092 X *c*.1096); T. 42 (1117), 79 (late 11th C.), 205 (mid-11th C.), 209 (2nd half 11th C.), 212 (1097), 256 (1052 X 1060), 269 (1064), 318 (1112), 320 (1121), 514 (*recte* 1073 X 1086). One *mansus* was also accepted earlier: T. 314 (996 X 1031), 421 (1031 X *c*.1050).

[11] B. 83 (1061 X 1076: rights of *vicaria*); SEL 72 (*c*.1060: unspecified share of rights in one church), 84[77] (*c*.1110: ½-*mansus* and censual renders); V. 82 (1108 X 1110: rights in 1 *mansus*, 12*d. cens* from another property, and 4*d.* in a third), 88 (1092 X *c*.1096: rights in 1 *mansus* and 6*d.* from an allodial property as well as a tithe and other rights), 193 (1111 X 1124: 2 *bordariae*, ½-*mansus*, and part of a tithe); T. 104 (1099: ½-*mansus* held from the abbey and 1*d.* from a dependant), 315 (1121: a tithe), 319 (1113: ¼ of 2 *mansi*); Uz. 242 (1100: rights of justice in 1 *mansus* specified as receipts in kind and ⅓ of the fines) 251 (1099: ¼ of 1 priest's fief itself held in fief by the donor), 446 (1062: 5*s.* fief-rent and possibly 4*d.*).

might be expected to give property which reflected their wealth and prestige. For example, the powerful and nobly born Peter of Malemort gave 'multos mansos' to Uzerche when his son Hugh became an oblate in 1072, in other words during a period when one-*mansus* or fractional entry gifts were becoming more common.[12] A member of the family of the lords of Rouffignac entered Vigeois in the third quarter of the eleventh century with four *mansi*.[13] Similarly, in 1091 Viscount Boso of Turenne gave to Tulle his son Ebles and half of an area of woodland, with rights in the remainder, which would serve as the site of an anticipated priory, complete with church, lodgings, outbuildings, and grayeyard.[14] Secondly, an entry gift might include property which the donor already held from the donee as a fief, or which was security for an existing loan.[15] A third variable arose from the common practice of making an oblation part of a compromise deal negotiated when a property dispute was resolved.[16] For example, around the middle of the eleventh century the abbot of Uzerche received a litigant's son and paid out 40*s.* in return for the surrender of rights in a valuable church.[17] Cathedral communities might be placed under similar pressures. Peter of Vic surrendered valuable rights of jurisdiction in return for guarantees that his son would be made a canon at Ste-Marie, Auch.[18] In 1097 Archbishop Raymond II of Auch protested at Viscount Peter of Gavarret's mistreatment of the church of Nogaro. In order to 'absolve' Peter, his widow and eldest son handed over another son to be trained as a canon.[19]

Given these variables, it is impossible to say that entry gifts were in every instance strictly equivalent in value to the property which the oblate might have hoped to receive had he stayed in the world. As institutions drawing income in cash and kind from many sources, religious communities enjoyed some latitude in agreeing to entry gifts; this flexibility was bound to increase progressively as the communities expanded their proprietary resources. Furthermore, it is possible that a fixed system of tariffs would have seemed to some to bear the taint of simony even at a time when the practice of requiring entry gifts was

[12] Uz. 111.

[13] V. 29. See ibid., 21 n. 2.

[14] T. 498.

[15] B. 121 (1031 × 1059); T. 269 (1064).

[16] Lynch, *Simoniacal Entry*, 12–13.

[17] Uz. 52⁴ (1036 × 1053). Cf. P. D. Du Buisson, *Historiae Monasterii S. Severi Libri X*, ed. J.-F. Pédegert and A. Lugat (Aire, 1876), ii. 188–9 (late 11th C.).

[18] Auch, 6 (1068 × 1096).

[19] Ibid. 28.

widely accepted.[20] At least some sons, had they not become oblates
would have been expected by their families either to remain single
anyway in order to diminish their long-term burden upon the patrimony
or to 'marry out' into other kindreds' sources of wealth. Some entry gift
clearly did represent substantial endowments. In the last example cited
from Auch, the church which accompanied Peter of Gavarret's son to Ste-
Marie was reckoned to have a redemption value of 300s. if the boy died
and his brothers wished to recover their father's rights; this figure must
have been set substantially lower than the site's true value, and it is
reasonable to suppose that the property could have contributed significantly to the comfortable existence of the younger son of a viscount. But
in contrast, and even allowing for lesser material expectations, it is hard
to see how Geoffrey of Barzolas could have lived from the 2s. 2d.
measures of corn, one-eighth of two *mansi*, three hens, and one-tenth of a
sow which accompanied him to Vigeois in about 1100;[21] or Arnald of
Rouffignac from the half-sack of corn taken from a tithe with which he
entered Uzerche.[22] There are also numerous examples of an oblate having
at least two,[23] three,[24] four,[25] or even five or more[26] brothers, which
suggest that in many instances the loss of one mouth to feed would not
have substantially improved the kindred's material prospects.[27]

On the basis of such examples, it is clear that oblation cannot be
explained adequately as a social utility reducible to questions of economic
loss or gain. (It is worth noting in addition that material considerations
alone cannot account for the great disparity between the numbers of male
and female oblates in this period.) To ask whether oblation disposed of
children is to miss the basic point that professed religious were not lost to
their kindreds forever when they entered the church. Had oblates grown
up to be absorbed into an anonymous mass of cowled figures with no
links to their past lives, the appeal of oblation to the laity would have

[20] See J. H. Lynch, 'Monastic Recruitment in the Eleventh and Twelfth Centuries: Some
Social and Economic Considerations', *American Benedictine Review*, 26 (1975), 431–3, 436,
446.

[21] V. 73 (1092 × 1110).

[22] Uz. 999 (1097 × 1108).

[23] B. 121 (1031 × 1059); Auch, 13 (1068 × 1096); St-M. 20 (c.1080), 23 (1073); V.
14 (1082 × 1091: a third brother was already a monk); T. 31 (996 × 1031), 426 (1070),
435 (1119), 498 (1091).

[24] LR 56[6] (early 11th C.), 66[39] (pre-1084); T. 78 (3rd qr. 11th C.), 85 (1110), 400
(1073 × 1086), 421 (1031 × c.1050).

[25] T. 269 and 269² (1064); V. 29 (3rd qr. 11th C.).

[26] T. 364 (n.d.); SEL 84[77] (early 12th C.).

[27] But cf. Lynch, *Simoniacal Entry*, 42 and table at 43.

been minimal. The importance which was attached to the persistence of old loyalties after entry into religion can be demonstrated by first considering the minority of recorded cases in which oblates did not have close family connections with those who placed them in a community. Some charters record instances of quasi-charitable oblations in which the donor was not intimately linked to the child. For example, in 1120 the viscountess of Baïgorry in south-western Gascony gave a mute boy to Sorde with the request that he be raised by the abbot; if he ever learned to talk he was to be trained to enter orders, but if his disability persisted he was to be guaranteed a home in the abbey for life in spite of the fact that, patently, he could not hope to become a fully functioning monk.[28] In 1112 Viscount Bernard of Comborn and his wife Petronilla of Latour presented a postulant to Tulle. Their charter specified no relationship between themselves and the entrant, and the boy's toponym is not recorded elsewhere, which suggests possibly that he was a minor dependant of the vicecomital household.[29] Other oblations were expressions of a lord's concern for his *fideles*. For example, Viscount Adhemar II of Limoges provided part of the entry gift from his allodial property, and consented to the alienation of the remainder, when the son of his *fidelis* Arbert of La Vallette entered Uzerche in 1071.[30] When Peter I of Pierrebuffière provided for the entry of Constantine of Meyrat into Uzerche in the early eleventh century, the oblate, who was described simply as *puer*, may have belonged to a family which served the lords of Pierrebuffière.[31] Laymen might also sponsor the entry into a community of adults who could make a contribution to the liturgical routine. In about 1100 Viscount Adhemar III of Limoges placed in Uzerche a kinsman who was a cleric; a generation earlier Adhemar's father Adhemar II had supported the entry of a *quidam clericus* into the same monastery.[32]

It is important to note that placements of this sort were not simply motivated by compassion or a disinterested concern to see that communities were staffed by priests. In the case noted above, the viscountess of Baïgorry placed the mute boy in Sorde with an entry gift of an entire church for the benefit of the souls of herself, her kinsfolk, and particularly her dead husband Garsia Loup.[33] Adhemar III of Limoges gave his cleric to Uzerche in order to lessen the burden of his own crimes.[34] In 1121 Viscount Archambald IV of Comborn placed in Tulle a monk called John of Thiviers (with whom no relationship was expressed) because he,

---

[28] S. 7.   [29] T. 318.   [30] Uz. 229.   [31] Ibid. 323 (1003 × 1036).
[32] Ibid. 166 (1068 × 1090), 358 (1097 × 1108). See also T. 507 (1117).
[33] S. 7.   [34] Uz. 358.

Archambald, had recently killed Amalvin of Belcastel, the brother o
the lord of Malemort.[35] It is also noteworthy that the placement o
dependants, vassal's kinsfolk, distant relatives, or clergy was largely
confined to the very highest levels of lay society, in particular vicecomita
families or powerful lords.[36] Such cases were exceptional, for the basis o
oblation was that it typically involved the giving of close blood-relatives

The operation of ties of kinship was axiomatic in the *Rule of S*
Benedict's provision for child oblation. The *Rule* required that a boy from a
noble family (as distinct from the *pauperiores*) should be presented by his
kinsfolk, who made the petition to be admitted to the community on the
boy's behalf and were competent to guarantee that he would not be
enticed away in the future.[37] Cartulary scribes did not usually trouble to
copy the written petitions which were presented when a child was offered,
but a rare survival from Tulle from 1099 demonstrates how the *Rule* was
followed in practice. The petition is in the name of Walter of Naves, who
offered a boy to the abbot. Holding a token of his oblation and the
petition, the boy's hand was wrapped in an altar cloth and placed over
relics. Walter swore on the boy's behalf that the latter would remain a
monk for life and learn to serve under the *Rule*. To ensure that the
petition remained 'firm', he stated that he had subscribed it himself, and
he proffered it to the witnesses present. Finally he gave a half-*mansu*
to the abbey. The oblate was the presenting party's own brother; the
petition added that Walter was acting on the advice of his mother and
three other brothers; and the half-*mansus* was property which he himsel
already held from the abbey. In this way the *Rule*'s requirement that an
oblate be given by his relatives found intimate and practical expression
through existing family ties to the monastery.[38]

In practice the *Rule*'s provision about kinship was usually understood in
the narrowest of terms. Although there are some examples of laymen
placing brothers, nephews, or cousins in religious communities,[39] the
great majority of cases involved the giving of sons.[40] The preambles o

[35] T. 320.
[36] Cf. OV iii. 126, 164.
[37] *La Règle de saint Benoît* (ch. 59), ed. and trans. A. de Vogüé and J. Neufville (Sources
chrétiennes, 181–3; Paris, 1972), ii. 632–4.
[38] T. 104. Walter may have been a monk, for he described himself as 'Ego frater
Galterius'.
[39] Uz. 242 (1100); T. 104 (1099), 316 (late 11th or early 12th C.), 319 (1113), 435
(1119); Auch, 10 (late 11th C.); St-Seurin, 37 (1st qr. 12th C.).
[40] e.g. B. 94 (1032 × 1060), 182 (1100 × 1108); A. 42 (pre-1140); SEL 72 (*c*.1060); V.
14 (1082 × 1091), 53 (1092 × 1096); T. 20 (1031 × 1060), 212 (1097), 259 (1109), 269

charters of oblation expressed in a theological idiom the sentiments of donors who wished to place their blood-kin in a religious community. In one charter from Beaulieu, for example, the donor was made to appeal to the hallowed example of biblical authority and the fitness of rendering one's offspring to God the Creator: the son who now entered the monastic community was likened to a young Israelite serving in the Temple of the Lord.[41] Oblates were treated, in effect, as living sacrifices.[42]

The common emphasis on the giving of offspring was reflected in cases in which the oblation of a brother was conceived as the vicarious act of the father from beyond the grave. For example, in the third quarter of the eleventh century two men gave their brother to Beaulieu in performance of the wishes of their father, who had recently died.[43] Towards the end of the century two brothers, Arnald of Tarsac and William, gave their brother Bernard to the Cluniac priory of St-Mont on the advice of their mother and uncle and for the soul of their father, suggesting that the father had so disposed before his death.[44]

The oblation of sons was prevalent because it met most satisfactorily both the kin-centred emotional loyalties of donors and the practical guarantees about property rights in the entry gift which religious communities required. This combination of needs can be seen in a dispute which involved the abbey of Sorde in the first quarter of the twelfth century. A woman named Guasen of St-Dos had failed to produce children from two marriages, and, anxious about the fate of her soul, asked the abbot of Sorde whether the abbey would receive her nephew with certain properties she had decided to donate. So typical was the convention that a donor give his or her own child that the scribe who noted Guasen's petition felt obliged to record that she was particularly attached to her nephew; in other words, it was accepted that he was a surrogate son. The abbot refused Guasen's request. It is significant that Guasen did not break off contact with Sorde after receiving the rejection, for on her death-bed she gave properties to the abbey, and was remembered by the monks thereafter for her virtue and generosity. The abbot's

---

(1064), 315 (1121), 426 (1070); Uz. 111 (1072), 154 (1068 × 1096); Auch, 13 (1068 × 1096), 32 (1073 × 1085); S. 77 (1118 × 1136); LR 65[28] (1st third 11th C.), 68[41] (1st half 11th C.); Ste-C. 38 (1124). For an analogous case involving the giving of a daughter, see Uz. 251 (1099).

[41] B. 83 (1061 × 1076); cf. ibid. 97 (1061 × 1076), 107 (3rd qr. 11th C.), 182 (1100 × 1108).
[42] Lynch, *Simoniacal Entry*, 40.
[43] B. 97 (1061 × 1076).
[44] St-M. 20 (*c*.1080).

reasons for turning down the petition became apparent later, when Guasen's brother William Bergo refused to consent to her final dispositions until his son—almost certainly either the same individual as Guasen's prospective oblate or one of his brothers—was received by Sorde. Under pressure from the boy's kindred, the abbot insisted that William Bergo provide an entry gift from his own resources. After some wrangling William Bergo finally agreed, handing over property which was considerably less extensive than Guasen's original grants, and part of which was, in fact, already pledged to the abbot. The abbot's misgivings about the oblation had rested, therefore, not on the size of the entry gift but on the threat to the property of Guasen's disgruntled relatives, men of note locally who feature in some of the abbey's most important documents and showed themselves capable of making trouble in other property disputes. The entry gift which the boy's father could grant directly was much more secure. Significantly, the boy's kinsmen appear to have been in agreement that it was fitting for William Bergo to provide for the oblation himself.[45]

Guasen had wished to place the oblate 'so that he might pray for her soul'. In other words, her strong attachment to the boy would, in due course, be reciprocated by his individual contribution to the abbey's communal acts of intercession. The belief that a close relative was particularly well suited to provide personalized spiritual services informs other documents. For example, in 1073 Garsia Aner of Couture, with his wife and sons, gave his son Peter to St-Mont in order to replace another son, Maurice, who had for reasons which are not recorded left the community. In the event of Peter predeceasing Maurice, it was decided that the latter should be readmitted to take his brother's place. This arrangement was agreed, it was stated, so that God's anger should not descend on Maurice or Garsia Aner.[46] In a remarkable case from 1110, Gerard of Chaunac, on the point of leaving for Jerusalem, petitioned

---

[45] S. 100 (1118 × 1136); cf. ibid. 79 (*c.*2nd qr. 12th C.), 81 (1120), 97 (1118 × 1136), 99 (1118 × 1132), 142 (mid-12th C.). The statement in S. 100 that Abbot William Martel of Sorde refused the father's petition 'rogatus a parentibus filii sui' strictly means that he conferred with the kinsmen of his own son. But from the context it is very probable that *sui* should have been rendered *eius*, i.e. of William Bergo. The fact that the relationship was stated to have been with the son, not the father, could mean either that the son was married or widowed and the complaining parties were his spouse's relatives, or that they were the son's maternal kindred. In either case it is difficult to see why they should have worried about the fate of an in-law's aunt's or in-law's sister's property unless they were generally interested in seeing that the boy's father made himself responsible for the entry gift.

[46] St-M. 23.

Tulle to accept his son. The boy, Archambald, had earlier been given to the Cluniacs of St-Sour, Terrasson, but the monks there had thrown him out. Abbot William of Tulle agreed to take the son on the strict understanding that his abbey's rights to the entry gift would not be challenged if the boy were reclaimed by another religious community. About to depart on a perilous journey, Gerard was clearly desperate to place his son somewhere in order to secure the personalized intercession he could provide.[47] In 1099 a man granted three churches to Ste-Croix, Bordeaux, on condition that one, at Carcans, be served by monks as a semi-independent priory, and that two of his sons be given the option of becoming monks there without the need of an entry gift. This opportunity was expressly denied other postulants, who were required to endow the priory from their patrimonies. It was further agreed that if either son declined to take the monastic habit but wished to serve Carcans as a secular cleric, the community of Ste-Croix would house and feed him from Carcans's resources. Thus the donor wished to complement the generous spiritual rewards which he received from Ste-Croix (he was granted confraternity with the monks) with the particularly powerful intercession of an offshoot community built around his children.[48]

Adult Conversion

Oblation was not the sole means of recruitment to religious communities. Benedictine monasteries and houses of regular canons also received adult converts. (The same was less true of communities of secular clergy, which were ill-suited to find employment for uneducated men who could not hope to enter the major orders, and where recruitment was probably conducted on a more overtly hereditary basis.) Adult conversion is a particularly significant institution because, in those observations upon the nature of the crusader's vocation which were noted earlier, Guibert of Nogent and Ralph of Caen treated entry into professed religion as the established expression of piety which most closely corresponded to crusading. The various forms which adult conversion assumed, therefore, and what the sources can reveal of converts' motives, help to establish the parallels with crusading as a means of attaining salvation.

[47] T. 85. The condition that Tulle's rights would be unimpaired 'si quis requiret eum [Archambald] atque per rectum posset habere' suggests that St-Sour was not the only community which Gerard had approached in the past.

[48] Ste-C. 84.

A recent thorough study of lay conversion at Cluny in the tenth and eleventh centuries has demonstrated that converts there were treated in many respects as full monks. In spite of the fact that converts, unlike oblates, might never acquire enough Latin to be able to participate fully in the *Opus Dei*, they had an important role in the liturgical routine, for example as thurifers, servers, and in formal processions, and they took their place alongside the other monks in chapter, the acid test of the brethren's status and relative seniority.[49] There is every reason to suppose that, on a more modest scale, the arrangements were very similar in the Cluniac and Benedictine communities in Gascony and the Bas-Limousin. Typically, when a man asked to enter a community, he asked 'that he might become a monk'. Within the limits imposed by his age, education, and intelligence, he expected to be recognized as an integrated member of the community, bound by the same vows and enjoying the same prospect of salvation as former oblates. Certainly he did not anticipate being subject to anything approaching the institutional distinction which was made in some twelfth-century reforming communities between the choir-monks and the often illiterate *conversi*, who assumed manual and administrative tasks.[50]

The integration of converts is demonstrated by two points. First, it was a corollary of those criticisms noted before, concerning former oblates' shortcomings, that religious communities could welcome and exploit adult converts' experience of the outside world, for example in administration, estate management, and liaison with secular authorities.[51] Second, there was no formal barrier to prevent converts (perhaps, but not necessarily, favoured by some earlier education) from entering major orders or attaining high office. Hildebert Grimoard, for example, was a *miles* based at the castle of Ségur who took the habit at Uzerche after receiving a near-fatal wound in battle. Recovering his health against all expectation, Hildebert respected his vows, remained at Uzerche, and rose through the major orders, becoming abbot in 1113. The author of

[49] W. Teske, 'Laien, Laienmönche und Laienbrüder in der Abtei Cluny: Ein Beitrag zum "Konversen-Problem"', *Frühmittelalterliche Studien*, 10 (1976), 255–7, 282–6, 290–309, 312–21. See also, N. Hunt, *Cluny under Saint Hugh 1049–1109* (London, 1967), 89–91; G. Constable, '"Famuli" and "Conversi" at Cluny: A Note on Statute 24 of Peter the Venerable', *RB* 83 (1973), 334–40; P. D. Johnson, *Prayer, Patronage, and Power: The Abbey of la Trinité, Vendôme, 1032–1187* (New York, 1981), 39–41.

[50] Hunt, *Cluny*, 91.

[51] Cf. C. Harper-Bill, 'The Piety of the Anglo-Norman Knightly Class', in R. A. Brown (ed.), *Proceedings of the Battle Conference on Anglo-Norman Studies*, ii. 1979 (Woodbridge, 1980), 68–9.

the short mid-twelfth-century chronicle of Uzerche remembered that Hildebert made a good abbot, was respected and pious, and applied himself to becoming expert in Scripture 'although he was not greatly educated in letters'. The author did not believe that Hildebert suffered by comparison with his learned predecessors, Gerald of Nontron (1068–97) and Gauzbert Malafaida (1097–1108), who had both been connected with the prestigious Cluniac community of St-Martial, Limoges. In fact he hinted strongly that Hildebert had been chosen, and proved able, to restore harmony within the community after the abbacy of Peter Bechada of Lastours (1108–13), an incompetent leader whose conduct had caused dissent among the monks, prompted an acrimonious complaint to the papal legate Bishop Gerald of Angoulême, and eventually forced his retirement. Hildebert was able to hold down the office of abbot for twenty years, and, significantly, it was his old age, not any handicap resulting inevitably from his late entry into monasticism, which finally led the monks to 'murmur' against him and to push him into dignified retreat in a dependent church.[52] Hildebert's experience was not unique. His contemporary Bernard III Grossus of Uxelles entered Cluny as an adult convert and in 1114 was elected abbot of St-Martial as the immediate successor of the venerable and very highly regarded Abbot Adhemar (1063–1114). In the event, Bernard proved less successful than Hildebert, for he soon accepted demotion to become prior of Cluny (an office which was itself no soft touch), but it is still very significant that he could have even been considered seriously for the highest office in a prestigious community with proud traditions.[53] Men such as Hildebert and Bernard formed a small minority of all adult converts, but their careers illustrate well the ability of professed religion to accommodate late arrivals.[54]

[52] 'Historia monasterii Usercensis', extr. *RHGF* 14. 339–40.

[53] Adhemar of Chabannes *et al.*, 'Commemoratio abbatum Lemovicensium basilice S. Marcialis, apostoli', ed. H. Duplès-Agier, *Chroniques de Saint-Martial de Limoges* (Paris, 1874), 10 (a very brief notice, suggesting that Bernard's impact upon St-Martial was slight); Geoffrey of Vigeois, 'Chronica' (ch. 18 [*recte* 38]), ed. P. Labbe, *Novae Bibliothecae Manuscriptorum Librorum* (Paris, 1657), ii. 299; Teske, 'Laien, Laienmönche', 10. 319, 320. Teske argues that Bernard must have been in major orders to have been considered for abbot.

[54] For other lay converts who attained high office, and sometimes acquired a formidable reputation, see H. Grundmann, 'Adelsbekehrungen im Hochmittelalter: *Conversi* und *nutriti* im Kloster', in J. Fleckenstein and K. Schmid (edd.), *Adel und Kirche: Gerd Tellenbach zum 65. Geburtstag dargebracht von Freunden und Schülern* (Freiburg, 1968), 325–45, esp. 326–7, 332–4. See also OV ii. 132 and, for a Gascon example, Rodrigo Jiménez de Rada, *Historia de Rebus Hispaniae sive Historia Gothica* (VI. 24), ed. J. Fernández Valverde (Corpus Christianorum, Continuatio Mediaeualis, 72; Turnhout, 1987), 206.

Modern treatment of adult conversion has tended to concentrate upon the experiences of founders of reforming communities or famous ecclesiastical leaders, for example Romuald of Ravenna, Herluin of Bec, Hugh of Cluny, and Bernard of Clairvaux. Men such as these are well served by Lives which devoted passages to the timing and conduct of their conversions and which, allowing for the conventions of hagiography, enable the historian to speculate about the psychological processes at work.[55] It is important to remember, however, that this sort of celebrated holy man was a rare individual. He was usually canonized (though not necessarily soon after his death); in other words, he came to be treated officially by the church as someone who had exhibited such particular virtue and devotion that he was believed to have entered Heaven. It is therefore reasonable to wonder whether the experiences of the small élite of conspicuously holy religious were different from those of the majority of less exalted converts; or, conversely, whether the holy men's conversions did share, in some accentuated form, the same basic motivations as the experiences of the rank and file of adult converts. In order to examine this question, it is useful to make a provisional distinction between converts who pioneered new forms of religious life or were among the first members of new communities, and those who entered long-established religious houses and thereby expressed the desire to conform to traditional practices.

An excellent example of the pioneering convert in action is provided by the foundation in 1079 of the abbey of La Sauve-Majeure in the woods of Entre-Deux-Mers fifteen miles east of Bordeaux. The inspirational figure behind the creation of La Sauve was St Gerard of Corbie, who was not himself an adult convert. Born around 1020 into the middling nobility of Picardy, Gerard was made a monk of Corbie as a boy, and rose to succeed his brother as abbot of St-Vincent, Laon in about 1074. Attempting to reform St-Vincent in accordance with his rigorist tastes, Gerard ran into fierce resistance from its monks and was dismissed by the local bishop. Under the influence of Arnulf of Pamele, the future bishop of Soissons, Gerard was installed as abbot of St-Médard, Soissons sometime in the mid-1070s, only to run into difficulties once more and be replaced by a candidate supported by King Philip I. This final setback seems to have persuaded Gerard to take to the road on a sort of pilgrimage of discovery in search of an opportunity to put into effect his belief in reformed

---

[55] e.g. A. Murray, *Reason and Society in the Middle Ages*, rev. edn. (Oxford, 1985), 350–82, esp. 374–9.

Benedictinism with an eremitical slant. Arriving in Aquitaine in 1079, he attracted the interest of Duke Guy Geoffrey and was granted land on which to settle. With the active support of local nobles and papal legates, La Sauve became a conspicuous success under Gerard as its first abbot. He died on 5 April 1095.[56]

Gerard's Life was composed in *c.*1140 by a monk of La Sauve who drew upon the recollections of Peter of Amboise, the seventh abbot and Gerard's former chaplain, and other men who had been influenced by Gerard in their younger days.[57] Naturally Gerard took centre-stage in the Life, which in conventional hagiographical terms praised his noble birth, his precocious aptitude for monastic discipline as an oblate, and the strictness of his routine as a young adult. With the benefit of the hindsight of Gerard's achievements at La Sauve, the Life's author did not dwell on the failures at St-Vincent and St-Médard, but he was also concerned to establish that Gerard was not a cloistered ineffectual. He praised, for example, Gerard's success in supervising the restoration of the abbey church at Corbie in the 1050s and 1060s.[58] Thus the principal thrust of the Life was that it had been Gerard's personal qualities as developed in a claustral environment, and his great experience of monasticism, which had made the foundation of La Sauve possible.[59] To this extent the Life celebrated an exercise in monastic renewal by an insider, someone who had been in the church for nearly all his life and had never had direct personal experience of the lay convert's problems.

Gerard had not struck out for Aquitaine alone, however. At early stages of his quest he attracted to himself various companions, whose numbers meant that by the time of his arrival at La Sauve he already had the nucleus of a community of monks. It is noteworthy that Gerard's Life devoted a long passage to these original followers, giving their names and potted accounts of their previous careers. The inner circle of Gerard's following comprised three men: Martin, a monk from St-Vincent who went on to become abbot of St-Denys-du-Mont; Gerard's nephew Aleran, later third abbot of La Sauve (1102–6); and Ebroin, a former knight who

---

[56] Gerard of Corbie, 'Notitia de fundatione monasterii Silvae-majoris', *RHGF* 14. 45–6; 'Vita S. Geraldi abbatis' (chs. 1, 3), *AASS* (Apr.) 1. 414, 419–20; G. M. Oury, 'Gérard de Corbie avant son arrivée a la Sauve-Majeure', *RB* 90 (1980), 306–14.

[57] 'Vita S. Geraldi', 414–23; see esp. (ch. 4), 421, 422.

[58] Ibid. (chs. 1, 2, 3), 414–15, 417, 419. For the dating, Oury, 'Gérard de Corbie', 309–10.

[59] Cf. G. M. Oury, 'La Spiritualité du fondateur de La Sauve-Majeure, saint Gérard (v.1020–1095)', *Revue historique de Bordeaux et du Département de la Gironde*, NS 29 (1982), 5–19.

had probably already become a recluse by the time that he attached himself to the group.[60] More significant in the present context, however, was an outer circle of five laymen who took the habit at La Sauve. These five had decided together to leave the world and had sought the advice of an anchorite (probably Ebroin or Arnulf of Pamele), who commended them to Gerard. Gerard's party assembled at St-Denis, and worshipped there, at Ste-Croix, Orleans, and St-Martin, Tours, before reaching Aquitaine. After arriving at La Sauve the five laymen journeyed on to Compostela, then returned to Gerard and became monks.[61]

The laymen's stories have three important features. In the first place, the five, despite being of differing social status, were linked by the fact that they had been active arms-bearers. The most important of them was Berlegius of Noyon, a royal vassal, who was described in the Life as noble and rich. Probably alone of the five, he had the advantage of some childhood education, no doubt because his family had originally intended him to enter the church. Guy of Laon, a vassal of the bishop of Laon, was remembered as being of lesser noble stock. The third layman, Tezzo, was a youth who, the author of the Life maintained, had attached great importance to the reputation he might win as a warrior. The final two, Walter of Laon and Lither of Laon, were knights. The author of the Life recognized that the laymen had been active warriors. For example, it was said of Berlegius, who was expert in the use of arms, that he had been unable to refrain from warfare as an adolescent when he became aware of its attraction for others of his age; it was quite some time before he felt the urge to convert. Lither, like Tezzo, was immersed in ideas of honour and prestige: conscious of the need to stand out amongst his peers, he directed his energies towards ever greater feats of *militia*. In other words, there was no suggestion that these men were failures in the world. Nor had they experienced distaste for military activity throughout their adult lives. The author of the Life implicitly recognized this fact when he recorded that, during their secular careers, the five had been able to exhibit some of those qualities which they took into their monastic vocation, such as charity, prudence, temperance, discretion, and *amicitia*, which may be interpreted as the ability to function as part of a group.[62]

---

[60] 'Vita S. Geraldi' (ch. 3), 420. Ebroin's exact status is unclear. The Life records simply that 'de militari habitu egrediens, Dei amore inclusus, usque ad mortem fortiter pugnavit', which could refer to his spiritual career after his arrival at La Sauve. The Life adds (ibid.), however, that the other followers of Gerard (for whom see below) were still in the lay habit in contrast to the first three, described as 'monachi jam facti'. This suggests that Ebroin had become a professed religious of some sort before he joined Gerard.

[61] Ibid. 419, 420.   [62] Ibid. 420.

The second important point about these men's conversion is that, with the partial exception of Berlegius, they were not formally learned but still capable of making an important contribution to the foundation of La Sauve. The Life stated that when they had been recommended to follow Gerard they were effectively acquiring 'the rudiments of faith'. They had never read Scripture, but they exhibited great enthusiasm to obey its commands. In performance of the prophecy in Isaiah 2: 4 ('and they shall beat their swords into ploughshares, and their spears into pruning hooks') they became prodigious clearers of the woods on the site of their monastery, whereas earlier they had been vigorous warriors. In addition the Life stated that Berlegius had been notably careful with money when a layman, suggesting that his experience was exploited by the community. In short, it was the laymen, much more than Gerard (who was then about 60) and the professed followers, who supplied the raw energy which gave La Sauve its early momentum.[63]

The final point is that, although the Life does not dwell at length on the motives for the laymen's conversions, some broad themes can be partly reconstructed from a few general remarks. It was believed that the laymen entered into a state of penitence, for they had first approached the hermit who introduced them to Gerard in order to make a full confession of their sins.[64] Berlegius's experience of conversion, which receives the most detailed treatment, is described as a second baptism and 'pure', or total, confession.[65] The sins which may have weighed on the men's minds are only expressed in vague terms, and there is no suggestion that any of the five converted after a specific unnerving misdeed or accident. The account of how Berlegius was touched by the right hand of God and realized that he had not been leading a good life immediately follows a passage which treats his secular career in very general terms. As monks, the Life asserts, the former laymen refrained from their past 'illicit acts' and strove to suppress their 'desires of the flesh'. Thus there is the suggestion, unremarkable in itself, that they consciously renounced violence and sexual activity. The fact that they chose to live on bread and water, and the contrast made in the Life between their former luxury and their ragged monastic habits, point to a revulsion from conspicuous consumption and trappings of status.[66] In sum, the Life drew on stock themes of monastic conversion, such as penitence and self-abasement, which were common currency (particularly so under the influence of twelfth-century reformed monastic ideals). Writing with an insider's

---

[63] Ibid. 419–20.   [64] Ibid. 419.   [65] Ibid. 420.   [66] Ibid.

perspective, the author of the Life treated the reasons for the men's conversions as largely self-explanatory.

Yet even if we fail to gain precise insights into the five laymen's states of mind, the Life provides important evidence for the manner in which the act of conversion was considered in the round. It is in this context that it is possible to isolate parallels with the crusading vocation. In the first place, the laymen's conversion was associated with pilgrimage. On one level the laymen's journey to Compostela was a device to assimilate their vocation with that of St Gerard himself. The Life devoted a good deal of space to two pilgrimages which Gerard had undertaken in his middle years: in *c.*1050 to Rome, Montecassino, and Montegargano in search for a cure of a chronic and debilitating illness; and probably in the 1060s to Jerusalem. By describing these journeys in some detail (and by omitting references to Gerard's failures in other periods of his life), the Life effectively proposed that the pilgrimages were significant phases in Gerard's spiritual development.[67] On a less exalted level the same applied to the knights. It is also noteworthy that the five insisted on fulfilling their vows to carry on to the shrine of St James as laymen, even though they had already committed themselves to serve Gerard. The pilgrimage was not some way of postponing the decision to take the habit: it was treated as an extension of the whole group's journey to the south-west, and as anticipating and complementing the laymen's conversions.[68] Further, and more fundamentally, the Life described the group's, and particularly the lay element's, vocation in language which is redolent of crusade terminology. It was said of the small band setting out from northern France that they abandoned the world and intended to follow in the path of Christ. Berlegius of Noyon renounced all his possessions in obedience to the instruction of the Gospels, and the five followed Christ by turning their backs on their goods, lands, and kinsfolk. The knight Ebroin had earlier abandoned the world for love of God, and the laymen attached themselves to Gerard for the same reason. These observations amount, in fact, to a fair summation of some of the more commonly cited scriptural bases of crusading.[69]

A new foundation such as La Sauve, which was supported by the secular and ecclesiastical authorities and which was not colonized by a

---

[67] Ibid. (chs. 2, 3), 416–17, 418–19.

[68] Ibid. (ch. 3), 420. For the whole group travelling from northern France like pilgrims, see ibid. 419.

[69] Ibid. 419, 420. See J. S. C. Riley-Smith, 'Crusading as an Act of Love', *History*, 65 (1980), 177–80.

mother house, offered ample scope for adult recruitment. There survive in La Sauve's *Grand Cartulaire* records of no fewer than seven arms-bearers who took the habit during St Gerard's abbacy, an average of nearly one every two years.[70] The needs of the *Opus Dei* meant that some clerics were encouraged to enter. A priest from near Civrac entered La Sauve with the consent of his family and gave vines and an orchard.[71] Gerard's successor as abbot, Achelm (1095–1102), was formerly archdeacon of the cathedral of St-André, Bordeaux.[72] It was also already possible during Gerard's lifetime for women to take the veil at La Sauve, though it is not known what arrangements were made for them and where.[73] The adult lay male convert was, however, the staple of La Sauve's early recruitment. A policy to attract men of this sort was both necessary for the community's survival and growth, and in harmony with the ideals of Gerard's first lay followers.

The preambles of some of the lay converts' charters survive in the cartulary. Raymond William, for example, was made to declare that he had become mindful of the evanescent nature of earthly existence, and alarmed by the 'terrifying places' which awaited mankind after death: he had therefore resolved to put aside his carnal desires by placing himself under God's yoke.[74] Another convert, a *miles* named Arnald, expatiated on the need to flee the world: time was pressing, and the shades (*tenebre*) were fast approaching; fearful of a sinner's death, he had begun to reflect on his past actions and to count up the punishments which his crimes merited; with a contrite heart, therefore, he now sought refuge with St Gerard.[75]

It is striking that the preambles' terminology is similar to statements made in Gerard's Life. Converts spoke of dissatisfaction with the world; of the generality of sins expressed by 'physical desire' or 'my deeds'; of the fear of spiritual punishment; and of the need to find a refuge in God's service. Even when such unspecific terms were used the parallelism with crusading was still evident. Thus a *miles* named Joscelin Paucus who became a monk rejected the vanities of the world, took up his cross, and followed Christ. This act was accounted as entering the service of the

---

[70] LSM, pp. 16, 58, 96, 142 (twice), 144, 155, 406.
[71] Ibid. 165.
[72] Ibid. 112.
[73] Ibid. 67.
[74] Ibid. 142. Raymond William's status can be inferred from the fact that his sons subsequently disputed his entry gift at the ducal court in Bordeaux.
[75] Ibid. 144.

Lord.[76] Another *miles* who converted during Gerard's abbacy or shortly thereafter left his wife 'for the Lord's sake' (though he made provision for her maintenance for the rest of her life).[77] The actions of these early converts demonstate that some arms-bearers were attracted by the example of St Gerard and his followers. In other words, there was a period while La Sauve was still a recent foundation when outsiders could hope to participate in the raw enthusiasm for conversion displayed by Gerard's small group, and it was appropriate to describe their change of lifestyle in the same bold terms.

It is worth noting, however, how quickly traditions of family, lordship, and location came to influence conversions. One, and possibly two, of the earliest converts came from distant Noyon and were most probably connected in some way with Berlegius.[78] To judge from the witness list of his charter, the convert Raymond William was a vassal of Auger of Rions, an important local lord on whose allodial land La Sauve had been built and who featured prominently in the abbey's earliest documents.[79] The convert Arnald William of Escoussans was the brother of another important lord of Entre-Deux-Mers, Bernard, who had received the benefit of the abbey's prayers in return for the surrender of claims against Auger of Rions's foundation grant; Bernard, like Auger, was a vassal of the abbey's leading lay supporter, William Amanieu of Benauges.[80] By the second quarter of the twelfth century, one of Arnald William's kinsmen had become a monk at La Sauve.[81] The convert Raymond Bernard, a *miles*, was probably William Amanieu of Benauges's vassal.[82] A *miles* who converted under Abbot Achelm had previously made modest gifts to the abbey during Gerard's rule.[83] Already during Gerard's abbacy entry into La Sauve was becoming institutionalized: at least one oblate was received, and the abbey was accepting donations in return for the deferred option of conversion at some unspecified point in the future.[84]

Within one generation, then, adult conversion at La Sauve began to

[76] Ibid. 142, 155.

[77] Ibid. 406. The entry is undated but is connected with the immediately preceding notice concerning a convert from the same region and dated 1079 × 1095.

[78] Ibid. 406.

[79] Ibid. 144.

[80] Ibid. 4–5, 16. For William Amanieu's importance to La Sauve, see ibid. 4, 5, 6–7, 10, 14, 15, 16.

[81] Ibid. 19.

[82] Ibid. 58.

[83] Ibid. 177. The notice makes it clear that the convert was not received *ad succurrendum*.

[84] Ibid. 50, 78. Cf. ibid. 100 for an oblation dated 1102 × 1106.

assume characteristics which we shall see were typical of the second type of conversion mentioned earlier: that involving entry into an established and well-known monastic tradition. One way to approach this second type is to study cases which may seem at first glance to form an intermediate group between the pioneering and traditional forms of conversion, sharing some of the features of both.

From the third quarter of the eleventh century there was a remarkable increase in the number of religious communities in southern Gascony which were founded or reformed by prestigious outside institutions, most notably Cluny and, to a slightly lesser extent, St-Victor, Marseilles. Southern Gascony, much more than Entre-Deux-Mers (which had the wilderness to attract experiments such as that at La Sauve) or the Bas-Limousin (where monastic settlement was already quite dense), saw the harnessing of novel, external influences to respect for traditional religious practices. For example, in 1064 the brothers Bishop Heraclius of Tarbes and Count Bernard II of Bigorre granted to Cluny the family monastery of St-Lizier, which, their charter stated, was an ancient community now in steep decline.[85] St-Orens, Auch, began to emerge from the shadow of its neighbouring cathedral in 1068, when it was granted to Cluny by Count Aymeric of Fezensac.[86] Ste-Foi, Morlaas, was a very recent foundation given to Cluny by Viscount Centulle IV of Béarn in 1079.[87] In the following year Centulle, as the husband of the heiress of the count of Bigorre, granted St-Savin, Lavedan, 'now almost wholly lost to monastic observance', to St-Victor with a remit to eject any obstructive monks and to introduce brethren from Marseilles in order that the mother community's customs could be reproduced there.[88] In 1087 Centulle similarly granted the abbey of St-Sever-de-Rustan near Tarbes to St-Victor with the same sweeping instructions. His widow confirmed St-Victor's rights in the abbey in 1091.[89]

At least from the point of view of the institutions invited to introduce reform, these Gascon communities were in such a parlous state that they needed monks from outside in order to bring them up to the *Rule*'s

---

[85] C. iv. 3402.

[86] Ibid. 3414. For tension between St-Orens and the archbishop of Auch, see also Auch, 162 (one of 14 documents from St-Orens appended to the *Cartulaire noir* of Ste-Marie, Auch) and H. E. J. Cowdrey, *The Cluniacs and the Gregorian Reform* (Oxford, 1970), 97–101.

[87] *Cartulaire de Sainte-Foi de Morlàas*, ed. L. Cadier (Collection de pièces rares ou inédites concernant le Béarn, 1; Pau, 1884), 1.

[88] *Cartulaire de l'abbaye de Saint-Victor de Marseille*, ed. M. Guérard (Collection des cartulaires de France, 8–9; Paris, 1857), i. 483; cf. ibid. ii. 841 (1081).

[89] Ibid. i. 484; ii. 818.

standards.[90] Once a community was reorganized and had gained in prestige, however, it was open to it to attract local converts who would help to secure the support of the region's arms-bearers. This process can be seen at work in the period's best documented instance of monastic reform in southern Gascony, the granting of St-Mont in Armagnac to Cluny.

St-Mont was founded in 1036 by a minor lord, Raymond of St-Mont. (Later Cluniac tradition had it that Raymond had become alarmed by a plague which was decimating the local populace, and decided on his foundation after an unnerving dream.) Anticipating the resistance of his mother and brother (correctly, as events turned out), Raymond sought the support of the local prince, Count Bernard II Tumapaler of Armagnac. Bernard was not at first enthusiastic but eventually agreed to act as guarantor for Raymond's donations. Exactly how and why he did so are unclear. A notice recording St-Mont's early history, written after its grant to Cluny and (for reasons which will become obvious) very biased towards Bernard, states that Raymond reneged on a vow he had made to become a monk of his own foundation and ran away to live with disapproving kinsfolk, thereby forcing Bernard to step in and fulfil his votive obligation to defend the community. After five years Raymond returned, negotiated a property deal to provide for his family from the lands he had earlier granted to St-Mont, and entered the community.[91]

Count Bernard became vigorous in his support of St-Mont, most notably in helping it to resist the jurisdictional claims of Archbishop Austinde of Auch;[92] and it was he who granted the community as a dependent priory to Cluny sometime before 1060.[93] A few years after the introduction of the Cluniacs to St-Mont, Bernard decided to become a monk. In the present state of the documentation, the chronology of the events leading up to Bernard's conversion is very confused. Consequently the reasons for his actions cannot be reconstructed with complete confidence. Bernard's case is worth considering in some detail, however, because it points to the existence of political and domestic pressures

---

[90] See also C. iv. 3471 (1074), 3630 (1088).

[91] St-M. 1; see also Auch, 14. By 1062 Bernard was claiming full credit for St-Mont's creation: St-M. 7 at pp. 15–17.

[92] St-M. 7 at pp. 17–18; Auch, 14.

[93] St-M., pp. xiii–xiv and no. 1 at p. 6. Cowdrey, *Cluniacs*, 79 follows the dating of 1055 which de Jaurgain proposed (St-M., p. 8 n. 1) on the basis of a scribal error of *feria* IV[a] for VI[a]. The absolute terminal dates are *c.*1050 (the accession of Archbishop Austinde) and King Henry I's death in 1060.

which could supplement the sort of religious responses found in charters' preambles and hagiographical texts.

The clock began to run on Bernard's conversion in 1032 with the death without direct male heirs of the last Mitarrid duke of Gascony, Sancho William. As a result of the succession disputes which ensued, the ducal power-centre in Bordeaux became detached from southern Gascony. In the north of the duchy members of the ducal line of Aquitaine made gradual progress in asserting their overlordship, but found themselves impeded by their own succession problems as each of Duke William V's elder sons died in turn. Odo, William V's son by Brisca, the daughter of Duke William Sancho of Gascony, was attempting to consolidate his power in Bordeaux when he was recalled to Poitou by the death without a son of William VI in 1038. He then had no opportunity to resume his efforts in Gascony before his own death in the following year. Poitevin influence in northern Gascony seems to have been suspended until 1044, when an expedition under William V's youngest son Guy Geoffrey succeeded in reducing the Bordelais. Guy Geoffrey lacked Odo's direct dynastic interest in Gascony, for his mother was not Brisca but Agnes of Burgundy, by then the wife of Count Geoffrey Martel of Anjou. But Odo had established the precedent for a Poitevin cadet seeking to carve out an area of authority from the deep south-west. In 1044 Guy Geoffrey's elder brother, Duke William VII Aigret, was himself just emerging from his minority and there was no reason to suppose that he would not produce sons in due course. (In the event, Guy Geoffrey succeeded him in 1058.) Guy Geoffrey attempted to consolidate his entry into Gascon affairs by marrying Odo's widow, Countess Aina of Bordeaux; and, with a Poitevin installed as archbishop of Bordeaux by 1045, his position there was secure. At this stage, however, Poitevin authority made little progress beyond the Landes into southern Gascony.[94]

The position in southern Gascony is more confused, but it seems that Bernard II of Armagnac very gradually emerged as the leading claimant to the Mitarrid succession (at least to the title of count of Gascony)[95] by reason of the rights transmitted by his mother, William Sancho's

[94] For the sequence of events described in this paragraph, A. Richard, *Histoire des comtes de Poitou 778–1204* (Paris, 1903), i. 268–70; J. Martindale, 'The Origins of the Duchy of Aquitaine and the Government of the Counts of Poitou (902–1137)', D.Phil. thesis (Oxford, 1965), 97–104; R. Mussot-Goulard, *Les Princes de Gascogne* (Marsolan, 1982), 189–90.

[95] W. Kienast, *Der Herzogstitel in Frankreich und Deutschland (9. bis 12. Jahrhundert)* (Munich, 1968), 273.

daughter Adalaïs. Significantly, Bernard won the support of his kindred, the viscounts of Béarn and Lomagne, and the witness lists of his charters demonstrate that he managed to forge alliances with many of the lords of south-central Gascony, including the counts of Fezensac, to whom he was also related. In this way an equilibrium emerged between north and south, only to be destroyed in 1058 when Guy Geoffrey succeeded William VII in Poitou and Aquitaine, and the prospect emerged of Poitevin resources bearing down on Gascony.[96]

It is against this background of increased Poitevin pressure after 1058 that Bernard's conversion should be set. The preamble of a later charter from St-Mont recorded that Bernard met Abbot Hugh of Cluny and travelled with him to Toulouse. There he made his dispositions to his wife and sons before proceeding to Cluny, where he took the habit.[97] These events almost certainly took place in 1062.[98] It has often been supposed that Bernard's conversion was an act of weary surrender after he had been crushed in battle by Guy Geoffrey. The sole piece of evidence for this belief is a seventeenth-century précis of a lost document from the abbey of St-Sever which states, after recording the subscription by Guy Geoffrey of a charter drawn up at the monastery of St-Jean de la Castelle, that nearby the duke, at the head of large forces, triumphed over his enemies.[99] The fragment is dated 7 May 1073, but the synchronisms in the dating protocol of epact, lunar phase, indiction, and day do not match; it also incorrectly refers to Pope Alexander II as alive (though on 7 May 1073 he had been dead for only a little more than a fortnight). Various dates have been suggested for the original document in an effort to fit it into the confused history of Gascony in this period.[100] It would seem that, in fact, too much has been read into the St-Sever fragment, which is no more than a late and highly corrupt tradition and makes no mention of Bernard of Armagnac. Even if we accept its date as May 1063, as have some scholars, we have seen that Bernard had probably already left Gascony by then.[101]

[96] Mussot-Goulard, *Princes*, 191–4, 195.

[97] St-M. 68.

[98] H. Diener, 'Das Itinerar des Abtes Hugo von Cluny', in G. Tellenbach (ed.), *Neue Forschungen über Cluny und die Cluniacenser* (Freiburg, 1959), 360, 399. See St-M. 7 at p. 19.

[99] P. de Marca, *Histoire de Béarn* (Paris, 1640), 280.

[100] P. Tucoo-Chala, *La Vicomté de Béarn et le problème de sa souveraineté des origines à 1620* (Bordeaux, 1961), 33 n. 12 decides for 1061 or 1062. Richard, *Histoire des comtes*, i. 290–2, followed by Mussot-Goulard, *Princes*, 204 n. 130, chooses 1063, but Richard misdates the siege of Barbastro. See the comments of Martindale, 'Origins of the Duchy', 106–7 and n. 74.

[101] Mussot-Goulard, *Princes*, 206 points out that Bernard subscribed a charter of November 1064 as Count of Armagnac (C. iv. 3402) and argues that he took the habit

There is therefore little reason to posit a single, cataclysmic reverse as the reason behind Bernard's conversion. It is more realistic to regard the background to it as Guy Geoffrey's growing pressure on southern Gascony. The precise stages by which Guy Geoffrey intruded his authority into the deep south cannot be traced with certainty. In 1061 and at least the early part of 1062 he would have been unable to intervene directly in Gascony because he was engaged in a long and difficult campaign wresting the Saintonge from Angevin control.[102] But by 1064 he was sufficiently confident of his position in the south and the security of the Pyrenean passes to contemplate participation in the Barbastro campaign. It is therefore probable that Poitevin pressure on southern Gascony had mounted in the intervening period. It was remembered at St-Mont that Bernard had gone to Toulouse 'unknown to his opponents', a phrase which suggests that he had felt himself on the verge of defeat.[103] In 1062 Bernard was in his mid-fifties and had two adult sons.[104] He could therefore be tolerably confident that the county of Armagnac would be secure in his family under any new Poitevin dispensation.

Such may have been Bernard's calculation. It would be mistaken, however, to treat his conversion as nothing more than a decorous retirement. In time he returned to Gascony as a monk, witnessed a number of St-Mont's charters, and supported the decision of his kinsman Centulle IV of Béarn to cede Ste-Foi, Morlaas to Cluny.[105] He lived on until about 1080. Thus Bernard was prepared and able to continue as an influential figure in the Gascon scene, and his behaviour was not that of someone who had become handicapped by a grave loss of status. It is reasonable to suppose that his conversion and support of progressive monastic reform in fact did much to salvage his prestige. Bernard's active career after he became a monk helps to put his political misfortunes into context. His difficulties complemented rather than supplanted the ideas contained in the observation of a scribe of St-Mont that Bernard left the world animated by the recollection of his sins and the judgement awaiting the good and evil.[106] His record of support for St-Mont and Cluny suggests that such sentiments were genuine.

---

shortly thereafter. Her argument is unconvincing, however, for the subscription may only mean that Bernard had not made his full profession by then or that the title was a scribe's honorific gloss.

[102] *La Chronique de Saint-Maixent 751–1140*, ed. and trans. J. Verdon (Les classiques de l'histoire de France au Moyen Âge, 33; Paris, 1979), 134–6.

[103] St-M. 68.

[104] Mussot-Goulard, *Princes*, 191.

[105] *Cartulaire de Sainte-Foi*, I; St-M. 31, 52, 73.

[106] St-M. 68 (amending 'futuram discussionem').

Bernard's case is a reminder that conversion might be prompted by personal or political changes of fortune of the sort that would seldom be mentioned in charters or hagiographical texts. Equally, however, it would be wrong to regard conversion as nothing more than an escape mechanism for individuals beset by crisis.[107] Charters usually pass over the background circumstances of conversions, but they are valuable evidence for the importance of other factors such as family tradition and ties of lordship. It was believed at St-Mont, for example, that Bernard's conversion directly inspired some of his followers (*sui*) to imitate his example. One such was Fort Brasc of Montagnan, a *miles* of Count Bernard who supported his lord when the latter defended St-Mont against Archbishop Austinde, and who himself gave property to the monks there. Fort Brasc's uncle was Garsia Brasc of Lannux, whose family had supported St-Mont since before its grant to Cluny and featured among the list of important local lords who swore to protect the community's jurisdictional rights in *c*.1055.[108] Fort Brasc entered Cluny alongside Bernard of Armagnac, returned to St-Mont, and vigorously defended the priory's claim to property which, having originally been granted to the monks by him, had been disputed by his brother during his absence.[109] Were it not for the unusual circumstance of a major prince setting an example at a time of political turmoil, Fort Brasc's conversion would be indistinguishable from other cases in which lords and *milites* entered long-established religious communities. His and Bernard of Armagnac's experiences bridge the impulses of pioneer converts and the ideas of men who entered old communities. We can now turn to the latter.

The appeal to arms-bearers of traditional forms of religious life can be demonstrated by a number of examples. It was common for an old man to convert in an act which capped a long career of support for a religious community. In the late eleventh and early twelfth century, for instance, Manuald Trenchelion and his brothers had very close ties to the Limousin abbey of Vigeois, acting as benefactors, consenting to the grants by others of property in which they had rights, and standing surety for a secured loan.[110] Manuald was also a benefactor of Uzerche, however, and it was there that he took the habit in 1113 × 1133, by which time he had an adult son and daughter and possibly as many as three grandsons.[111]

---

[107] But see OV iii. 104.   [108] St-M. 1 (at p. 7), 7 (at p. 16), 44, 47.
[109] Ibid. 68. See also 87[7].
[110] V. 23 (1084 × 1096), 42 (*c*.1080), 104 (1092 × 1110).
[111] Uz. 556 (1113 × 1133), 263 (1113 × 1133), 532 (1113 × 1133), 548 (early 12th C.).

Arnald Rufus of Nontron was the brother of Abbot Gerald of Uzerche. He gave that abbey valuable properties on a number of occasions over nearly fifty years, and became a monk there as a very old man in 1108 × 1113. By this time Uzerche had became firmly established as the intercessory centre of the lords of Nontron. For example, in about 1090 Arnald granted the monastery his rights in the church of St-Angel in order that his father's, his mother's, and his own anniversary be celebrated by the monks 'with honour' (the provision makes it highly likely that the mother and father were buried at Uzerche). Gerald's sister Aina gave a son as an oblate to Uzerche during her brother's abbacy, and she took the veil there herself under Gerald's successor Gauzbert. Arnald's widow Petronilla, a daughter of the southern Limousin lord Peter of Malemort, gave a son to Uzerche, and the oblate's three brothers confirmed the entry-gift on the day their father was buried at the abbey. In the interim Petronilla had also taken the veil at Uzerche.[112] Arnald's conversion was thus one notable point in a sequence of actions which had intimately linked his family's spiritual prospects to the monks of its favoured abbey.

Other examples further illustrate the importance of family tradition. When Ranulf of La Roche became a monk at Tulle, he gave four *mansi* and also confirmed the abbey in possession of properties which his mother had given earlier.[113] When Bernard of L'Echamel converted at the same abbey in 1117 he granted his interest in a church in which a quarter of a century before the monks had received, in return for two oblations, sundry other rights from a family-group to which Bernard was probably related by marriage.[114] In a number of cases men became monks in a community in which relatives were already brethren. For example, when Amelius of Charieyras converted at Vigeois, it was noted that one of his many sons was a monk there; this monk Peter also witnessed a later gift by some of his brothers, at least one of whom went on to be buried at the abbey.[115] Peter of Tournemire (who was married to a sister of Peter of Malemort) became a monk at Tulle in about 1077, sometime after a son had become an oblate there. He later witnessed an important donation

---

[112] Ibid. 290 (1081 × 1096 and 1108 × 1113: the date given of 1066 cannot be correct), 291 (1066 or later), 292 (1072), 293 (1113 × 1133), 294 (1068 × 1096), 295 (1072), 297 (1068 × 1096 and 1097 × 1108), 298 (1108 × 1133 and 1135 × 1149).

[113] T. 415 (probably 1st half 11th C.).

[114] Ibid. 413 (*c.*1092 × 1097), 414.

[115] V. 280 (1124 × 1164), 303 (1111 × 1124), 315 (1145 or 1146); see also ibid. 118 (1111 × 1113), 140 (1092 × 1110), 172 (1108 × 1111), 259 (1124 × 1164).

made by his sons and grandsons, and lived on to, probably, 1109, when one of his sons, about to depart for Jerusalem, confirmed in his hands all his family's past grants. Before then, in 1086, Peter's daughter had taken the veil at Tulle, her entry gift comprising property which her father had held of the abbey in fief. So close were the family ties with Tulle that the son leaving for Jerusalem stipulated that if his pregnant wife gave birth to a son, the boy was to be given as an oblate with an entry gift of a whole *villa*; if she bore a daughter, the mother or another son might take vows there in return for the same entry gift.[116] In about 1100 a married man with at least one son converted at Beaulieu, where his brother was a monk.[117] In the late eleventh century Guarin of Tiurans entered La Réole after his son had taken the habit. When sometime later another of Guarin's sons lay dying he gave one of his own sons as an oblate.[118] Geoffrey of Favars entered Tulle in 1107, about twenty years after his wife Emma (a daughter of Peter of Malemort) had taken the veil there.[119] Peter of Beaumont became a monk at Tulle a year after witnessing the entry gift of Gerard of Murat; the two men were very probably related.[120] Gerard of Murat's own connections with Tulle may have dated back to the oblation of his brother Bernard. Gerard gave Tulle lands about a decade before his conversion, and his entry into the abbey cemented his family's loyalty to it. One of his sons, Archambald, took the habit there *ad succurrendum* in 1110. A daughter, Petronilla, gave property to the monks on her death in 1124, by which time one of her sons was a monk there. Another daughter, Rixendis, buried her own daughter in the abbey.[121] Around the middle of the twelfth century Gerald Malafaida of Noailles entered Vigeois in return for confirming properties which had long before been granted for the oblation of his uncle, Gauzbert. In this instance Gerald's choice of community could not have been clear-cut. Gauzbert became abbot of Uzerche in 1097, and it was there that one of his brothers, also called Gerald, chose to convert. A third brother was a canon of St-Martin, Brive. Clearly, however, the family's connection with Vigeois persisted and influenced the younger Gerald's conversion.[122]

---

[116] T. 79 (pre-1077), 322 (1077 × 1084), 329 (*c.*1077), 388 (1086), 465 (possibly 1109).

[117] B. 123–4 (1097 × 1108).

[118] LR 73[46], 74[47].

[119] T. 171, 196 (1084 × 1091).

[120] Ibid. 160 (1091), 161 (1092). See also ibid. 407 (1110).

[121] Ibid. 157 (3rd qr. 11th C.), 159 (1073 × 1084), 162 (1110), 163 (1124), 122.

[122] V. 322 (1124 × 1164); Uz. 52[8] (*c.*1100). See also Geoffrey of Vigeois, 'Chronica' (ch. 35), 298.

Reception *ad succurrendum*

If a layman did not enter a religious community while he was still healthy, he might still choose to receive the habit shortly before death. This was reception *ad succurrendum*.[123] As in the case of oblation, there is some evidence that purists were becoming suspicious of this practice by the later eleventh century. In a well-known passage in his Life of Herluin of Bec, for example, Gilbert Crispin contrasted his hero, who had invited the opprobrium of his peers by deciding to convert in early adulthood, with an aggressive and hedonistic knight named Ralph Pinellus, who often taunted Herluin with the promise that he would become a monk 'when he had grown weary of arms and had had his fill of the pleasure of the world'.[124] Ralph could, of course, have postponed that decision until he was on the point of death. It is important, however, to place Gilbert's meaning in context. He was praising Herluin by attacking Ralph's particular antipathy towards, and willingness to abuse, conversion; he was not dismissing the institution of reception *ad succurrendum* as such. Similarly, an Anglo-Norman tractate from the early twelfth century which explored the practice of monasteries receiving the ill and dying emphasized that intention was the validating factor: if an ill man became a monk fully prepared to respect his vows in the event of his recovery, his action would benefit his soul, whereas mere 'bet-hedging' would not.[125] Doubts about a dying convert's purity of motive were bound to be stimulated by men who miscalculated their chances of survival, recovered, and then came to regret their vows.[126] Peter Damian fulminated against bishops who encouraged such men to return to the world, albeit under orders to live moderately.[127] Anselm of Canterbury urged a convert who was wavering, now that he had regained his health, to respect his vows

---

[123] For a sociological analysis covering the entire medieval period, see W. Brückner, 'Sterben im Mönchsgewand: Zum Funktionswandel einer Totenkleidsitte', in H. F. Foltin *et al.* (edd.), *Kontakte und Grenzen: Probleme der Volks-, Kultur- und Sozialforschung. Festschrift für Gerhard Heilfurth zum 60. Geburtstag* (Göttingen, 1969), 259–77. See also L. Gougaud, *Devotional and Ascetic Practices in the Middle Ages*, trans. G. C. Bateman (London, 1927), 131–45. For a fine example of the practice, see OV iii. 192–8.

[124] 'Vita Domni Herluini Abbatis Beccensis', ed. J. A. Robinson, *Gilbert Crispin Abbot of Westminster* (Notes and Documents relating to Westminster Abbey, 3; Cambridge, 1911), 94–5.

[125] J. Leclercq, 'La Vêture "ad succurrendum" d'après le moine Raoul', *Analecta Monastica*[3] (Studia Anselmiana, 37; Rome, 1955), 161–2.

[126] See e.g. the cases of two lapsed converts in OV ii. 44–6.

[127] 'Opusculum decimum sextum' (chs. 1–2, 8), PL 145. 366–9, 375–7.

and use his talents for the benefit of his new brethren.[128] To pre-empt recidivism there was a convention that the postulant's illness should be as far advanced as possible before formal reception took place.[129] Such caution, however, only reflected the entrenched belief that converts *ad succurrendum* were accounted full monks sharing the same spiritual prospects as their healthy brethren, and that to decide otherwise was tantamount to prejudicing God's mercy.[130] We cannot tell from the surviving evidence whether or how thoroughly religious communities screened ill or dying postulants. There was often a powerful impulse, however, to give them the benefit of the doubt.[131] After all, as the career of Hildebert Grimoard at Uzerche demonstrates, laymen who took vows at moments of personal crisis might go on to make a valuable contribution to the church.[132]

It is usually impossible to gauge from the charters exactly how ill individual postulants were at the moment of conversion. Inevitably documents drawn up after the event would often record converts' conditions in such terms as 'ad obitum suum', 'in extremis positus', 'in extremo vite sue positus', 'in sua ultima infirmitate', or similar phrases.[133] We should not distinguish too rigorously between such expressly terminal cases and instances of conversions by persons described simply as 'in infirmitate'.[134] When in 1086, for example, a *miles* named Peter of Veix converted at Uzerche 'long exhausted by old age', he was still sufficiently active to make his own entry gift before the altar, but he must have been expected to die soon afterwards.[135] Although there was always a danger that religious communities might become obliged to become virtual rest-homes for the chronically ill but stubbornly alive,

---

[128] (Ep. 335), *Opera Omnia*, ed. F. S. Schmitt (Rome, 1940; Edinburgh, 1946–61), v. 271–2.

[129] Ibid. (Ep. 325) 256–7; C. M. Figueras, 'Acerco del rito de la profesión monástica medieval "ad succurrendum"', *Liturgica*, 2 (1958), 369–72.

[130] See the 12th-C. tractate in A. Wilmart, 'Les Ouvrages d'un moine du Bec: un débat sur la profession monastique au XIIᵉ siècle', *RB* 44 (1932), 36. Also Leclercq, 'Vêture', 162–4. Cf. the story of the dead robber-lord saved from demons' clutches by St Benedict in *Chronica monasterii Casinensis* (III. 40), ed. H. Hoffmann (MGH SS 34; Hanover, 1980), 417–18.

[131] For monastic criticism of bishops who discouraged conversion by the dying, see Hamelin of St Albans, 'Liber de monachatu', extr. *Thesaurus Novus Anecdotorum*, ed. E. Martène and U. Durand (Paris, 1717), v. 1456.

[132] Cf. OV ii. 132.

[133] T. 162 (1110), 293 (1112), 401 (1096), 407 (1110), 430 (1103); V. 158 (*c*.1100); Uz. 88 (*c*.1090).

[134] T. 192 (1119); Uz. 704 (1097 × 1108).

[135] Uz. 406, 411.

whether a late convert entered at the moment of death or some short time before was a less important issue than the appeal of dying in the habit.

Consequently there is evidence that laymen might plan ahead for their late conversions, or at least not wait until their final illness was so advanced that they could not control their actions. When Adhemar Robert, 'in infirmitate positus', entered Uzerche, his entry gift was witnessed by his wife, son, brother, and probably his lord, Viscount Bernard of Comborn.[136] When in 1103 Robert of Pandrigne entered Tulle just before he died, his final dispositions were witnessed by six sons and one daughter.[137] Family gatherings of this sort would have been impossible without reasonable advance warning. Furthermore, it was not strictly necessary to be able to reach the religious community in order to be received. When, for example, Gerbergis, the mother of Viscount Raymond of Turenne, fell terminally ill, she took the veil before being carried to Tulle, where she died.[138]

It would be wrong to regard most receptions *ad succurrendum* as occasions of scrambled panic. Rather, they might be the subject of considerable forethought. This is demonstrated by cases which reveal ill or dying converts reasserting or improving their ties to religious communities. When Guy of Malemort became a monk at Uzerche as he lay dying, he confirmed the gift of a church which had been made by his mother, and insisted that no heir or other person could reclaim it by means of rights transmitted through himself; this suggests that Guy's kindred had harboured doubts about the original grant of the church.[139] Around 1140 two brothers asked the canons of Aureil to receive as security for a large loan property which was threatened by the exactions of local *milites*. Later one brother became a canon just before he died, granting to Aureil in alms his share of the pledged property.[140] In 1124 the son of the lord of a benefactor of Ste-Croix, Bordeaux, who had resisted the donation of a tithe, reached an agreement which included the promise that he might be received as a monk 'in life and likewise in death'.[141] Two brothers converted *ad succurrendum* at Tulle within a year of one another.[142] A man's conversion was believed to strengthen his

---

[136] Uz. 704 (1097 X 1108); cf. ibid. 490, which establishes the identity of Adhemar Robert's brother.

[137] T. 430.

[138] Ibid. 504 (1103).

[139] Uz. 88 (*c*.1090); see ibid. 87, 89–90.

[140] A. 80. For the date, see the similarity between the witness-list of ibid. 82 and the donors in ibid. 8, dated by a later hand to 1137.

[141] Ste-C. 38.   [142] T. 162, 407 (1110).

family's bonds with the monastery and so increase the likelihood of future endowment. When Viscount Ebles I of Ventadour became a monk as he lay dying at Tulle in 1096, he gave two *mansi* and ordered his wife Almodis and two sons to give two further properties. When Ebles died, his wife and sons did as he had demanded. Thirteen years later one of the sons, Ebles II, gave more property, adjacent to the land given earlier, for the souls of his father, mother, and kindred.[143] Helias of Cornil was received *ad succurrendum* at Tulle in 1112 with the gift of one *bordaria*. Eight years later Peter Gerard of Cornil, almost certainly a kinsman, gave the monks vines and woodland in the same location before he died.[144]

Reception *ad succurrendum* was thus one means of regulating the ties between religious communities and local arms-bearers. Far from being simply a bolt-hole for the inveterately secular, the practice helps to demonstrate the variety of contacts which could adapt to the various stages of an individual's or kindred's life-cycle. The adaptability of lay–religious contacts is further illustrated by a practice which was, strictly speaking, distinct from conversion but in reality conceived as closely associated with it: burial.

Burial

If a layman did not enter professed religion at any stage of his life it still lay open to him to arrange for his burial at a monastery or community of canons. For men and women in the eleventh and twelfth centuries burial was much more than a question of the commodious disposal of bodies. The location of one's final resting-place, and the liturgical rituals which might accompany interment, occupied men's minds greatly, and sometimes burdensome obligations were placed upon religious or kinsfolk to ensure that a dead individual's burial wishes were respected.

Not the least beneficial effect of conversion *ad succurrendum* was burial according to the rites reserved for professed brethren.[145] Many laymen seem to have treated death-bed conversion and burial as closely equivalent

---

[143] Ibid. 401–2.

[144] Ibid. 293–4.

[145] For the procedures surrounding a monk's death at Cluny, which were very influential, see e.g. *Liber Tramitis Aevi Odilonis Abbatis* (ch. 33), ed. P. Dinter (Corpus Consuetudinum Monasticarum, 10; Siegburg, 1980), 272–8; Ulrich of Cluny, 'Antiquiores consuetudines Cluniacensis monasterii' (III. 28–9), PL 149. 770–5. For the impact of Cluniac customs in the Limousin, see J.-L. Lemaître, *Mourir à Saint-Martial: la commémoration des morts et les obituaires à Saint-Martial de Limoges du XIᵉ au XIIIᵉ siècle* (Paris, 1989), 171–93.

acts. In the early twelfth century, for example, Rainald Hugh, a *miles* of Comborn, his wife, and brother arranged that they should be buried at Vigeois or alternatively might take the habit whenever they so wished.[146] When Peter Rigald of Soularue gave various properties to Tulle in 1121, he was promised in exchange the right to be buried at the abbey without the need for further endowment, or, if he so decided, to become a monk there.[147] In 1114 a man converted at Tulle at the same time as he buried his wife at the abbey.[148] When in 1091 Gerald of Roussanes took the habit at Tulle during his last illness, his entry gift was accounted to be a donation for his burial.[149]

At St-Mont there were two distinct areas of consecrated ground for the bodies of professed brethren and the laity, a practice which was in keeping with the arrangements made at the mother house at Cluny, and there is no reason to suppose that a similar formal division was not observed by other communities in Gascony and the Limousin.[150] Recent research has demonstrated, however, that the liturgical provision for favoured laymen's burials at Cluny was not dissimilar from that provided for monks; and quasi-regular burial must have been meant in requests found in some charters that an individual be buried 'with honour'.[151] Burial was thus regarded as an effective way of approximating one's chances of salvation to those of the spiritual élite.

In some instances a lay person might be particularly anxious to be laid to rest in a spot which accorded reverence and prestige. The very pious Engalsias of Malemort, for example, was buried between the cloister and the church of Arnac, which she and her husband Guy Niger of Lastours had founded in *c.* 1015.[152] But interment within a church or chapterhouse was inevitably restricted to a small élite of founders and princes.[153] For the majority of laymen a final resting-place in the lay persons' cemetery sufficed to satisfy the cardinal goal of burial: that proximity to the scene of intercession helped to fix one's memory indelibly in the minds of the

---

[146] V. 159 (1111 × 1124). For Rainald Hugh's status, ibid. 156.

[147] T. 384.

[148] Ibid. 393.

[149] Ibid. 448.

[150] St-M. 87¹ (*c.*3rd qr. 11th C.); D. Poeck, 'Laienbegräbnisse in Cluny', *Frühmittelalterliche Studien*, 15 (1981), 75–6.

[151] B. 126 (955 × 985); T. 328 (1086 × 1091); Poeck, 'Laienbegräbnisse', 76–8. For an example from a community of canons, see St-Seurin, 76 (1126).

[152] Geoffrey of Vigeois, 'Chronica' (ch. 6), 281.

[153] See e.g. Adhemar, *Chronique* (ch. 66), 190–3; *Historia Pontificum et Comitum Engolismensium* (ch. 26), ed. J. Boussard (Paris, 1957), 21.

intercessors. Thus one benefactor of Tulle asked 'that he be buried at Tulle and [his name] entered in the *Rule*'.[154] In the first half of the twelfth century a man buried his wife at Aureil with the request that the produce of the land he gave be used to feed the canons on the dead woman's anniversary.[155] Burial was an efficacious way of concentrating religious' spiritual services upon a given soul. Viscount Ranulf of Aubusson, for example, hoped for his brother Viscount Rainald, buried at Uzerche, that the Lord might deign to release his soul from the pains of Hell.[156] Brothers who wished to reserve the choice of burial at Beaulieu or elsewhere required that, if they were not buried at that abbey, the monks should still celebrate one hundred Masses for them. Similarly, the brothers Peter and Boso of Courson asked that when they came to be buried at Tulle, fifty Masses should be sung for them by the monks and five by Bishop Bernard of Cahors.[157] In Gascony the occasional grant of formal absolutions associated with graveyards helped to reinforce the view that fitting burial was one of the optimal routes to salvation.[158]

The importance of burial to the laity was reflected in the amount of resources diverted towards it. Consequently burial rights could become the subject of bitter disputes between religious communities.[159] Between the first half of the eleventh century and the early years of the twelfth, for example, the monastery of St-Orens and the cathedral of Ste-Marie, Auch were engaged in an uncharitable wrangle over two rival cemeteries. The dispute proved so intractable that it had to be referred more than once to the papal *curia*.[160] Similarly, in the final quarter of the eleventh century St-Seurin and St-André, Bordeaux became embroiled in a dispute over burial rights which had to be submitted to the abbot of the third important local community, Ste-Croix.[161]

Obviously competition for burial rights in urban settings was apt to be particularly intense if not carefully controlled, but it is noteworthy that

[154] T. 367 (1104).

[155] A. 143 (*c.*1100 × 1140).

[156] Uz. 119 (*c.*1020 × 1030).

[157] B. 99 (1061 × 1108); T. 308 (probably to be dated by the episcopate of Bishop Bernard III (1047–54) on the basis of the brothers' appearance in Uz. 441 (1030), 842 (1048), 1195 (1055)).

[158] St-M. 6 (1073); Auch, 24 (*c.*1046). For this form of absolution, which was distinct from absolution associated with penance, see Cowdrey, *Cluniacs*, 125–6.

[159] For Anglo-Norman exx. see B. Golding, 'Anglo-Norman Knightly Burials', in C. Harper-Bill and R. Harvey (edd.), *The Ideals and Practice of Medieval Knighthood: Papers from the First and Second Strawberry Hill Conferences* (Woodbridge, 1986), 35–6, 37–8.

[160] Cowdrey, *Cluniacs*, 97–101.

[161] St-Seurin, 16, 17.

relations between rural houses, too, could be soured by disagreements of this sort. We are fortunate to possess records from both sides concerning a dispute over burials between Vigeois and Uzerche. Trouble arose over the monasteries' respective claims to bury the inhabitants of a local settlement (Uzerche and Vigeois are only four miles apart) when a peasant fell ill and summoned a priest from Uzerche, who administered the last rites. When the priest returned to fetch the corpse, men acting for Vigeois chased him off with violence, killed one of Uzerche's servants, and hijacked the body, which was taken for burial at Vigeois. According to the Vigeois version of events, Gauzbert, archdeacon of Limoges, was asked to mediate, and he forced an agreement between Abbots Gerald of Uzerche and Adhemar of St-Martial (who was then acting abbot of Vigeois). Uzerche, however, continued to nurse its grievance and reopened the case before Bishop Eustorgius (1106–37) more than twenty years later. The bishop heard various witnesses' accounts and ordered that the body be exhumed and taken to Uzerche for reinterment.[162] The community at Uzerche appears to have been particularly jealous of its burial rights. In the early 1090s it agreed with the clergy of Souillac that if any inhabitant of the parish centred on the castle of Turenne felt himself dying and wished to guarantee burial at Uzerche, he had simply to make his way to the abbey's dependency at Gondres (about one mile away) and expire there. By 1122 this arrangement had broken down, and a further agreement had to be negotiated.[163]

The two cases involving Uzerche reveal that in the years either side of 1100 religious communities were seeking to establish that burial rights attached to their propertied interest in given areas. The Uzerche case against Vigeois had been that the dwelling from which the dead man came was in its parish—which Vigeois also claimed—and within its *jus* and *potestas*. A further case, from southern Gascony, also demonstrates how communities attempted to assert burial rights over their localities. In

---

[162] V. 84; Uz. 79. The Vigeois document, which purports to have been nearly contemporaneous with the original dispute, may have been produced later in connection with the appeal to Bishop Eustorgius. If its contents are to be taken as accurate, they suggest that the dispute flared up in or shortly after 1082, that is after the grant of Vigeois to St-Martial by the lords of Bré and before the election of the first Cluniac abbot, Gerald of Lestrade. The Cluniac connection may well have been the stimulus which goaded Uzerche into vindicating its rights, for it could no longer take for granted its superior prestige over its hitherto modest near-neighbour. The Uzerchois version, however, suggests that when the original dispute erupted Abbot Adhemar of St-Martial was still able to act as a neutral observer.

[163] Uz. 78.

about the 1120s an inhabitant of the valley of the Gave d'Azun (a tributary valley leading into the Gave de Pau from the high Pyrenees) was buried at Marsous, about seven miles over awkward terrain from St-Savin, Lavedan. The monks of St-Savin protested to Count Centulle II of Bigorre. Their contention that the man had been buried 'contrary to ecclesiastical law and with no clergy present' was probably hyperbole, but their indignation reveals that they wished to establish their community as the burial-centre for the remote valleys nearby. Count Centulle supported the monks and ordered that the corpse, by now badly decomposed, be exhumed and taken to St-Savin.[164] Similar sentiments to those of the monks of St-Savin informed the attempt by Tulle in 1105 to enshrine the idea that it had a right to bury the dead of certain areas. In that year the abbey obtained from Pope Paschal II a bull which confirmed, *inter alia*, its right to bury the *milites* from twenty locations within an approximately fifteen-mile radius, as well as the viscounts of Comborn and Ventadour.[165]

By about 1100, then, choice of burial site was beginning to become circumscribed by constraints determined by parochial organization and religious communities' tenurial power. Many of the twenty locations recorded in Tulle's bull correspond to the toponyms of kindreds with recorded links to the abbey and, presumably, members buried there. In other words, Tulle was attempting to convert habits of allegiance into formal duties. The macabre stories of exhumations in some of the above examples reveal also that communities were keenly aware of the value of precedent in securing a string of subsequent burials. It is important, however, not to exaggerate the speed with which choice of burial site was narrowed down. The disputes between religious communities were not collusive codifications of abstract rights, but were rather sparked off by particular events which served to expose problems which might otherwise have long gone unforeseen. In the dispute between St-Seurin and St-André, for example, trouble only arose when a visiting southern Gascon lord, a man with no obvious ties to either community, happened to die in Bordeaux.[166] Typically of medieval institutions, religious communities groped their way very gradually from particular cases to general statements of right. In the meantime nobles and knights could use their wealth and mobility to exercise choice in their burial sites and to build up family traditions of burial at certain religious houses.

---

[164] *Cartulaire de Saint-Savin*, 17. Cf. ibid. 4.
[165] T. 3 at p. 13.   [166] St-Seurin, 16.

The importance which laymen attached to burial in a religious community is demonstrated by a number of further points. In the first place, monks and canons might use requests for burial to settle or forestall property disputes. For example, when Peter Bernard of Tulle and his wife buried their son at the local abbey they surrendered their rights in tithes which the monks already claimed.[167] When Stephen of Castres gave a vine to Vigeois he had to offer guarantees against the anticipated opposition of his nephew Helias. Later Helias was killed, and Stephen encouraged Helias's son to confirm the gift of the vine and add some renders in return for the father's burial.[168] When Gerald of Voutezac's body was borne to Vigeois for burial the monks raised the matter of his tenure of a *mansus* which they claimed. Gerald's lord, Viscount Bernard of Comborn, negotiated a deal between Abbot Peter and Gerald's children whereby the latter were able to retain the usufruct of the property in return for modest renders and the promise that they too would be buried at the abbey.[169]

Secondly, the burial of an individual was understood to regularize his kindred's existing ties with a religious community. When in 1116 Rigald and Stephen of Bouchiat surrendered their claim to property which had previously been contested by their father, they did so for their father's burial at Tulle and in order that their brother Bernard would be received as an oblate.[170] In or shortly before 1068 a *miles* granted land near a church which he had already given to St-Mont in return for the burial of his small son.[171] Similarly, in return for future burial 'with honour' at Tulle, two brothers, Gerard and Andrew of Torenx, granted their rights in a *villa* where the monastery was extending its interest in order to develop a newly built church.[172] It was sometimes appreciated that burial in a satellite church was the most effective way to seal a kindred's loyalty. In the third quarter of the eleventh century, for example, Dodo of Bernède granted his local church to St-Mont in return for burial there, not at the mother priory.[173] Throughout the latter half of the eleventh century the monastery of Uzerche was involved in a series of agreements which slowly unravelled the web of laymen's rights in the church of Exandon, which had originally been held by the viscounts of Limoges, the counts of La Marche, and a network of vassals including the lords of Pierrebuffière and Lastours. By late century the most important out-

---

[167] T. 164 (1084 × 1091).   [168] V. 234 (1111 × 1124 and 1124 × 1164).
[169] Ibid. 169 (1096).   [170] T. 35, 379.   [171] St-M. 30.
[172] T. 328 (1086 × 1091). See ibid. 322–7.   [173] St-M. 11.

standing lay claim was a *malatolta* levied on the church's lands by Gerald
Malafaida of Noailles. When Gerald left for the East, possibly on the
First Crusade, he not only surrendered part of his claim but was also
made to swear that he would be buried at Exandon. His brother Gerbert
did the same sometime later. Family connections suggested a choice of
burial sites. One of Gerald and Gerbert's brothers, Ranulf, was a canon of
St-Martin, Brive. At the time of Gerald's vow another brother, Gauzbert,
was sacrist of Vigeois and soon afterwards became the abbot of Uzerche.
Yet it was accepted that the brothers' burial at a lesser site was a
particularly effective way to quell a long-standing and troublesome
dispute.[174]

A third index of the significance attached to burial is the often considerable trouble taken to transport bodies to their final resting-place.
When benefactors of religious communities were granted rights of future
burial, the monks or canons must have been understood to assume some
corresponding duties.[175] Although it might fall to a dead man's kinsmen
to bring his body for burial,[176] religious themselves often performed this
task. In the early twelfth century Bernard William of Lanne was mortally
wounded and sent for a monk from Sorde to administer the last rites.
When Bernard William died, the monk arranged for his body to be borne
by boat approximately ten miles down the Gave d'Oloron to the abbey.
(In a bizarre development, the dead man's kinsmen then stole the body
from Sorde, evidently for fear of their enemies, and asked the monks to
escort them to their local church where the remains might be better
protected.)[177] In contrast, the kinsmen of another dead man asked the
abbot of Sorde to bury him in their local church, not at the abbey as he
had requested, because they could not carry him to the monastery. (A
journey of about fifteen miles was involved, but up the River Adour.)[178]
When the wife of Arnald Ainard of Laleugue was buried at Mormès, not
St-Mont (ten miles distant), her husband felt that he had to justify his
violation of her dying instructions by pleading that winter snows had
made the journey impracticable.[179] When Loup of Beaulieu died, his
brother, who was a monk at the priory of La Réole in the Bazadais, was
deputed to fetch the corpse for burial.[180]

---

[174] Uz. 52[7-8]. For Exandon's previous history, ibid. 52[1-6, 9].
[175] e.g. V. 93 (1092 × 1110), 116 (1073 × 1086), 159 (1111 × 1124), 258
(1124 × 1137); T. 308 (1037 × 1054), 384 (1121).
[176] e.g. T. 441 (1117); LR 71[44] (*c*.1090).   [177] S. 31 (*c*.1105 × *c*.1119).
[178] Ibid. 114 (*c*.1119 × 1136).   [179] St-M. 15 (later 11th C.).
[180] LR 80[54] (early 12th C.).

A final indication of the seriousness with which burial was viewed is that lay men and women often attached spiritual and emotional importance to being buried next to their kinsfolk.[181] (This practice was undoubtedly encouraged by the notion that members of a family would be summoned to Judgement together.) In one late but clear example, the widow of a man who had been buried in a cemetery belonging to the abbey of Ste-Croix, Bordeaux granted lands to the monks to build a new basilica on condition that she, her sons, and her second husband might lie 'in some private place' next to him.[182] The daughter of Berald of Cabre placed upon the monks of Beaulieu the duty to fetch her body, wherever she might happen to die, so that she could lie beside her father.[183] By the mid-twelfth century Aureil was treated by *milites* from Noblat and Laron as their families' mausoleum.[184] The fullest account of a princely burial from the twelfth-century Limousin describes how Viscount Boso II of Turenne was killed in battle in June 1143 and borne to Tulle, where he was buried the following December. (No reason is stated for the delay, but perhaps he had died excommunicated.) At an impressive gathering attended by the viscount of Limoges (Boso's brother-in-law), the viscounts of Comborn and Ventadour, other nobles, and five abbots (including Boso's uncle Ebles of Tulle) no fewer than eleven *mansi* were granted to the abbey so that Boso might lie next to his father Raymond (a first crusader) close to the doors of the abbey church.[185]

From birth to death religious centres embraced the pious aspirations of local arms-bearers. Traditional norms of recruitment and benefaction were not immutably fixed, for otherwise new foundations such as La Sauve, or attempts at reform such as those which took place in many southern Gascon houses, could not have succeeded. Yet even at new or revitalized communities the rhythms of lay–religious contacts would typically, within one or two generations, settle down into conformity with established patterns. This was so because of the motive force of the kin-centred lay spirituality which dominated the period. Kinship bonds were not in themselves the sole motivating influence behind laymen's relations with religious communties; they were the framework for the expression of pious values. In other words, the fact that laymen addressed spiritual concerns through sentiments as self-justifying and intimate as their family's solidarity and prestige demonstrates that those concerns were

[181] e.g. B. 57 (882).   [182] Ste-C. 128 (1149).
[183] B. 41 (1097 × 1108).   [184] A. 4, 6, 45, 83, 143.   [185] T. 490.

deep-rooted and sincerely held. It is, of course, unreasonable to posit oblates' kinsmen, ageing postulants, or even all young, adult converts being in some advanced stage of pious fervour when they approached a religious community, for such a heightened energetic state would have been impossible to sustain over generations and around every religious centre. So, when the monks of Tulle observed that one dying man became a monk 'with great devotion', perhaps they believed that his particular religious fervour raised him above the normal run of converts *ad succurrendum*.[186] On the other hand, it is fundamentally important not to interpret the usual absence from the charters of references to religious ecstasis as evidence for a lack of genuine piety on the part of nobles and knights. Doubtless there were men like Ralph Pinellus who took a robustly independent view of the religious life; and throughout this study, of course, families which did not give to or enter religious communities, and so leave no charters, are the great unknown. But there is no evidence for a non-religious lay counter-culture—even Pinellus, after all, said that he might become a monk one day—only for inter-mittent irreligion. One abbey's wicked oppressor, moreover, might be the next monastery's active benefactor. From the values of the moderately generous donors (who were probably nearer the lower end of the scale of piety than is often supposed), through those of conspicuous benefactors, families with members in the habit, and adult converts, to the spirituality of conspicuous holy men, there stretched a continuum of belief. When we consider that the crusade vocation was substantially modelled on features of adult conversion, and that this was simply one element in an integrated system of contacts between the laity and professed religion (albeit the one which usually demanded the greatest personal commitment from the individual), we can begin to see where crusade enthusiasm was rooted. To explore this matter further, we must now consider the sorts of beliefs at work when laymen supported local religious communities.

[186] Ibid. 92 at p. 71 (*c.*1115).

4

Aristocratic Piety in the Eleventh Century

ALTHOUGH much must remain obscure, the religious ideas of arms-bearers around the time of the First Crusade may be retrieved, in some significant measure, from narrative and documentary sources, especially the latter. Narratives such as chronicles and hagiographical texts were, broadly speaking, written by educated churchmen for consumption by their peers. But in so far as these sources often noted the actions of laymen in relation to the church and its ideas—the First Crusade chronicles are a case in point—they are useful evidence for lay piety. Charters do not justify themselves so readily. Because they were formularistic and written in a language in which very few laymen had received any instruction (although there is a case for arguing that intelligent Occitan-speakers could have followed simple spoken Latin), it is tempting to dismiss the non-dispositive and non-attestative elements of charters as, at best, the putting of words into disinterested laymen's mouths or, at worst, empty verbiage. To scholars who regard charters principally as so many units of socio-economic and genealogical data such an approach may be attractive. It is also misguided, for two points suggest that these documents are valuable sources for the religious ideas of arms-bearers.

First, while charters from our period tended to be very similar in form and determined by customary phraseology, their content was not so tightly restricted by conventions that scribes could not sometimes introduce interesting but strictly extraneous details. It was noted, for example, that one donor to a religious community had been struck down with leprosy;[1] another's brother was on the point of death;[2] a third had been wounded in battle and had come to the monastery to receive medical treatment.[3] Generically more common situations involved donors suffering

---

[1] LR 80[55] (early 12th C.). See also P. de Marca, *Histoire de Béarn* (Paris, 1640), 281–2 (*c*.1060).

[2] St-Seurin, 70 (1123).

[3] T. 88 (1014 × 1022). See also St-M. 53 (1062 × 1085); Uz. 124 (mid-12th C.); T. 382 (1113); LR 77[51] (late 11th C.). For monks attending an ill benefactor, see also *Les Miracles de saint Benoît* (III. 6), ed. E. de Certain (Paris, 1858), 143.

from a serious illness or on their death-bed,[4] or embarking on a hazardous undertaking such as a pilgrimage.[5] Such explicit examples appear in only a minority of the surviving documents, which means that we can only speculate about the backgrounds to the bare recitations of facts in the many more tersely worded charters. But the fact that material of this sort could be recorded—always a possibility when charters were not so much title-deeds as records of those formal actions, donations for example, which themselves had legal consequences[6]—demonstrates that religious communities took an interest in, and were responsive to, benefactors' personal circumstances. An interest in circumstances, moreover, pre-supposed some appreciation of underlying ideas and motivations. In this light, it is impossible to propose a model of arms-bearers' piety which involves professed religious spoon-feeding sophisticated ideas to the uncomprehending.

The value of charters as evidence for lay piety also emerges from the evidence provided in the previous chapter that arms-bearers attached importance to, and might be influenced by, the presence of close relatives in religious communities. Such interaction between laymen and professed religious, however closely related, would have been impossible if there had been no shared terms of reference to enable a dialogue to take place. It is reasonable to argue of course that what, say, an abbot might tell his brother, a local lord, by way of vernacular religious instruction would not correspond completely to the ideas expressed in that lord's charters. Furthermore, it is an inevitable consequence of the nature of the evidence that the documents highlight the conventions and beliefs surrounding benefaction of the church far more than other forms of pious behaviour. Yet even though charters were not suitable vehicles for potted catechisms containing everything a pious layman was expected to believe, they drew upon funds of images and ideas which were used so regularly that they seem to reflect a good deal of what passed in discussions between the laity and religious.[7]

---

[4] Uz. 421 at p. 241 (*c*.1100), 1214 (*c*.1086 × 1097); V. 54 (1092), 273 (1124 × 1164), 304 (1124 × 1137); T. 142 (1117), 272 (1115); St-Seurin, 79 (1128), 85 (1131); A. 66 (*c*.1100 × 1140).

[5] See below, Ch. 5.

[6] For the way in which actions were understood to have legal force in this period, see S. D. White, *Custom, Kinship, and Gifts to Saints: The* Laudatio Parentum *in Western France 1050–1150* (Chapel Hill, NC, 1988), 70–3, 82.

[7] See the comments of R. Mortimer, 'Religious and Secular Motives for some English Monastic Foundations', in D. Baker (ed.), *Religious Motivation: Biographical and Sociological Problems for the Church Historian* (Studies in Church History, 15; Oxford, 1978), 77.

The Purposes of Supporting Religious Communities

It was seen in the previous chapter that there was a widespread belief that entry into the religious life required a gift. Oblations, conversions, or requests for burial, however, form the background to only a minority of the charters recorded in the cartularies. Taken as a whole, the charters demonstrate that, irrespective of the exact circumstances behind any given transaction, the normative type of formal contact between the arms-bearing laity and religious communities was the grant—usually by gift but also possibly by sale, lease, or pledge—of property, typically but not invariably land or appurtenant rights. It requires only the most cursory examination of the cartularies to reveal that this sort of transaction was very common, involving at least the larger religious communities in contacts with laymen many times each year. This recurrent pattern of behaviour, it is reasonable to suppose, was informed by a corresponding pattern of entrenched ideas and influences which motivated lay benefaction of religious institutions. Thus, although it is quite impossible in every instance to enter the mind of an individual benefactor in order to analyse his exact thoughts, we possess enough information to reconstruct a generalized picture to which at least most specific cases approximate.

It is important to note that it is unnecessary for our purposes to search back into early Christian history in order to find the roots of the beliefs evidenced by the charters. This approach is not intended to evade the issue of historical causation, but rather attempts to share the perspectives of laymen who may not have been intellectually equipped—nor have felt the need—to know, for example, the theological origins of the Communion of Saints or the complete scriptural basis of alms-giving, but whose beliefs about the value of giving to religious communities were none the less sincere and entrenched.

The basic premiss informing grants to religious bodies was expressed in about 1127 by Peter the Venerable, the abbot of Cluny, to Bernard of Clairvaux. Peter's comments formed part of a lengthy defence of the Cluniac observance against the criticisms of progressive twelfth-century reformers; but what he wrote applies equally well to eleventh-century conditions and to all religious institutions, Cluniac or otherwise, which encouraged donations from the laity. Monks, Peter argued, received the gifts of the faithful and in return the donors participated in the *bona* performed by the brethren; that is to say, they shared in the spiritual merit earned by prayers, fasts, and good works.[8] The verb which Peter

---

[8] Peter the Venerable, *The Letters* (no. 28, ch. 19), ed. G. Constable (Harvard Historical Studies, 78; Cambridge, Mass., 1967), i. 84. See also OV iii. 260–4.

used to express the faithful's receipt of the spiritual benefits was *recompensare*. He was thereby evoking the widely held medieval notion that all gifts, between laymen as well as to religious bodies, were not occasions of unilateral and disinterested largess but rather merited a counter-gift, some quid pro quo which cemented the ties between the two receiving parties. This notion of reciprocity has led some scholars to treat benefaction of religious communities as symptomatic of a gift- or exchange-culture in which social patterns and individual or collective status were constantly reinforced by the formal giving and receiving of property.[9] As a tool of analysis such an approach is perfectly valid, as far as it goes. To rely on it alone, however, robs the contacts revealed by the charters of any religious content: donors' ideas become, literally, soulless. When lords and knights gave their property to a religious community, they were not, on the level of consciousness, simply obeying social imperatives. They knew what they wanted; and what they wanted was that the grant would do them good.

It was believed, for example, that securing the support through benefaction of religious communities and the saints whom they represented availed the faithful in this life. On one simple level this notion was consistent with the practice at pilgrimage shrines of giving *ex voto* offerings as tokens of assistance received.[10] Even modest churches received offerings which were accounted among their most important appurtenances when they changed hands.[11] A series of documents from Gascony demonstrates that temporal benefits were also believed to flow from larger-scale benefaction. In about 1091 Viscount Gaston IV of Béarn confirmed the gifts made by his father Centulle to the Cluniacs of Ste-Foi, Morlaas (Centulle's own foundation), and added further properties and rights. It was stated that Gaston did this for the souls of himself, his wife, and children, and 'so that God may help us in this world in all our needs, and in the future grant us eternal life'. In 1101 Gaston granted Ste-Foi jurisdictional rights in Morlaas 'for the salvation of my soul and body, of my wife, and all my kinsfolk [*parentes*], and so that Almighty God may give me prosperity in all things and free me from the clutches

---

[9] See White, *Custom*, 19–20, 27–8, 75, 156–63, 165–7; B. H. Rosenwein, *To be the Neighbor of Saint Peter: The Social Meaning of Cluny's Property 909–1049* (Ithaca, NY, 1989), 44–5, 48, 110–11, 121–2, 124–43, 202–3. For an earlier period see also L. K. Little, *Religious Poverty and the Profit Economy in Medieval Europe* (London, 1978), 4–6.

[10] P.-A. Sigal, *L'Homme et le miracle dans la France médiévale (XIᵉ–XIIᵉ siècle)* (Paris, 1985), 86–107.

[11] e.g. T. 177 (1060 × 1073), 411 (*c.*1095), 523 (1086 or 1087).

of my enemies as long as I live, and grant me a perpetual inheritance with him in Heaven'. When Gaston's son Centulle V confirmed his grandfather's and father's grants in 1131 he recognized that by their past benefactions they had 'enjoyed favourable successes in this life', adding that by reaffirming his family's ties to Ste-Foi he might merit the same *prosperitas*.[12] Other documents demonstrate similar beliefs. In about 1090, for example, Count Bernard III of Armagnac made over rights to St-Mont so that the Lord might have mercy on him in this world and the next.[13] In about 1075 Viscount Fedac of Corneillan recalled that some years before, when pursued by unidentified enemies 'without any hope of escape', he had granted his whole honour to St-Mont in order that he would be protected and granted victory by means of the intercession of St John (the monastery's patron) with God.[14] A spectacular example of how the support of God and the saints, mediated by religious, was believed to benefit a layman comes from an entry in the cartulary of Sorde which relates events of the late eleventh or early twelfth century. A man named Arnald Sancho was waging a vendetta against William Malfara, who had betrayed and killed his brother. Arnald Sancho came to pray at the abbey of Sorde and granted properties to the monks 'so that God might give him the traitor'. He later caught up with William, cut off his hands, nostrils, tongue, and genitals, and presented the 'spoils', William's full knightly armour, to Sorde. To the extent that they were prepared to receive Arnald's offering and thought it worthy to record his actions, the monks of Sorde made themselves a party to what appears to have been nothing short of ritual dismemberment. They did so because they shared Arnald's belief that God and St John, Sorde's patron, had granted him vengeance.[15] Like the begetting of heirs, success in war was a pre-occupation of arms-bearers which could be accommodated by contacts with religious communities.

Important as was the wish for success in this life, however, many more charters reveal that benefactors of religious communities were anxious about their fate after they died. As the above examples from Ste-Foi demonstrate, succour in this life and the next world were not treated

---

[12] *Cartulaire de Sainte-Foi de Morlàas*, ed. L. Cadier (Collection de pièces rares ou inédites concernant le Béarn, 1; Pau, 1884), 2, 3, 5.

[13] St-M. 88 (1085 × 1096). See also ibid. 55 (1077 × 1085).

[14] Ibid. 73. The dating clause places the charter between 1068 and 1085, but Fedac's evocation of St John alone and not St Peter also establishes that the document recalled events before St-Mont was granted to Cluny in *c.*1055, possibly—Fedac gave woodland 'ad edificandum monasterium cum officinis suis'—soon after St-Mont's foundation in 1036.

[15] S. 32.

as alternative benefits but rather part of a continuum which was not interrupted by death.[16] The basis of the belief that religious could help the faithful after they died was the provision of intercession. We have seen that laymen considered burial in religious communities important because it guaranteed liturgical commemoration. This was one expression of a wider belief in the value of regular prayer by holy men. Leaving Gascony for Jerusalem in the late eleventh century, Peter of Vic asked the canons of Auch to establish clergy in his local church so that they might pray 'all the time' for him and his family.[17] A man who ended a long-running dispute over a church with St-Mont commended himself 'to God, St John, and the prayers of those serving God there'.[18] A benefactress of St-Seurin, Bordeaux in about 1030 hoped that she would become 'orationum . . . particeps'.[19] Similarly, an early twelfth-century benefactor of Ste-Croix, Bordeaux believed that he was joining the souls of himself and his parents to the monks' prayers.[20] The psalmody lay at the core of the intercession which could be provided, but there is also evidence of widespread belief in the efficacy of all the liturgical and extra-liturgical acts of religious communities as facets of one service. Thus, when in 1107 a *miles* surrendered to St-Mont rights in some peasants which he had been contesting, his charter evoked the Masses, Psalms, alms, and 'other good things' which made up the monastic routine and could now operate in his favour.[21]

The word used of St-Mont's spiritual services in this last charter was *beneficia*. This term became part of the vocabulary which expressed laymen's confraternity with religious communities, an institution which developed from and complemented the practice whereby religious bodies built up prayer associations between themselves or with individual clerics in order to ensure greater intercession, especially for dead members.[22]

---

[16] See also J. de Jaurgain, *La Vasconie* (Pau, 1898–1902), i. 396–7.

[17] Auch, 6 (1068 X 1096).

[18] St-M. 25 (mid-11th C.).

[19] St-Seurin, 10.

[20] Ste-C. 115 (1132 X 1138).

[21] St-M. 41.

[22] H. E. J. Cowdrey, 'Unions and Confraternity with Cluny', *JEH* 16 (1965), 154–7. For particular exx., J.-L. Lemaître, 'Les Confraternités de La Sauve-Majeure', *Revue historique de Bordeaux et du Département de la Gironde*, NS 28 (1981), 5–34, esp. 11–12; id., *Mourir à Saint-Martial: la commemoration des morts et les obituaires à Saint-Martial de Limoges du XIe au XIIIe siècle* (Paris, 1989), 365–76; A. Sohn, *Der Abbatiat Ademars von Saint-Martial de Limoges (1063–1114): Ein Beitrag zur Geschichte des cluniacensischen Klösterverbandes* (Beiträge zur Geschichte des alten Mönchtums und des Benediktinertums, 37; Münster, 1989), 142–50, 171 ff.

At Cluny lay confraternity, though not a new idea, was particularly developed under Abbot Hugh (1049–1109).[23] The Gascon and Limousin evidence likewise points to the later eleventh century as the period when this institution became more widespread, in Cluniac monasteries influenced by the mother-house but also in non-Cluniac Benedictine establishments and in communities of canons. For example, when Guy, Gerard, and Gulpher of Lastours agreed to grants to Beaulieu for their father's soul a few years before the abbey's cession to Cluny in 1076, it was provided that three Masses would be celebrated immediately for their father, followed by one by each ordained monk. In addition the father's name was to be recorded in the *Rule*, and his anniversary celebrated in the years ahead.[24] A generation later, after Cluniac influence had penetrated Beaulieu, Gerald of Chanac granted rights *post obitum* in a church in order that his parents' anniversaries would begin to be celebrated, his own death would be commemorated by three hundred Masses and anniversaries, and he and his kindred would receive 'the whole benefit of the monastery'.[25] That 'the benefit of the monastery' and analogous phrases were shorthand for particularly close bonds is further demonstrated by examples dating from between *c.*1050 and *c.*1150 from the cathedral community of Auch,[26] the canons of St-Seurin, Bordeaux,[27] the non-Cluniac Benedictines of Ste-Croix, Bordeaux,[28] and the Cluniacs of St-Martial and (from 1082) of Vigeois.[29] In a rare instance of more explicit language being used, two men confirmed the entry gift of their brother to St-Mont in return for livestock and money to the total value of 40*s.* as well as the grant in chapter of *societas*.[30]

In this last example from St-Mont it was recorded that the brothers had previously objected to the entry gift so vehemently that they had

---

[23] Cowdrey, 'Unions and Confraternity', 157–9, 160–2.

[24] B. 14 (1062 × 1072).

[25] Ibid. 181 (1097 × 1108).

[26] Auch, 72 (1143): a woman commends herself to Ste-Marie in order to be 'tam spiritualium quam actualium bonorum illius ecclesie particeps'.

[27] St-Seurin, 38 (1st third 12th C.): 'beneficii ecclesie . . . particeps'; 89 (*c.*1120 × 1143): 'in consortio canonicorum' and 'de beneficio . . . participem'; 90 (1144): a surrender of property in return for a fief and to be 'particeps de beneficio ecclesie, tam in animis quam in corpore'.

[28] Ste-C. 84 (1099): 'participem bonorum Sancte-Crucis'.

[29] Lemaître, *Mourir à Saint-Martial*, Annexe 2, no. 21 (1063 × 1114): 'hujus loci beneficium omne'; V. 93 (1092 × 1096): 'participem . . . totius beneficii quod in monasterio Vosiensi actum fuerit'; 139 (*c.*1095): 'et in vita et post mortem . . . in consortia Sancti Petri'.

[30] St-M(Sam). 60 (*c.*1070); see also St-M. 87[6]. For a further example, Lemaître, *Mourir à Saint-Martial*, Annexe 2, no. 1 (1063 × 1079).

threatened the monks with death. Such initial hostility, and the fact that money changed hands when an agreement was eventually struck, reveal that confraternity might be accorded at the conclusion of disputes. In about 1075 two brothers surrendered rights in *mansi* given to Tulle by a third party in return for an ounce of gold and the benefit of the monastery.[31] The same spiritual concession was granted when around 1100 a man surrendered *post obitum* his rights in property which had earlier been given to Vigeois.[32] Confraternity was a special favour granted by religious communities, sometimes after hard bargaining, which solemnly guaranteed laymen that they would feature in their intercession and good works. Although evidence from Cluny reveals that by the third quarter of the eleventh century confraternity was granted in a special ceremony which was formally distinct from the simple donation of property, it would be misleading to speak of a discrete institution or a clear two-tier system of association whereby religious communities distinguished *confratres* from mere benefactors; laymen entered confraternity with a religious community but did not join a confraternity which established bonds between its members. The total numbers involved were probably quite small.[33] Furthermore, it is impossible to see how, from a layman's perspective, the benefits of confraternity in the eleventh century would have been qualitatively different from, for example, the wish expressed by a benefactor of Beaulieu in 866 that, when he died, the community would daily sing five Psalms for his soul and annually commemorate his death with a Mass; or the duty imposed upon the same community in 885 to celebrate a donor's anniversary by supplying the refectory with the produce of the property given.[34] Confraternity was simply one way of accommodating the widespread concern to benefit from the provision of intercession and association with good works. This concern might be so strong that confraternity could elide into conversion.[35] When in about 1075 Robert of Rouffignac surrendered a claim to property which had been given to Tulle by his brother, he asked that a Mass be celebrated for his father, mother, and another brother, that his name be recorded in the abbey's martyrology, and that he share in the

---

[31] T. 177.

[32] V. 264 (1092 × 1110); cf. ibid. 265 (1124 × 1164).

[33] Cowdrey, 'Unions and Confraternity', 157, 162; Lemaître, *Mourir à Saint-Martial*, 376–7.

[34] B. 3, 55; see also ibid. 166.

[35] See A. 199 (1106 × 1137). Cf. Lemaître, *Mourir à Saint-Martial*, 362–3 and Annexe 2, no. 1 (1063 × 1079).

benefits of Tulle 'as if he were a monk'.[36] As well as being a specific request for confraternity, Robert's demands are a particularly clear expression of the broader notion that the spiritual services of religious communities could be directed towards laymen's benefit as an extension of the ability of the brethren to attract spiritual merit to themselves.

Evidence thus abounds that many laymen wished to associate themselves as closely as possible with the merits of religious. This much is clear. It is more difficult, however, to find a coherent body of evidence which would explain why laymen felt that they needed to rely on religious communities in this way: in other words, what sort of sins they committed and how often. Penitentials, handbooks for confessors, list many of the sins which a confessor could expect to hear; and by comparing the frequency with which certain sins are mentioned, and the nature of the penances recommended for each sin, it might be possible to draw up a list of faults very approximately graded by the severity with which they were viewed by the church.[37] Such an exercise would not reveal, however, whether laymen committed some sins more than others, had their own scale of values about the relative seriousness of various faults, or indeed were even aware that some actions were sinful. The problem is compounded by the fact that penitentials' prescriptions for specific sins varied.

The charter evidence is also problematical, for it throws up random instances of sinful behaviour from which no clear picture emerges. Centulle IV of Béarn-Bigorre granted Ste-Foi, Morlaas to Cluny because, it was recorded, he had been persuaded that he had married his first wife within the prohibited degrees, and wished St Peter to intercede on his behalf. This impediment, however, had not prevented him from the consanguineous marriage in the first place; and it may be suspected that the political attractiveness of his second bride (Beatrice, the heiress of the count of Bigorre), and the fact that he already had an heir (the future crusader Gaston IV), weighed on Centulle's mind at least as much as the church's fulminations.[38] A small but not insignificant number of documents recorded that a donor was guilty of homicide or assault.[39] In

---

[36] T. 180.

[37] For the difficulties involved in such an exercise, see C. Vogel, *Les 'libri paenitentiales'* (Typologie des sources du Moyen Âge occidental, 27; Turnhout, 1978), 104–7.

[38] *Cartulaire de Sainte-Foi*, 1. See also Gregory VII, *Das Register* (VI. 20), ed. E. Caspar, 2nd edn. (MGH Epistolae Selectae, 4–5; Berlin, 1955), ii. 431–2.

[39] Uz. 355 (early 11th C.); Auch, 53 (1068 × 1096); T. 320 (1121), 346 (*c*. 1100 × 1111); A. 185 (mid-12th C.); Ste-C. 112 (1132 × 1138).

the case mentioned above which involved the vengeful Arnald Sancho, however, the protagonists had patently not believed that homicide was inexcusable in all circumstances. Given the popularity in our period of the vendetta, the relatively few charters expressly referring to violent behaviour cannot support any claim that there was a growing revulsion towards bloodshed. The morality of each incidence of violence seems rather to have been judged on its merits by means of secular as well as Christian sets of values: sets of values which, from an arms-bearer's perspective, were often complementary.

A more substantial body of evidence reveals that the mistreatment of religious communities or their property could be construed as sinful. The preamble of an Aureil charter from the early twelfth century, for example, noted that many important persons withheld many properties from the church through ignorance; in recent times the learned and devout statements of popes and other religious authorities had made it clear that this was a sin.[40] In this instance the donors were giving a half-church and a castle chapel which, as far as the records can reveal, had not been granted to Aureil before. The preamble thus expressed the general Gregorian programme of extracting churches from lay control.[41] Such a generalized appeal to principle was exceptional, however, it not being a common technique of reformers working with laymen at the grass-roots level to bombard them with statements that their control of churches was in itself sinful.[42] It was quite a different matter, however, with regard to churches and other forms of property in which a religious community claimed an existing interest. In the first decade of the twelfth century members of a family surrendered rights in a *villa* claimed by Vigeois, recognizing that they had held it unjustly, by fraud, and to the endangerment of their souls.[43] A later notice recalling a surrender made to Tulle by Archambald I of Comborn in 970 used the language of sin to explain his unjust possession of a property belonging to the monks.[44] Similarly, a document from Aureil of *c*.1130 described as an injury to God and St John the unwarranted retention of modest renders by the son of the donor.[45]

Obviously, religious communities had a vested interest in portraying as sinners those who contested or disrupted their property and rights.

[40] A. 133 (*recte* 1107?).

[41] See also C. iv. 3630 (1088), a Gascon charter.

[42] But cf. H. Dormeier, *Montecassino und die Laien im 11. und 12. Jahrhundert* (MGH Schriften, 27; Stuttgart, 1979), 58–62.

[43] V. 133 (1106 × 1108). See also T. 523 (1086 or 1087).

[44] T. 345.

[45] A. 85.

The charters are consequently bound to reveal a disproportionately large number of cases involving this particular type of sinful conduct. But, even though these examples do not take us far forward in assessing the quantitive incidence of laymen's various sins, they throw valuable light on how sins were perceived qualitatively, for they demonstrate laymen acting upon religious communities' definition of their faults; if two laymen contested a piece of property between themselves, the eventual loser would not have thought of himself as a sinner. In other words, the language used of property disputes involving religious institutions reveals a value-system in which laymen gauged their sinfulness by using the example of religious as a yardstick. Thus many more charters evoked a benefactor's awareness of sin in general terms than picked out specific misdeeds. For example, in about 1100 Imbert of La Gardelle granted properties to Beaulieu because he was fearful of his past transgressions.[46] In the third quarter of the tenth century Count Odo of Fezensac was portrayed in a charter of Ste-Marie, Auch as mindful of how great was the 'bundle' (*sarcina*) of his sins; more than a century later, when Odo's descendant Count Astanove was about to leave on the First Crusade, his charter evoked his 'innumerable and daily' excesses.[47] In about 1100 Azivera, the daughter of Viscount Odo of Lomagne and mother of Count Bernard III of Armagnac, was described in a notice of St-Mont as 'remembering all her own faults and the sins of her kinsfolk'.[48] In other documents sins were lumped together as 'magnitudes', 'excesses', or synonymous terms.[49] When the *miles* Raymond of Liviniac granted a church to La Réole, he stated that he was aware of the 'heap' (*congeries*) of his errors; a contemporary who gave a half-church to the same community in 1098 or 1099 described himself in his charter as terrified by the 'enormity' of his sins.[50] A charter from St-Seurin, Bordeaux from the early eleventh century developed the idea of the sinfulness of lay life by introducing the pitfalls of youth: a woman granting a mill declared that in her younger days she had committed many faults, and the memory of them now made her feel wretched.[51]

In all these examples the benefactors were presented as weighed down by a burden of sinfulness. Charters were not ideal vehicles for blow-

---

[46] B. 40 (1097 × 1108); cf. ibid. 42 (1097 × 1108).
[47] Auch, 54 (*c*.960), 57 (1097).
[48] St-M. 88. Cf. St-Seurin, 23 (*c*.1080).
[49] T. 109 (923 × 933); St-M(Sam). 23 (1062 × 1085); A. 132 (1147 × 1189).
[50] LR 78[52] (*c*.1083 × *c*.1100), 82[57].
[51] St-Seurin, 10 (*c*.1030).

by-blow accounts of a donor's past faults. For reasons of deference, discretion, and gratitude, moreover, religious communities may not have wished to berate donors with the darker aspects of their past lives, or at least to make them a matter of formal record. On the other hand, recourse was made so regularly and openly to generalized evocations of sins that it becomes apparent that laymen shared a sense of sinfulness which flowed from their immersion in the world and was focused by the example set by monks and canons, men who conspicuously avoided the taints of daily life. In this light, it is perhaps an irritation but not an insurmountable problem that it is impossible to know whether lord X worried about the slaying of an enemy more than having sexual relations with his wife during the prohibited phases of her physiological cycle or the liturgical calendar;[52] or whether *miles* Y agonized more about his technically consanguineous marriage to a distant cousin than about his rough treatment of some peasants. Religious stood apart (or at least were supposed to) from the full range of sins and vices: in particular, sex, violence, pride, and avarice.[53] When laymen supported religious communities they were bearing witness, even if only implicitly, to widespread dissatisfaction with the lay condition. Immediately we have isolated one reason why the appeal of the crusade, a meritorious exercise proposed to laymen *qua* laymen, was so powerful. We must now ask whether the links between the crusade and the relationship between religious communities and the laity ran deeper than this.

The First Crusade 'Indulgence' and the Significance of Penance

It is the central argument of this study that the first crusaders' pious motivations, the main element among all the reasons why men went on the First Crusade, were moulded by contacts with religious communities. It has just been seen that laymen supported ecclesiastical bodies for their own good, particularly their spiritual welfare. Participation on crusade was motivated by the same concern. In the broadest terms, therefore, one

[52] For restrictions on sexual activity, see J.-L. Flandrin, *Un temps pour embrasser: aux origines de la morale sexuelle occidentale (VIᵉ–XIᵉ siècle)* (Paris, 1983), esp. 8–54, 91–114, 128–43; J. A. Brundage, *Law, Sex, and Christian Society in Medieval Europe* (Chicago, 1987), esp. 154–61 and flow-chart at 162.

[53] See L. K. Little, 'Pride goes before Avarice: Social Change and the Vices in Latin Christendom', *AHR* 76 (1971), 16–49, esp. 20–6, 31–9.

might possibly argue that benefaction of churches and crusading were no more than parallel pursuits, tending to the same aim of salvation but formally distinct one from the other. In fact it becomes evident on close examination of the evidence that the two activities were intimately, even organically, linked. To examine how this could be so, it is first necessary to understand precisely what Pope Urban II offered the faithful when he launched the First Crusade.

As it was recorded on behalf of Bishop Lambert of Arras, the second canon of the Council of Clermont (the 'indulgence') was predicated on the supposition that the faithful would respond to the prospect of spiritual rewards couched in the language of penance: 'Whoever for devotion alone . . . shall set out for Jerusalem to liberate the church of God, to him may that journey be reckoned in place of all penance.'[54] There is further evidence for an emphasis upon penance in crusade preaching. In September 1096 Urban II wrote to the clergy and people of Bologna proposing the crusade as a 'total penance for sins for which they shall have made true and perfect confession'.[55] It should be noted that the terminology of Urban's few surviving pronouncements concerning the crusaders' spiritual rewards is not entirely consistent. In December 1095 he wrote from Limoges to the faithful of Flanders proposing the crusade 'for the remission of all sins' and making no explicit mention of penitential acts.[56] The apparent inconsistency, however, was only the result of the imprecision of the formula *remissio peccatorum*, not of muddled thinking on the pope's part. What Urban stated was that participation on the crusade was a satisfactory penance, a task so arduous that it would expunge the consequences of all confessed sins.[57]

The nature of Urban's promise has generated a good deal of controversy, much of it coloured by the hindsight of how the mature crusade indulgence slowly developed in the twelfth century and became fixed in the thirteenth. To express matters in their simplest terms, what Urban did not offer was the later medieval indulgence, the extra-sacramental remission by the church, by means of the inexhaustible fund of good built

---

[54] *The Councils of Urban II*, i. *Decreta Claromontensia*, ed. R. Somerville (Annuarium Historiae Conciliorum, Suppl., 1; Amsterdam, 1972), 74.

[55] *Die Kreuzzugsbriefe aus den Jahren 1088–1100*, ed. H. Hagenmeyer (Innsbruck, 1901), no. 3, p. 137.

[56] Ibid., no. 2, p. 136.

[57] N. Paulus, *Geschichte des Ablasses im Mittelalter vom Ursprunge bis zur Mitte des 14. Jahrhunderts* (Paderborn, 1922–3), i. 17–18, 196; B. Poschmann, *Der Ablass im Licht der Bussgeschichte* (Theophaneia, 4; Bonn, 1948), 54–5; J. S. C. Riley-Smith, *The First Crusade and the Idea of Crusading* (London, 1986), 27–9.

up in the Treasury of Merits, of the temporal punishment due for sins.[58] This developed idea presupposed a clear distinction between the punishment which attached to sin—in this world and in the afterlife prior to entry into Heaven—and the guilt of sin, which could be forgiven through sacramental absolution. The later indulgence also rested on the practice which gradually emerged in the eleventh and twelfth centuries whereby a confessor would absolve a penitent after confession but before the performance of the penance he had enjoined. This sequence in part reflected the extension into pastoral routine of concerns expressed by twelfth-century theologians that proper emphasis should be placed on the penitent's intention, not his mechanical actions, and further that no human act, however onerous or time-consuming, could even begin to match the offence done to God by sinful conduct: however assiduously the faithful performed their penances, there would always be a balance of punishment in this world and the next (that is, in 'Purgatory', the substantive used from the twelfth century to label the 'middle place' between Heaven and Hell).[59] The core of the penitential act by this stage was consequently the sinner's contrition, expressed in confession, and the confessor's mediation of Grace through absolution. In effect, the sinner was throwing himself on God's mercy. The performance of the penance was not an empty gesture. It served as an earnest of intent to demonstrate the penitent's submission to the church's discipline,[60] and it was a step in the right direction towards easing the punishment of sins. But it was not the cardinal event in the penitential sequence.

Before this system developed, the sequence of events involved in confession had been subtly but significantly different. The sinner made confession, whereupon the confessor, quite possibly using a penitential as a guide, ordered that he perform a suitable penance. The penitent duly completed what was required, anything from a fast to years on wandering pilgrimage.[61] Finally, the confessor, satisfied that the penance had been

---

[58] See E. Jombart, 'Indulgences', *DDC* 5 (1953), 1331–3.

[59] A. Angenendt, 'Theologie und Liturgie der mittelalterlichen Toten-Memoria', in K. Schmid and J. Wollasch (edd.), *Memoria: Der geschichtliche Zeugniswert des liturgischen Gedenkens im Mittelalter* (Münstersche Mittelalter-Schriften, 48; Munich, 1984), 152–4; K. Müller, 'Der Umschwung in der Lehre von der Busse während des 12. Jahrhunderts', in *Theologische Abhandlungen Carl von Weizsäcker zu seinem siebzigsten Geburtstage 11. December 1892 gewidmet* (Freiburg, 1892), 297–309.

[60] For early statements of this idea, see P. Anciaux, *La Théologie du Sacrement de Pénitence au XIIe siècle* (Universitas Catholica Lovaniensis, Dissertationes ad Gradum Magistri in Facultate Theologica vel in Facultate Iuris Canonici consequendum conscriptae[2], 41; Louvain, 1949), 52–3.

[61] Fasting was the staple penitential act for laymen: Angenendt, 'Theologie und Liturgie', 142–3.

acquitted, granted absolution. In this system the granting of sacramental absolution was tantamount to the church stating that the penitent had erased the sins he had confessed (deliberately concealing sins from the confessor was itself sinful); he had 'satisfied' the debt which his sins had created. Pursuing the rationale of this system to its extremes, it would have been technically possible for an individual to go straight to Heaven provided that he confessed every sin which he ever committed, performed every penance completely, and dropped dead at the instant of sacramental absolution. In such circumstances he would have been in the same position as an infant who died immediately after being cleansed of Original Sin by baptism. It was, however, supremely difficult for anyone, and virtually impossible for a layman, to keep an accurate tally of every sin and maintain the necessarily heroic levels of spiritual cleanliness over a lifetime. If anyone managed it and did so publicly, then he might become recognized as a saint—and the church canonized very few laymen indeed before the thirteenth century.[62] The motive force behind the system of satisfactory penances was not, therefore, that it made people believe that there was a short cut to eternal salvation, but rather that it encouraged the faithful to do as much as they could to limit the pains which they would suffer after death.[63] Urban's proposal to the first crusaders belonged to this earlier system of ideas.

Or, to be more precise, it was an extension of it. It is impossible to fix upon a precise date when the old penitential discipline gave way to the new, given that the problem was not the subject of one definitive piece of legislation, and that considerable regional variation in practice must have resulted from the progressiveness or otherwise of local senior clergy. What is reasonably clear, however, is that the First Crusade fell somewhere within the period of transition.[64] Much of the uncertainty which has surrounded the First Crusade 'indulgence' stems from the vagueness which surrounds the changes in the church's practice, with scholars attempting to project Pope Urban's meaning forward in time to conform

[62] See A. Vauchez, *La Sainteté en Occident aux derniers siècles du Moyen Âge d'après les procès de canonisation et les documents hagiographiques* (Bibliothèque des Écoles Françaises d'Athènes et de Rome, 241; Rome, 1981), 310–14.

[63] The best account of penitential discipline before the 12th C., although it over-emphasizes the importance of absolution relative to that of performance, is Müller, 'Umschwung', 290–2. See also C. Vogel, *Le Pécheur et la pénitence au Moyen Âge* (Paris, 1969), 27–8, 31. For the basis of penitential equivalence, see Angenendt, 'Theologie und Liturgie', 118–26, 134–7.

[64] See B. Poschmann, *Die abendländische Kirchenbuße im frühen Mittelalter* (Breslauer Studien zur historischen Theologie, 16; Breslau, 1930), 189–97; Anciaux, *Théologie*, 27–8, 43–51.

with later penitential observance.[65] The issue becomes much clearer if we consider that the First Crusade 'indulgence', far from being obscured by the uncertainty surrounding penitential discipline, was the direct result of it. In other words, Urban was drawing upon the old idea that penances could be satisfactory, but also addressing emerging anxiety that normal penitential forms were unattractive or fell short of appeasing God. (Laymen no doubt rationalized the unattractiveness of penances as their inefficacy.) The crusade was in effect a 'super-satisfaction'. The concerns to which Urban responded did not only come from intellectuals worried about the quality of sin, but also from the ordinary faithful. Their quarrel was not with the idea that penances could be satisfactory; rather, they felt that they had fallen behind in the performance of their penances so much that it would be impossible to acquit themselves before death. Professor Hans Mayer has attempted to rebut the argument that the First Crusade was a satisfactory penance by maintaining that no penance could have been as arduous an undertaking as participation on the crusade.[66] Of any single penance this is true. But this argument misses the fundamental point that the crusade was not a substitute for any one other penitential act but for all penances, both those outstanding before the crusade and those enjoined by potential crusaders' confessors shortly before departure. Thus, according to a Montecassino chronicler, who was very probably drawing on ideas which were voiced within Urban II's close circle, the pope suggested the crusade to certain penitent French princes who 'could not perform a fitting penance for their innumerable offences amongst their own people', not least because they felt that laying aside their arms during periods of penitential performance placed them at a disadvantage among their peers.[67]

It is therefore unnecessary to posit a situation in which the pope and senior clergy declared a limited indulgence comprising a commutation of

---

[65] See e.g. Paulus, *Geschichte des Ablasses*, i. 78. See also A. Gottlob, *Kreuzablass und Almosenablass: Eine Studie über die Frühzeit des Ablasswesens* (Kirchenrechtliche Abhandlungen, 30–1; Stuttgart, 1906), 71.

[66] *The Crusades*, trans. J. Gillingham, 2nd edn. (Oxford, 1988), 32.

[67] *Chronica monasterii Casinensis* (IV. 11), ed. H. Hoffmann (MGH SS 34; Hanover, 1980), 475. Hoffmann (ibid., pp. xxviii–xxx) argues that this passage was based upon the writings of Leo Marsicanus (d. 1115), a monk from Montecassino who entered the papal *curia* in the latter half of Urban II's pontificate and became cardinal bishop of Ostia under Paschal II. Leo is known to have written an 'ystoria peregrinorum', which is probably a lost account of the First Crusade and may have influenced the *Historia peregrinorum* composed at Montecassino in *c.*1140. See also H. E. J. Cowdrey, *The Age of Abbot Desiderius: Montecassino, the Papacy, and the Normans in the Eleventh and Early Twelfth Centuries* (Oxford, 1983), pp. xvii–xix, 26, 217–18.

penance, whereupon the popular preachers of the crusade and the faithful connived at misconstruing what was on offer as forgiveness of all sins.[68] At this stage there was no clear distinction between canonical punishment —the penance—and the temporal penalties in this world and the next created by sin. Pope Urban knew his audiences too well, and spent too long among them in 1095–6 at the head of a mobile *curia* drawing on local expertise, to have let the fundamentally important spiritual reward element of his crusade message grow out of control. For Urban and the faithful who responded to his appeal, *remissio peccatorum* was equivalent to *remissio poenitentiae* because of the super-satisfactory quality of the expedition. Thus Guibert of Nogent's description of the crusade, noted at the beginning of this study, as 'a new way of attaining salvation' recalled the novelty not only of the idea of an armed pilgrimage to Jerusalem, but also of the way in which the crusade opportunity cut through the existing norms of penitential discipline.

It follows from the foregoing that the crusade appeal was directed at laymen who were acquainted with the demands of penitential practice. It is now necessary to examine whether this familiarity was linked in any way to the relationship between the faithful and religious communities.

Charters and Penance

It is very difficult to know how often individual eleventh-century nobles and knights subjected themselves to penitential discipline, and still harder to generalize about the practices of whole social groups. Charters are not an ideal source, for, as has been seen, they usually recorded only one aspect of a layman's relations with the church, the formal grant of property, and did not consistently mention the circumstances behind a property transaction. The most informative documents deal with formal, public penances. Although penances of this sort must have formed a minority of all cases, it is useful to begin with them.

An interesting case from southern Gascony demonstrates some of the salient features of public penances. In about 1034 Count William of Astarac was persuaded by Archbishop Garsia II of Auch that he had

---

[68] Mayer, *Crusades*, 31–3. Mayer's argument about the loss of central control over the 'indulgence' message is consistent with his view that Jerusalem was not suggested as the crusade's goal by the pope but became so through popular enthusiasm (ibid. 9–11, 33). H. E. J. Cowdrey has established that this was singularly unlikely: 'Pope Urban II's Preaching of the First Crusade', *History*, 55 (1970), 177–88.

married within the prohibited degrees and that this placed him under the penitential discipline of the church.[69] The penance imposed upon him comprised abstinence from meat on Mondays and Wednesdays and from meat and wine on Fridays, the feeding of one hundred poor folk annually, the washing of twelve paupers' feet on Good Friday, and the distribution of alms. Further prescriptions dealt with periods of sexual abstinence.[70] The penance had three significant features. First, the possibility of commutation was allowed from the beginning. Abstinence from wine on Fridays was considered equivalent to giving 3d. to an unspecified number of the destitute, and forty days' fasting might be replaced by the giving of 5s. in alms over the same period. Archbishop Garsia's penitential ordinance expressed the penances in the plural, which suggests that he attempted to frame general rules for all those guilty of consanguinity in the light of a particular cause célèbre. An undated fragment appended to the penitential decree in the cartulary of Auch, however, suggests that bargaining and compromise were normal, at least when the penitent was a powerful figure.[71] Although no penitent is named in it, this fragment is in the singular. It refers to far longer periods of sexual abstinence (Lent, Advent, and five days per week) than the general ordinance, and its provision for the poor is less exacting (three paupers to be fed and clothed indefinitely). It most probably represents an early draft of Count William's penance before it was subjected to negotiation. This illustrates an important feature of penitential discipline: that penances, once enjoined, were not immutably fixed. A penitent might avail himself of a number of approved actions which were considered to be of equivalent value. This factor will be of fundamental importance when we come to consider the reasons why laymen gave properties and rights to religious communities.

The second important feature of Count William's penance is that it reveals some understanding of the difference between lesser and graver errors. This did not simply involve a contrast between venial and mortal sins, an idea which became more refined in the twelfth century and turned upon the intrinsic quality of the fault. Rather, a working distinction was used based on public notoriety.[72] Garsia's ordinance argued that,

[69] For the dating, Jaurgain, Vasconie, ii. 161.

[70] Auch, 41–2.

[71] Ibid. 43.

[72] Regino of Prüm, 'De ecclesiasticis disciplinis et religione christiana' (I. 289–92), PL 132. 245–6; 'De vera et falsa Poenitentia' (ch. 11), PL 40. 1123; Lanfranc, 'De celanda confessione', PL 150. 630–2; C. Vogel, 'Les Rites de la pénitence publique aux X\ue et XI\ue

just as serious wounds require dressing, so there was a correspondence between the gravity of faults and the means required to compensate for them; whatever had been committed openly, therefore, could only be remedied likewise.[73] In other words, Count William's actions belonged to a species of conduct beyond the scope of everyday, lesser faults which could be dealt with by private confessors (and consequently were much less likely to be put in writing). This serves as a warning that the surviving documents are likely to present a 'top-heavy' picture of laymen's sinful careers, and that it would be unwise to underestimate the capacity of laymen to become burdened by a long sequence of individually private, and often relatively minor, sins.

Thirdly, the imposition of this form of penance was reserved for bishops. A verbose account of Count William's dealings with the arch-bishop, in the name of 'Odo the Deacon', expounded the jurisdictional authority of Garsia's actions.[74] It recalled the Petrine Commission, the Power of the Keys, and the power to bind and loose.[75]

Similar features are apparent in the penance to which Viscount Adhemar II of Limoges submitted in 1073.[76] Adhemar had burned the *cité* of the cathedral of St-Étienne, pursued the fleeing clergy and populace, and allowed his troops to loot and kill. Not surprisingly, the canons considered this act a 'most serious crime' and 'great misdeed'. It is quite possible that a formal Lenten penance was involved.[77] Our source, a charter from St-Étienne, makes no reference to ashes (which were scattered over the penitent at the commencement of the Lenten penitential period), but it reveals that Adhemar went barefoot to the cathedral, prostrated himself before the main altar, and received penance. In the charter Adhemar was made to say that he approached St-Étienne 'as though I have sought refuge at the safe haven of salvation', a phrase which seems to recall the physical presentation of the penitent at the cathedral's west door. The circumstances behind Adhemar's act of violence are not made clear. 1073 witnessed a vacancy in the see of Limoges, and the turbulence may have been caused by Adhemar's attempts

---

siècles', in P. Gallais and Y.-J. Riou (edd.), *Mélanges offerts à René Crozet* (Poitiers, 1966), i. 138; Anciaux, *Théologie*, 44–6.

[73] Auch, 42.

[74] Odo may have been a member of the comital family of Astarac: see Jaurgain, *Vasconie*, ii. 158, 160.

[75] Auch, 41; cf. Poschmann, *Die abendländische Kirchenbuße*, 127–8, 198–201.

[76] SEL 80[73].

[77] Cf. *Le Pontifical romano-germanique du dixième siècle* (XCIX. 220–67), ed. C. Vogel and R. Elze (Studi e Testi, 226–7 and 269; Vatican City, 1963–72), ii. 58–71.

to impose his own candidate as bishop; it is perhaps significant that the next vacancy at Limoges, in 1086, sparked off serious disorder in the city. Unfortunately, the charter does not reveal which bishops were present. What is made clear, however, is that earlier Duke Guy Geoffrey of Aquitaine, described in the charter as Adhemar's lord, had imposed a settlement concerning Adhemar's rights in the *abbatia* of St-André in Limoges, which the canons of St-Étienne also claimed. Adhemar's act of contrition, therefore, served a symbolic, political purpose, which may explain the omission from the charter of any details of the penance imposed. The charter records a number of property transactions at some length, however. Adhemar gave to St-Étienne an allodial *mansus* and water rights in Limoges, and confirmed an earlier grant of the *abbatia* of St-André. On one level the transactions appear as a quite conventional act of benefaction, for Adhemar invoked the consent of his wife and sons, and symbolized the transfer of the *mansus* by joining two of his sons in placing a token on the high altar; the words *dono* and *donatio* were used of the whole process. On the other hand, in the charter the record of the ceremony of donation at the altar immediately follows the account of the prostration and receipt of penance. This suggests that a single con-tinuous event was involved. It appears that the donation of the *mansus*—carefully distinguished in the charter from the confirmation of the *abbatia*, which would seem to have been one of the problems at issue in the acts of violence—was considered part of the penitential act.

Similar cases point to the penitential character of some important acts of benefaction. In the case mentioned above, William of Astarac granted to the cathedral of Auch the church and adjacent settlement of Ste-Aurence, which, it was claimed, had been lost to Ste-Marie through the aggression of 'perverse men'.[78] Within eight years of his return from the First Crusade, Viscount Raymond of Turenne did penance for the burning with great loss of life of the castle of Clérans.[79] A public Lenten penance was imposed. Our source makes no express reference to episcopal involve-ment (Clérans lay in the diocese of Périgueux), but this must be under-stood in the light of Raymond's statement that 'as it was enjoined upon me, amongst other things I have remained within the monastery of Uzerche, with the servants of God, up to Easter Day'.[80] The monks of Uzerche exploited Raymond's stay amongst them to exact from him a

---

[78] Auch, 41–2.

[79] Uz. 266 (1100 × 1108).

[80] For the practice of penitential reclusion in a monastery, see Vogel, 'Rites', 141–3; Poschmann, *Die abendländische Kirchenbuße*, 118–20.

promise to protect the monastery's property at Gumont (seven miles north-west of Turenne) which was exposed to the depredations of aggressive neighbours, including, most probably, Raymond himself. The promise was solemnly confirmed on Easter Sunday. Raymond had presumably been absolved on the previous Thursday (the prescribed day for the absolution of Lenten penitents), so the link between the penance and benefaction was not as immediate as in the Auch and Limoges examples. But Raymond was made to declare that the lifelong protection of Gumont had been promised 'as a penance and for the salvation of my soul'. It would therefore seem that an arrangement had been reached between the bishop of Périgueux and the monks whereby Raymond was allowed to reckon a typical feature of the relationship between a powerful lord and a local monastery as part of his penance, or as a commutation of it.

Co-operation between Benedictine communities and the episcopate is more clearly documented in another case. In 1079 William Arnald of Vignoles performed a Lenten penance at St-Mont alongside his wife and son. He remained at the monastery to celebrate Easter and then granted the monks his full rights in the church of Aurions. At some time between Easter and the following October the gift was confirmed by William Arnald's lord, Centulle of Béarn-Bigorre, and by his local bishop, Bernard of Lescar. It is likely, therefore, that the donation was made by prior arrangement with Bishop Bernard.[81]

Express references to penitential acts are quite rare in monastic records. It is not mere chance that instances of formal penance occur more frequently in the charters of cathedral communities. Between 1068 and 1096, for example, William Arnald of Tremblade was excommunicated by Archbishop William I of Auch for the murder of a priest, and then restored to the cathedral of Ste-Marie, 'to alleviate the penance', land which had come to him through his father's marriage into the family of Archbishop Raymond I Coppa (1036–*c*.1050).[82] Between 1052 and 1073 a fratricide received penance on the orders of Bishop Itier of Limoges, and gave St-Étienne one *bordaria*.[83] At about the same time a priest, Robert of Mortemart, received penance at Limoges on Ash Wednesday from a canon of St-André and granted to the cathedral chapter his rights in his own church before leaving on pilgrimage to Compostela.[84]

The preponderance of references to formal penances in the records of

---

[81] St-M. 43.   [82] Auch, 53.   [83] SEL 49.
[84] Ibid. 60[56] (1052 × 1060).

cathedral communities reflects the power of bishops to excommunicate and absolve.[85] Relying upon a suspect privilege of Pope Benedict IX (1032–44), the Benedictines of Ste-Croix, Bordeaux claimed the power to bind and loose over any aggressor and considered themselves and their property immune from any interdict.[86] This document was clearly part of a struggle for jurisdiction with the archbishop and secular clergy of Bordeaux; and in the competition for patronage and prestige the close proximity of Ste-Croix to the cathedral of St-André and the community of canons of St-Seurin created problems for the monks similar to those faced by the Benedictine communities of Limoges, for example, but less pressing for rural monasteries. Bishops were generally very jealous of the right to excommunicate, and 'excommunications' pronounced by monks could simply anticipate or amplify episcopal acts.[87] Two late examples from the abbey of Sorde make this clear. According to one charter, a man was excommunicated for withholding the tithe he owed to the abbey, and then surrendered his claim at Dax cathedral, not at the monastery, in the presence of the bishop and canons. In return he received money, food, and clothing from the monks.[88] In about 1135 the abbot of Sorde approached Bishop William of Dax to complain about the seizure by a layman of some monastic property. The bishop pronounced excommunication, and the aggressor eventually released his claim before both bishop and abbot. Again the quitclaim came at a price, revealing how a bishop might act as mediator between a monastic community and litigants by virtue of his pastoral authority.[89]

Bishops' authority over sinners, therefore, was of great importance, particularly in cases of notorious sins which were more likely than most to become a matter of record. It would be mistaken, however, to treat monasteries as little more than adjuncts of episcopal control. In fact some charters suggest that ordained monks were asked to act as laymen's confessors, and that consequently acts of benefaction or surrender to

---

[85] See e.g. St-Seurin, 11 (989 × 1010), 15 (1073 × 1085), 26 (1088), 27 (possibly early 12th C.), 29 (1119), 64 (1102 × 1130); SEL 37[35] (*c.*1056), 65[61] (*c.*1100).

[86] Ste-C. 85. For this document, see A. Chauliac, *Histoire de l'abbaye Sainte-Croix de Bordeaux* (Archives de la France monastique, 9; Ligugé, 1910), 63–4.

[87] e.g. Mart. I. 36 (1027); cf. L. K. Little, 'La Morphologie des malédictions monastiques', *AESC* 34 (1979), 51–3. See the procedure against violators of St-Mont's *sauveté* established by St-M. 6 (written post-1109 of events in 1073). Cf. the 9th-C. formula of malediction from the Limousin printed in L. K. Little, 'Formules monastiques de malédiction aux IXe et Xe siècles', *RM* 58 (1975), 386.

[88] S. 64 (mid-12th C.).

[89] Ibid. 110.

monasteries might be linked to private penances. In about 1060, for example, Arnald Loup of Goron renounced his claims to a gift made to Sorde by one of his tenants. His charter is full of the language of penitence, describing how he came to the abbey 'making satisfaction to God' through prayer and fasting, and petitioned the monks to intercede on his behalf. He then received penance. What this involved is not stated, but we learn that Arnald Loup granted property to feed the monks and to provide alms for the poor. The charter records that this was done 'for the redemption of his relatives' souls'—a perfectly conventional formula in an act of donation—and 'for the absolution of his sins', a phrase which may point to a formally penitential element.[90]

The anathema clauses of some charters supposed that the act of submission by a future litigant would have an at least quasi-penitential character. The relevant formulas followed no strict rules, and not all referred directly to spiritual sanctions. Whenever appropriate the threat of secular authority might be invoked. A St-Mont charter of 1062, for example, envisaged that an aggressor would be subjected to the ban of the duke of Gascony and forced to pay 100 gold pounds; the amount was so enormous that it could only have had symbolic value, but even so the principle of pecuniary liability was thereby asserted.[91] Contemporary charters from the same priory referred to men 'forced by the laws' or judged 'before the princes of the land';[92] and a privilege of Pope Calixtus II (1119–24) (a document which is not above suspicion, but nevertheless reveals the monks' approach to these matters) postulated a formal procedure against aggressors whereby, after three warnings, they might be excommunicated and also deprived of their offices and lands, the latter sanction obviously requiring the co-operation of secular powers.[93]

Other formulas, however, clearly anticipated acts of submission which were voluntary, or at least formally distinct from secular penalties. An Auscitain charter of 1094 provided that a malefactor would be cursed and excommunicated 'until he shall come to make emendation and satisfaction'.[94] Similarly, an individual might be anathematized 'unless he shall repent' ('nisi penituerit').[95] The verb *penitere* did not always bear the connotation of submission to the sacramental act of penance, for it might be used simply to mean 'to regret' or 'to see the error of one's ways'. Thus a Limousin notice of *c.*1120 recorded that men in dispute with St-Étienne about a predecessor's donation first repented of the injury ('de injuria . . .

[90] Ibid. 39 (*c.*1050 × 1072); cf. ibid. 38.   [91] St-M. 7.

[92] Ibid. 11 (1063), 14 (1062 × 1085); see also ibid. 62 (1074).   [93] Ibid. 8.

[94] Auch, 39.   [95] St-M. 34 (1068 × 1085).

penituerunt'), and then approached the canons to arrange an agreement.[96]
Penitere might be synonymous with *resipiscere* ('to regain one's senses'), for
the two verbs were used in very similar constructions.[97] *Penitere*, however,
also carried a more precise meaning. In one charter from St-Mont the
phrase 'unless he shall do penance' ('nisi penitentiam egerit')[98] was used
in lieu of the more common 'unless he shall repent and make emends'
('nisi penituerit et emendaverit').[99] Similarly, the formula 'si non se
penituerit, vitam non habeat domini' supposed some formal act of submission to the church as intermediary between God and man rather than
a simple change of heart.[100] A Sorde charter from 1120 distinguished
clearly between the three actions required of an aggressor, who would
be damned unless he realized his error, made 'worthy emends', and
performed penance.[101] Furthermore, 'satisfaction', which might evoke the
performance of penance, could be distinguished from 'legal emends'.[102]
A grant of rights 'to make emends for the misdeeds' ('pro emendatione
malefactorum') might represent material compensation for past aggression,[103] but equally a gift 'in emendatione' might involve the gift of
extra property not at issue in a resolved dispute.[104] From these examples
it is apparent that the language of charters' anathema clauses was fluid;
different sanctions might be invoked in the acts of a single community
within a few years of one another and irrespective of any difference
between the types of transaction involved.[105] Nevertheless, the clauses
suggest that many gifts and confirmations made to religious communities
were linked to penances, even though most charters were so worded as to
represent an unsolicited act of alms-giving.

It is likely, in fact, that the distinction between voluntary good works
and enjoined acts was often blurred in practice, especially in the minds of
laymen burdened by unperformed penances. Moreover, although bishops
dominate the surviving references to penances, there is evidence that
monasteries could co-operate with them in matters of penitential dis-

---

[96] SEL 86[79].

[97] St-M. 43 (1079), 73 (*c.* 1080).

[98] Ibid. 88 (*c.* 1100).

[99] e.g. ibid. 79 (*c.* 1080 × 1085), 86 (1068 × 1085).

[100] Ibid. 59 (1077 × 1085). Cf. Little, 'Formules monastiques', 387 (9th-C. Limousin
malediction): 'si autem emendare noluerint deo et sancto marciali accipiant damnacionem'.

[101] S. 7. Cf. Little, 'Formules monastiques', 392 (mid-9th-C. curse from St-Wandrille in
Normandy).

[102] St-M. 2 (1068 × 1085).

[103] SEL 51[47] (1062 × 1073).

[104] Ibid. 112[93] (*c.* 1050).

[105] e.g. St-M. 3 (2nd half 11th C.), 14 (1062 × 1085), 36 (3rd qr. 11th C.).

cipline; that monks could act as confessors; and that benefaction of religious communities of all types was often treated as supplementing or even replacing the performance of penances. In this way the charters reveal a level of submission to penitential discipline which helps to explain the powerful attraction of the First Crusade 'indulgence'.

Laymen's Religious Ideas: The Charters

The foregoing has established that laymen's acts of benefaction were motivated to a significant extent by religious ideas. Furthermore, those ideas were more precise than mere notions of doing good or being charitable: they were often influenced and sometimes directly channelled by the demands of, and the conventions surrounding, the church's penitential discipline. It follows that it is necessary to examine the laity's religious ideas in order to understand how they were able to be expressed in this specific way.

It was widely accepted that when charters were drawn up some effort should be made to pronounce upon the religious motivations which lay behind property transactions. It is difficult to trace clear patterns in the movements of ideas through time and space, for charters' contents were habitually determined by custom and respect for ancient authority.[106] Typically, one of the first acts of newly elected heads of religious communities, especially reformers, was to order a review of the archives. Reacquaintance with old documents influenced the practices of *scriptoria*. Thus, for example, one formula often used in Beaulieu's heyday in the second half of the ninth century reappeared under the Cluniac reformer Abbot Gerald (1097–*c*.1120).[107] The sample of relevant documents is not large, for when charters were copied into cartularies, their preambles and maledictions were particularly vulnerable to corner-cutting. The later-twelfth-century cartulary of Vigeois, for example, contains virtually no preambles (though it is worth noting that the first dozen folios, which most probably preserved copies of the abbey's most elaborate privileges and charters, are lost).[108] For these reasons the following discussion is a composite account of charters' language using examples taken from a

---

[106] H. Fichtenau, *Arenga: Spätantike und Mittelalter im Spiegel von Urkundenformeln* (Mitteilungen des Instituts für Österreichische Geschichtsforschung, 18; Graz, 1957), 123.

[107] B. 42 (1097 × 1108), 43 (887), 52 (895), 63 (893), 153 (868). But cf. ibid. 114 (mid-11th C.).

[108] For an exceptional survival, V. 106 (1100 × 1108).

broad chronological and geographical range. Some communities were more progressive than others: the records of the small but rigidly Cluniac priory of St-Mont, for example, furnish a particularly wide range of material. Broadly speaking, however, the ideas of all the studied communities were remarkably similar and may thus be treated as a whole.

In the period *c.*970–*c.*1130 private charters from our three areas of south-western France varied somewhat in form. In particular the distinction between the *arenga* (a generalized evocation of religious principles) and the *narratio* (an account of the specific circumstances behind the transaction described in the document) might be rigidly observed or compressed into a general statement of a donor's pious motivations. The practice was virtually unknown, however, of composing purposefully obscure preambles in order to create a solemnly euphonic but meaningless effect.[109] The few unorthodox passages which survive appear to have been the product of exuberance or self-conscious erudition on the part of scribes. One monk from Uzerche, for example, who composed a charter passed by Aimilina, sister of Viscount Adhemar II of Limoges, evidently believed in his ability as a versifier and scholar of ancient literature when he came to evoke his benefactress's fear of the 'waters of Acheron' (a river in the lower world).[110] When the dying Raymond Fort of Sion made over modest gifts to St-Mont his sentiments were expressed in an introspective monologue: 'Oh, the anguish! What am I to answer? What should I, a wretch, say? Why does the barren tree loom over its land? I know what to do; I should take refuge . . . in the haven of penance and consolation.'[111]

It was common form for the charters which recorded a community's most important property transactions to contain the longest *arengae* and/or *narrationes*. For example, the document drawn up when Viscount Archambald III of Comborn established the priory of Meymac in 1086 contains an elaborate *arenga* built around the proposition that all Christians shared a single faith which should be expressed through support of the institutional church. The *narratio* explains how Archambald, motivated by a desire to increase the material resources of the church and to provide for his kinsmen's souls, petitioned Bishop Guy of Limoges for permission to erect a Benedictine priory at Meymac, a church which he held as a fief from the bishop and canons of St-Étienne.[112] An important

[109] But see Bibliothèque Nationale, Paris, MS lat. 12752, p. 357, a copy of a charter of 1023 from the Gascon monastery of Simorre. For the date, see Jaurgain, *Vasconie*, ii. 160.

[110] Uz. 87 (1073 × 1085).

[111] St-M. 50 (*c.*1070 × 1090).

[112] *GC* 2, Inst., cols. 183–4.

feature of the preamble is its lack of originality. Its ideas were arranged brick-like one upon the other, and we shall see that many of its elements appear, in isolation or in various combinations, in numerous other charters. From these elements it is possible to reconstruct, in bare outline, the popular ideology of lay benefaction of religious communities in the years before the First Crusade.

In Archambald's *arenga* the duty of all Christians to support the church was expanded into a statement of the particular duties of a major lord: whoever was acknowledged as notably powerful in this world should add to the church's material resources according to his means; and God's ministry was furthered by nobles' endowments, which made ornamentation and building-work possible. The references to secular authority and nobility served to emphasize that the élite of the regional aristocracy was held to have a special duty towards the church which flowed, in part, from birth and social position. It is important to note, however, that Archambald was not explicitly invested with responsibilities by virtue of an office granted *Dei gratia*, an idea which would have been applicable to only a small minority of laymen. Thus the premiss that the tenure of property created an obligation to give to religious houses could be extended generally to laymen of all social levels.

Ecclesiastics were in the potentially awkward position of trying to stimulate largess and protection from the major lords while not giving the impression that there was an automatic and direct correlation between the amount given in alms and the donor's chances of salvation. A crude compromise was reached which involved some sort of proportionality between an individual's means and what he might be expected to give away (although no document from this period attempted to reduce this idea into any sort of arithmetical equation). If, as seems quite likely, families worked out some rough formula for the proportion of their wealth which could be alienated in any generation, this would be impossible to reconstruct from the documentation as it survives. Probably for reasons of discretion, charters tended to address the issue in general terms. A notice from St-Mont, for example, records that Count Bernard II Tumapaler of Armagnac made a number of grants to the priory 'having a greater abundance of goods than all those living thereabouts'.[113] A charter of St-Seurin, Bordeaux, which probably dates from the late eleventh or early twelfth century, contains a particularly clear statement that only the amount given in alms, not the duty to give itself, was

---

[113] St-M. 14; cf. ibid. 60 (3rd qr. 11th C.).

determined by social status and available means: ancient authority established that every believer, whoever he might be, should endow the
church, to the extent that his wealth permitted, as an expression of the
love of God and for the remission of his sins.[114] Some *arengae* reveal that
it was considered suitable for even the most important laymen to invoke
such general precepts. A charter of Viscount Guy I of Limoges cited the
authority of St Paul that alms should be given to the church to the extent
that lawfully held property was available for the purpose.[115] Seventy years
later a charter passed by Guy's grandson Adhemar II used exactly the
same argument.[116] When Viscount Raymond I of Turenne granted two
bordariae to Vigeois, he invoked the duty of all Christians to make
offerings to God and the saints.[117] Count Hildebert of Périgueux was
made to utter similar ideas in an Uzerchois charter of 1109.[118] In
a document from St-Mont, Viscount Odo of Lomagne referred to the
need of every individual to redeem sins through alms 'according to his
means'.[119]

Statements such as these reveal that alms-giving was treated as a form
of gift which might merit the counter-gift of salvation. This belief was
often asserted and reinforced by the use of a number of frequently cited
scriptural passages.[120] The most common was Christ's admonition to the
Pharisees (Luke 11: 41): 'give for alms those things which are within;
and, behold, everything is clean for you.' In an Auch charter of 956 this
passage was interpreted to mean that Christ warns man to redeem his sins
in order to avoid damnation (*tartareum*).[121] It, and Christ's promise that
'Give, and it will be given to you',[122] could be used to justify the belief
that the church should encourage all the faithful to give from their own
resources.[123] Luke 11: 41 might also form the core of a statement that
provision for one's soul through property was imperative because death
was inevitable.[124] This precept could be accounted 'the foremost and

---

[114] St-Seurin, 27; see also ibid. 43 (1102 × 1130), 76 (1126).

[115] Uz. 462 (998 × 1003).

[116] Ibid. 1020 (1068).

[117] V. 106 (1100 × 1108).

[118] Uz. 517.

[119] St-M. 5 (mid-11th C.); see also ibid. 12 (*c*.1080).

[120] See A. Giry, *Manuel de diplomatique*, 2nd edn. (Paris, 1925), ii. 539; W. Jorden, *Das cluniazensische Totengedächtniswesen* (Münsterische Beiträge zur Theologie, 15; Münster, 1930), 48–50; C. B. Bouchard, *Sword, Miter, and Cloister: Nobility and the Church in Burgundy 980–1198* (Ithaca, NY, 1987), 226–7.

[121] Auch, 25.   [122] Luke 6: 38.

[123] Auch, 48 (early 11th C.); cf. T. 12 (*c*.930).

[124] T. 17 (*c*.1080), 46 (986), 288 (1060 × 1073), 511 (984).

greatest' of all the Lord's commands,[125] and the desire to obey it might be explicitly cited as a donor's motive for giving to a religious institution.[126] Another popular passage, with obvious connections with imagery of the afterlife, was 'Just as water extinguishes fire, so alms extinguish sin.' This dictum was not a verbatim scriptural borrowing, but it was a close paraphrase of Ecclesiasticus (Sirach) 3: 33.[127] It was sometimes used in conjunction with Luke 11: 41 to provide a rounded statement of the means and effects of alms-giving.[128] Similarly, Christ's command, illustrated by the parable of the unjust steward, to 'make friends for yourselves by means of unrighteous mammon, so that . . . they may receive you into the everlasting habitations',[129] was used to demonstrate that an individual was able to convert property into intercession.[130]

Other forms of preamble linked the duty of Christians to give alms to a pessimistic view of the world, usually in the form of general statements into which the individual donor's experience was sublimated. Pronouncements of this sort might be illustrated by scriptural allusions, but they were not all directly biblical in inspiration, and might express, in a Christianized form, ideas which dated back to the Stoics and other writers of pagan Antiquity.[131] For example, it was common for documents to dwell upon the fragility of the human condition or of the world. In 975 a husband and wife giving to Beaulieu were made to link the fate of the world with the scheme of resurrection: 'Nos . . . consideravimus casum fragilitatis hujus saeculi, ultimamque corporis efflationem.'[132] The first element in this phrase might equally well stand alone as the expression of a donor's religious motivation,[133] or it might preface a statement of hope of future reward.[134] A charter of Ste-Croix from *c.*1130 in the name of a senior cleric forcefully developed the theme of human frailty in allusions to the Fall and the manner in which misery on Earth anticipated suffering in Hell.[135] Laymen's charters could share the same sentiments and imagery.

---

[125] Ibid. 396 (930).
[126] Ste-C. 7 (1126 × 1137); Sol. fo. 40^r–v (*c.*1060 × 1080).
[127] Sirach 3: 30 in the RSV.
[128] T. 396 (930), 399 (*c.*1055 × 1060); Mart. I. 36 (1027).
[129] Luke 16: 9; cf. Matt. 6: 24; Luke 16: 11–13.
[130] St-M(Sam). 66 (final third 11th C.); Uz. 455 (1072); cf. *Chartes et documents pour servir à l'histoire de l'abbaye de Charroux*, ed. P. de Monsabert (Archives historiques du Poitou, 39; Poitiers, 1910), 6 (1079).
[131] Fichtenau, *Arenga*, 124–9.
[132] B. 75.   [133] Ibid. 158 (889).
[134] Ibid. 17 (879 × 884); Ste-C. 115 (1132 × 1138).   [135] Ste-C. 97.

It was common practice to describe the world as growing old,[136] a belief which might be justified by appealing to 'increasing ruinations' or 'certain signs'.[137] A doctrinally more precise proposition was that the end of the world was near.[138] Although the sample of extant preambles is too small to admit precise chronological analysis, there is little evidence to suppose that statements of the imminent end of the world were most frequent in the years immediately before the millenia of the Nativity or Resurrection (1033).[139] Taken out of context, some charters might support the idea that apocalyptic fears were particularly common at those times and were being transmitted by monks to the laity. In about 1000, for example, Viscount Rainald of Aubusson was made to say in a Tulle charter that the duty to give to the church was now particularly pressing because the world was growing old and the Day of Judgement was at hand.[140] But similar ideas had been expressed at Tulle and nearby at Beaulieu before and after 900.[141] In 967 it had been believed that the end was approaching 'quickly'.[142] And in the latter half of the eleventh century apocalyptic language of equal urgency could be revived for particularly solemn charters.[143]

Rather than commit themselves to an exact forecast for the end of the world, the composers of charters often chose to appeal to the more open-ended and flexible belief that the material creation was transitory. Such an approach had the great merit of being able to appeal to every generation by locking into the potent ideas concerning the correct uses of property. A practice common in the ninth and tenth centuries which survived into the eleventh was to imagine the donor contrasting this world and the next: 'It behoves everyone to cross over from the earthly to the heavenly, from the perishable to the eternal.'[144] From this it was a small step to introduce the role of benefaction: one ought to acquire eternal life by means of transitory things.[145] A charter of St-Mont, probably dating

---

[136] St-M(Sam). 66 (final third 11th C.); B. 150 (984), 166 (885).

[137] B. 3 (866); cf. Marculf, 'Formulae' (II. 3), *MGH Legum Sectio*, 5. 74; 'Formulae Turonenses vulgo Sirmondicae dictae' (ch. 1), *MGH Legum Sectio*, 5. 135.

[138] T. 12 (*c*.930); B. 46 (878); Fichtenau, *Arenga*, 128.

[139] Giry, *Manuel de diplomatique*, ii. 543–4. For a contrary view, based upon charters from northern and western Aquitaine, see D. F. Callahan, 'The Peace of God and the Cult of the Saints in Aquitaine in the Tenth and Eleventh Centuries', *HR* 14 (1987), 451 and n. 33.   [140] T. 350.

[141] Ibid. 12 (*c*.930); B. 55 (885), 57 (882), 66 (927).

[142] B. 73.

[143] Sol. fo. 35ʳ (*c*.1070); B. 14 (1062 × 1072).

[144] B. 42 (1100 × 1108), 43 (887), 52 (895), 58 (943), 69 (909), 114 (1032 × 1060).

[145] Ibid. 55 (885); cf. ibid. 46 (878); SEL 18[16] (914).

from the final third of the eleventh century, vividly made a similar point: everything which grows must age, and water in a stream cannot flow back towards its source; likewise a fine appearance is quickly marred by even a mild fever, and worldly honour swiftly passes from one individual to another; every Christian should, therefore, consider how to dispose of the property granted to him by God.[146] A statement of doctrine which expanded upon the transitory nature of creation, and which members of the arms-bearing classes would have readily understood, was that no individual could have foreknowledge of the date of his death.[147] From this it was a short step to statements of the idea that man must be constantly vigilant[148] against the possibility of sudden death[149] lest he be 'unprepared'.[150] Providing for one's soul while there was still time was therefore a pressing need.[151] It also required no feat of the imagination for laymen to grasp that they could not take their property into the next life.[152]

Many charters made it clear that a donor's concern for his spiritual welfare should be focused upon the Last Judgement, which, after the Resurrection, was the cardinal event in God's scheme of salvation. As has been seen, the redactors of charters could not be precise about when Judgement would occur. There can be no doubt, however, that the laity was encouraged to fit this event into a chronological sequence derived from its own experience of this world. Judgement lay 'in the future';[153] it would take place within one day;[154] on that day all would tremble.[155] Extrapolation into the next life of everyday experience was also encouraged by using the language of the court. On the *dies judicii*[156] the faithful would be summoned before a *tribunal*[157] where Christ would appear as *judex*.[158] Belief in the Day of Judgement required of the laity some

---

[146] St-M(Sam). 66.

[147] Uz. 77 (1108).

[148] B. 150 (984).

[149] Ibid. 42 (1100 × 1108), 52 (895), 58 (943), 69 (909), 130 (885). Cf. P. Ariès, *The Hour of Our Death*, trans. H. Weaver (London, 1983), 10–13; Lemaître, *Mourir à Saint-Martial*, 158–61.

[150] B. 3 (866), 19 (860); cf. Marculf, 'Formulae' (II. 2, 4), 74, 76; 'Formulae Salicae Merkelianae' (ch. 1), *MGH Legum Sectio*, 5. 241.

[151] SEL 9[7] (1010 × 1014); Uz. 1021 (1097 × 1108).

[152] Uz. 466 (*c.*1060 × 1086).

[153] St-M. 68 (*c.*1070), 71 (4th qr. 11th C.).

[154] Auch, 55 (*c.*920); Uz. 53 (1019), 475 (1001); Mart. I. 36 (1027); B. 150 (984).

[155] Sol. fo. 25ᵛ (1031 × 1060); St-M. 67 (1062), 88 (1085 × 1096).

[156] Uz. 61 (1019), 174 (1001).

[157] B. 17 (879 × 884), 44 (928), 55 (885); Uz. 315 (1036); Auch, 25 (956).

[158] B. 125 (1061 × 1108); Ste-C. 97 (*c.*1130).

appreciation of the distinction between Christ's roles as Redeemer and
Judge. An unusually descriptive preamble in a charter from St-Mont
stated that whenever man sinned he provoked Christ to anger. He had
been born, suffered, and died for us. Consequently he shall appear severe
and terrifying at the end of the world, whereas once he had chosen to be
meek and gentle; the more men redeemed their sins now, the less
forbidding he would seem to them later when they came to be judged.[159]

Judgement was considered a single process in which all the faithful
would be involved simultaneously. In its most basic form, it was the
occasion when the good and evil would be divided.[160] Each man would
be judged individually according to his works,[161] and each soul would
receive a form of retribution peculiar to it.[162] Nevertheless, the belief in a
single event also reinforced the idea that all individuals would be judged
together, particularly in family groups.[163]

So much was clear. The belief that a kindred group spanning several
generations would be judged together, however, raises the question of
what laymen believed had happened—or, rather, was happening—to
their dead relatives. Understanding this is fundamental to the whole
question of lay piety. In particular, the appeal of the First Crusade
'indulgence' can only be understood once we know why eleventh-century
laymen were afraid of what would happen to them when they died. It is
noteworthy, therefore, that regarding this point the precision of the
charters' language breaks down somewhat, and statements can be adduced
to support a number of apparently contradictory views of the afterlife.

To begin with the most straightforward element of the charters' treatment of life after death, there was an obvious contrast between Heaven
and Hell. This basic division was commonly expressed in the coupling of
the *poena* and *felicitas/gloria* awaiting respectively the reprobate and the
just.[164] Many biblical metaphors for Heaven were well suited to appeal to
the lay aristocracy, and from their frequent use in charters it appears that
the laity was encouraged to apply them literally to its mental picture
of Paradise. Most broadly, Heaven was a kingdom.[165] Perhaps more

---

[159] St-M. 60 (3rd qr. 11th C.: amending 'cum' in the last phrase to 'eum').

[160] Ibid. 68 (c.1070), 77 (1089 × c.1120).

[161] SEL 56[52] (c.1060); St-M. 31 (1077 × 1085), 65 (2nd half 11th C.); B. 28 (c.945).
Cf. Rom. 2: 5–11.

[162] St-M. 66 (c.1055 × 1062).

[163] B. 147 (916); St-M. 5 (c.1050 × 1060), 88 (1085 × 1096); Auch, 25 (956).

[164] SEL 9[7] (1010 × 1014); St-M. 52 (1062 × c.1080).

[165] SEL 29[27] (1014 × 1021). Biblical references to the heavenly kingdom are legion.
See e.g. from the Gospels, Matt. 3: 2, 4: 17, 5: 3, 5: 10, 7: 21, 18: 1 (Vulgate: 'regnum

pertinently in areas where royal authority was weak or non-existent, it was also a heavenly homeland.[166] Some care was evidently taken in the choice of language. For example, the vassal of the lords of Lastours who gave his enfeoffed rights in a church to Beaulieu would have responded favourably to the monks' evocation of the 'hall of the upper city'.[167] Describing Heaven as an inheritance must have had a particularly strong hold upon nobles' and knights' imaginations;[168] so too appeals, sometimes built around Luke 16: 9, to use property in order to make God, Christ, or the saints one's heirs.[169] The belief that alms-givers laid up treasures for themselves in Heaven,[170] and the prospect of receiving a reward, possibly multiplied many times over, for services to the Lord, made good sense if understood in the light of secular social bonds.[171]

In the *arengae* and *narrationes* Hell was described more in terms of what befell its inhabitants than as a collection of topographical features. (We shall see presently that malediction clauses tended to be more graphic.) Hell (*Gehenna*) was above all a place of torments[172] and punishments (*poenae*)[173] where the damned suffered.[174] In some documents the threat of *Gehenna* was expressly located in time after the Day of Judgement. For example, the *arenga* of a charter from St-Mont stated: 'All men should reflect on the fearsome Judgement Day and how they might then be safe without any fear of infernal *Gehenna*.'[175] The formula presupposed the faithful entering Hell from an unspecified but distinct other, or 'middle', place. This was at variance, however, with ideas expressed in a charter from Tulle of 1059 or 1060, in which Archambald III of Comborn, with his mother and brother, confirmed or gave lands to the monks for the soul

---

caelorum'); Mark 4: 26, 10: 23–5; Luke 9: 60, 13: 20; John 3: 5 (Vulgate: 'regnum Dei', the predominant term in the subsequent books of the NT).

[166] B. 17 (879 × 884).

[167] Ibid. 14 (1062 × 1072).

[168] SEL 9[7] (1010 × 1014), 29[27] (1014 × 1021); *Cartulaire de Sainte-Foi*, 3 (1101). See Mark 10: 17; Luke 10: 25, 18: 18; Eph. 5: 5; Heb. 9: 15.

[169] *Chartes, chroniques et mémoriaux pour servir à l'histoire de la Marche et du Limousin*, ed. A. Leroux and A. Bosvieux (Tulle, 1886), charte 7 (1027), 11 (1082).

[170] St-Seurin, 35 (11th C.); *Chartes, chroniques*, charte 7 (1027). See Matt. 13: 44, 19: 21; Mark 10: 21; Luke 18: 22 (these three last were important crusading passages).

[171] St-Seurin, 11 (later account of events 989 × 1010); Ste-C. 80 (1043). Cf. Matt. 19: 29 (another crusading passage: e.g. Fulcher of Chartres, *Historia Hierosolymitana* (1095–1127) (I. 6), ed. H. Hagenmeyer (Heidelberg, 1913), 163; S. 81 (1120)).

[172] St-M. 5 (c.1050 × 1060), 12 (1062 × 1085), 72 (2nd half 11th C.).

[173] Ibid. 64 (1068 × 1085).

[174] SEL 139[119] (1092).

[175] St-M. 67 (1062: the dating is the editor's; the charter falls in the periods 1060 × 1062 or 1085 × 1108). See also Mart. I. 36 (1027).

of his father Archambald II, who had recently been killed and was now being buried at the monastery. The gift was made 'so that the Good Lord might deign to snatch it [the soul] from the fire of *Gehenna*'.[176] A decade later Viscount Aymeric III of Rochechouart anxiously approached the monks of Uzerche to resolve a long-standing problem which was compromising his family's spiritual prospects. In 1019 his grandfather Aymeric I Ostafrancs (a cadet of the vicecomital line of Limoges) had granted the monastery half the church of Nieul. In due course Aymeric I's son Aymeric II had given the other half but then reclaimed the property. Later Aymeric III, anxious that his father had met an 'unfortunate death', became worried by the repossession, 'knowing that without doubt his [Aymeric II's] soul was being crucified in Hell [*in inferno*] because of what he had unjustly seized'.[177] In these instances laymen were clearly motivated by feelings that dead kinsmen were suffering literally as the donation was made, not simply that the threat of future agonies after Judgement hung over them.

It is therefore important to ascertain what laymen believed about the topography and time-scale of the afterlife. The most consistently graphic evidence in the charters is found in the clauses of malediction which appear in documents from throughout our period. Typically, dire spiritual harm was threatened to anyone who might dispute or violently undo the property transaction evidenced by the document. The practice of invoking spiritual penalties in this way began to be queried in the eleventh century, and gradually died out in the twelfth. Initially, however, the criticism was voiced mainly by zealous reformers such as Peter Damian and Stephen of Muret, who argued that pronouncements of malediction by monks (who lacked the apostolic authority of bishops) were tantamount to prejudicing God's infinite mercy.[178] Traditionalists continued to appreciate the value of maledictions and, as far as the often truncated documents can reveal, few Limousin or Gascon religious communities in the eleventh century did not make regular use of them.

It is conventional wisdom that malediction clauses were most common in areas where, and at times when, public authority was weak.[179]

---

[176] T. 399. The witness clause of this charter presents problems, probably because two transactions, the former taking place in 1014 × 1022, were conflated by the scribe. There is no reason, however, to question the authenticity of the *pro anima* clause. For a similar construction, *Chartes, chroniques*, charte 1 (954).

[177] Uz. 53–5.

[178] Little, 'Formules', 384–5; id., 'Morphologie', 55–6.

[179] A. de Boüard, *Manuel de diplomatique française et pontificale* (Paris, 1929–48), i. 280; Little, 'Formules', 385; id., 'Morphologie', 47.

Although taking a very broad overview this appears to be true, there are some instances of charters threatening both spiritual and secular sanctions.[180] The maledictions were, therefore, treated as more than a means of last resort to intimidate potential lay aggressors who resisted public authority. The curses complemented the positive spiritual ends which a charter expressed, and by their normal position at the foot of the document they balanced the solemnity of the preambles. Nor were they an idle threat. Some surviving charters end with the words 'Fiat' or 'Amen';[181] undoubtedly this was a feature of many more documents before they were abridged by cartulary copyists. The anathema clause thus often recorded a pronouncement made at the time of the transaction. It might be read out whenever the charter was used in subsequent disputes, and it presaged the actions, including the ceremony of malediction or the more formal *clamor*, which the religious community might feel itself forced to take against a dangerous aggressor. It was clearly in the community's interest, therefore, to impress upon its benefactors the thrust of the curses' meaning. Consequently it is legitimate to treat the maledictions as further evidence of the religious ideas to which laymen were exposed.

A few clauses developed themes which were common in preambles. It was wished of an aggressor, for example, that 'he may not become an heir in the Lord's kingdom'.[182] He might forfeit his share in the kingdom of Heaven, and never experience eternal life with the Lord.[183] The hopeful sentiments of the preamble might even be turned on their head, as in the formula 'may he have a kingdom with the Devil and his angels in Hell'.[184] The great majority of clauses, however, dealt directly with what would happen to the accursed (a further indication that they reflected attempts to mould popular ideas). Consequently the maledictions tended to be more forthcoming than the other elements of charters on the subject of the dark side of the afterlife. They did not dwell on any middle place between Heaven and Hell, for the threat of a period of punishment or purgation followed by the possibility of salvation had scant deterrent

---

[180] e.g. SEL 116[97] (1052 × 1060); Ste-C. 35 (1137).

[181] S. 25 (4th qr. 11th C.); St-M. 13: 'Respondete Amen.' (late 11th C.), 30 (1068), 43: 'Amen dicant cuncti qui hanc audierint legere.' (1079), 59 (1077 × 1085), 63 (c.1090), 73: 'Amen, amen, dicant omnes qui hoc legere audierint.' (c.1077 × 1081); *Cartulaire de Sainte-Foi*, 3: 'Amen. Amen. Amen. Fiat. Fiat. Fiat.' (1101).

[182] Sol. fo. 34ᵛ (11th C.).

[183] Mart. I. 32 (1029); St-M. 59 (later account of events 1077 × 1085); Sol. fo. 39ʳ⁻ᵛ (mid-11th C.).

[184] St-M. 19 (1085 × 1110).

value. But they provide important details about how laymen were invited to imagine Hell.

Hell was considered a place of unspeakable pain, where one resided with the Devil and his demons.[185] One word used of its agonies, *cruciatus*, was redolent of Christ's suffering on the Cross,[186] with the important difference that the pain lasted forever.[187] Understood literally, some charters reinforced the belief that Hell was located beneath the Earth, for the souls of the damned were 'submerged'.[188] Entry into Hell meant joining a small group of named sufferers whose individual fates carried a didactic meaning. The inhabitant of Hell *par excellence* was Judas. The reason for his presence among the damned might be considered so obvious that it required no explanation.[189] Often, however, he was described, on the basis of the Gospels, as the man who had 'betrayed' or 'sold' the Lord.[190] Alternatively, he was simply 'the traitor'.[191] Judas, of course, warranted his frequent mention in the charters because of his role in the story of Christ's Passion—Christ had said that it would have been better had he never been born—but references to lordship and betrayal were also clearly intended to impress the lay aristocracy by appealing to themes with which they were familiar from everyday life. A more subtle, but no less pertinent, message lay behind the inclusion in Hell of Sapphira and Ananias, for they had been killed by the Holy Spirit for trying to mislead St Peter about the disposal of their property in alms.[192] Dathan and Abiram's presence among the damned both helped to place Hell beneath men's feet and encouraged obedience. They had

---

[185] Ibid. 19 (1085 × 1110), 43 (1079), 64 (1068 × 1085); Sol. fo. 38[r-v] (c.1070 × 1090).

[186] Auch, 89 (c.1050 × 1068).

[187] St-M. 67 (1060 × 1062 or 1085 × 1108), 88 (c.1100); cf. Matt. 25: 41; Mark 9: 42-3.

[188] Sol. fos. 36[v]-37[r] (c.1060 × 1090), 38[r-v] (c.1070 × 1090); SEL 116[97] (1052 × 1060). For popular belief in a subterranean Hell, and some clerical resistance to the idea, see P. Dinzelbacher, *Vision und Visionsliteratur im Mittelalter* (Monographien zur Geschichte des Mittelalters, 23; Stuttgart, 1981), 91-6. See also A. J. Gurevich, *Medieval Popular Culture: Problems of Belief and Perception*, trans. J. M. Bak and P. A. Hollingsworth (Cambridge, 1988), 130-1.

[189] Sol. fo. 27[v] (1105); Auch, 25 (956); St-M. 34 (1068 × 1085), 62 (1074).

[190] Sol. fo. 26[r] (c.1087 × 1090); *Chartes, chroniques*, charte 1 (954); T. 46 (986); S. 7 (1120); St-M. 79 (c.1080 × 1085). Cf. Marculf, 'Formulae' (II. 3), 76. For the scriptural authority, Matt. 10: 4, 26: 25, 27: 3; Mark 3: 19; John 12: 4, 18: 2, 18: 5.

[191] Auch, 11 (c.1040); SEL 116[97] (1052 × 1060); St-M. 11 (1063), 30 (1068), 43 (1079).

[192] S. 36 (c.1100 × 1118); *GC* 2, Inst. cols. 183-4 (1085 or 1086); Auch, 41 (c.1034). See Acts 5: 1-11.

been absorbed by the earth for leading rebellion against Aaron and Moses (for whom read the divinely constituted authority of the church); poignantly for nobles and knights, Dathan's and Abiram's kinsfolk, households, and property had shared the same fate.[193] Caiaphas and Annas suffered for having been the high priests who persecuted Christ and the Apostles.[194] Korah (the less frequently cited companion of Dathan and Abiram), Pharaoh, Nero, Pilate, and Simon Magus were all occasionally mentioned in malediction clauses. The relevance of the last two in particular to the conduct of an eleventh-century lord is immediately apparent.[195]

The charters reveal a body of ideas which exploited vivid imagery, deep emotions such as fear, and analogies with mundane experience. All these elements point to the conclusion that the ideas were intended to be communicated to, and readily grasped by, laymen.[196] It has also been seen, however, that the meaning of charters could tend to obscurity with regard to the question of the precise fate of the soul immediately after death. It is therefore useful to turn for assistance to other sources.

Laymen's Religious Ideas: Some Narrative Evidence

Accounts of visions in narrative sources help to clarify the sometimes confusing picture of the afterlife conveyed by the charters. The Limousin Life of the Blessed Geoffrey of Le Chalard (d. 1125),[197] for example, is a particularly rich mine of information, for it contains descriptions of seven visions, of which five have a bearing upon the fate of the soul after death. As a source the Life is not without problems. Written at the very end of our period,[198] it was the product of the progressive milieu of reforming

---

[193] Auch, 25 (956), 41 (*c.*1034), 48 (early 11th C.); S. 7 (1120), 25 (4th qr. 11th C.); SEL 116[97] (1052 × 1060); St-M. 3 (early 12th C.), 73 (*c.*1077 × 1081). See Num. 16: 1–33, esp. 32–3.

[194] St-M. 13 (late 11th C.). See Luke 3: 2; John 18: 13, 18: 24; Acts 4: 5–21.

[195] e.g. T. 46 (986); Auch, 48 (early 11th C.); Little, 'Formules', 392. For a forceful denunciation of simony put into the mouth of Viscount Adhemar II of Limoges, see C. de Lasteyrie, *L'Abbaye de Saint-Martial de Limoges: étude historique, économique et archéologique, précédée de recherches nouvelles sur la vie du saint* (Paris, 1901), pièce just. no. 6, pp. 426–7 (1062).

[196] See J. Avril, 'Observance monastique et spiritualité dans les préambules des actes (Xᵉ–XIIIᵉ siècle)', *Revue d'histoire ecclésiastique*, 85 (1990), 5, 17–23, 28.

[197] 'Vita Beati Gaufredi', ed. A. Bosvieux, *Mémoires de la Société des Sciences Naturelles et Archéologiques de la Creuse*, 3 (1862), 75–119, with commentary, 120–60.

[198] J. Becquet, 'Les Chanoines réguliers du Chalard (Haute-Vienne)', *BSAHL* 98 (1971), 155 places the Life between 1135 and 1169. It is probable that the canons were preparing

canons regular and, in its surviving form at least, was intended for a clerical audience. In addition, the Life's anonymous author clearly had an axe to grind, for the visions served to establish the historical basis of the liturgical practices observed at Le Chalard. The two years after Geoffrey's death were a period of some uncertainty within the priory, possibly as a result of tensions between the evangelical, eremitical, and coenobitical strains in Geoffrey's own observance which had been unresolved while the charismatic founder of the community was still alive.[199] The Life's author welcomed the accession of Prior Gerard in 1127 as an affirmation of what he considered to be Geoffrey's true dispensation, a regimen which gave prominence to a lengthy psalmody, the Office of the Dead, and daily Masses for individual benefactors. The visions helped to illustrate the efficacy of this policy.[200]

The visions are of more general interest, however, because they both amplify ideas which are found in the charters and draw upon the same assumptions concerning the afterlife. This is well illustrated by one vision in which Peter Brunus, an inhabitant of Limoges known for his personal piety and support of Le Chalard, appeared to Geoffrey himself in a dream.[201] Peter had just died. When asked by Geoffrey what he expected would happen to him, Peter made two points: he had no need to fear, because he already knew that he would receive eternal life on the Day of Judgement (located clearly in time by the use of the future tense); and in the mean time (*interim*—another unambiguous expression of sequence) he would suffer unspeakable agonies (*poenae*). (This expressed an ancient theme of Christian thought, that the pains experienced before Judgement were far greater than anything possible on Earth.)[202] Peter's suffering was particularly related to a specific set of circumstances in this life. In recent years his son-in-law had tried to restrain his habit of annually giving in alms property worth 1,000*s*.[203] The Life's unfavourable description of the son-in-law as someone who gravely resented Peter's generosity places the vision in the same thought-world as numerous charters which record laymen trying to claw back some of an older kinsman's donations to a religious community. Also redolent of ideas revealed by the charters is

---

for Geoffrey's canonization before the community's first papal privilege, granted in 1150: ibid. 162.

[199] See ibid. 161–2.
[200] See 'Vita Beati Gaufredi' (ch. 9), 106.
[201] Ibid. (ch. 7), 101–2.
[202] J. Le Goff, *The Birth of Purgatory*, trans. A. Goldhammer (London, 1984), 84, 99.
[203] 'Vita Beati Gaufredi' (ch. 7), 102–3.

Peter's instruction to his son, communicated through Geoffrey, that he must grant his sister her share of the patrimony: only in this way would communal tenure, the root cause of potential ill-feeling within the family, be ended. The amount Peter is said to have given in alms is almost certainly exaggerated, but the precision with which it is recorded and the regularity with which it was dispensed are both significant. Peter told his son to make good the balance lost through the son-in-law's meanness, otherwise they—Peter and the son—were both threatened with 'what must be feared' (*metuenda*). The son dutifully did what was necessary and dispensed a satisfactory amount of alms.[204] Thus Peter believed that there was a direct connection between his alms and the succour he could expect after death (his fate at Judgement, remember, was not at issue). Unfortunately, however, the account of the vision does not state precisely how and where Peter would suffer, nor whether the *poenae* might cease altogether once the particular problem of the alms had been resolved.

These issues were addressed a little more directly in a second vision, another dream in which Geoffrey spoke to two dead sons of Viscount Adhemar III of Limoges named Guy and Helias.[205] Unlike the case of Peter Brunus, there is no suggestion that either brother had died in the immediate past, though they would not have been dead for more than about ten years.[206] Like Peter Brunus, however, the two men were easily recognized and visible in bodily form. Their appearance could leave no doubt that they had been suffering greatly; their limbs were charred 'as if roasted in a fierce blaze'.[207] (For the duration of the vision, however, they seem to have passed out of the place of pain.) Geoffrey observed that, from his outward form, Guy seemed to have been suffering less, and he expressed surprise since Guy was the elder brother—an interesting clue that sinfulness could be conceived as an aggregation of sins committed through time. Guy's explanation of his relative comfort throws light on an important element of the individual's fate after death: the proportionality between earthly misbehaviour and subsequent pain. Whereas Helias was experiencing agonies for having committed adultery, he had been more restrained during his life and less given over to acts of *libido* and other shameful conduct.[208] This statement should not necessarily

---

[204] Ibid. 103.
[205] Ibid. (ch. 9), 105–6.
[206] See Geoffrey of Vigeois, 'Chronica' (ch. 37), 298–9.
[207] 'Vita Beati Gaufredi' (ch. 9), 106.
[208] Ibid.

be treated as evidence of laymen's attitudes towards the sinfulness of sexual activity, since Guy's utterance truly reflected Geoffrey's own particular horror of physical defilement, a point made elsewhere in the Life.[209] (This was possibly why the author emphasized the vague term *libido*, which camouflaged any hierarchy of specific sexual or marital delicts which could be matched by a corresponding scale of punishments; the fact that Guy sinned 'less' simply resumed the idea of aggregated individual sins.) What is clear, however, is that, although the brothers had still to be judged, they shared Peter Brunus's conviction that they would enter Heaven: 'we shall rise amongst those who will be at God's right hand.'[210] This reveals that the dead had some sure means of learning their eternal fate, rather than that salvation was simply held out to them as a possibility. Even so, the purpose of the brothers' visit was to solicit prayers from Geoffrey because the suffrage of such a renowned holy man was particularly effective in making their *poenae* more tolerable. It was not only the sanctity of the intercessor, moreover, which the brothers considered important. They asked Geoffrey to pray and celebrate Mass for them daily. This story thus clearly reflects the belief revealed by the charters in the value of association with regular intercession as well as with holiness in itself.

Unfortunately this vision is scarcely more explicit than that featuring Peter Brunus about the setting of the sufferings of the dead, though both stories reveal that some form of middle place was involved. Moreover, the stories only partly explain the exact purpose of the agonies. They could not have been probative, for otherwise the dead men would not have known that they would enter Heaven. Were they, then, purgative or punitive? Was this a distinction which would have meant anything to laymen? It would appear that the dead men of the visions were not in Purgatory, if by this is understood the doctrinally distinct place of spiritual cleansing which came to be defined by churchmen from the twelfth century onwards.[211] Nevertheless, the two dreams point to a belief in an intermediate region of the next world which closely resembled Hell but through which passed men and women who would ultimately be saved.

This is made clear by another of Geoffrey's visions which contains a traditional picture of the afterlife. In a dream Geoffrey found himself on a

---

[209] See ibid. (ch. 11), 112.

[210] Ibid. (ch. 9), 106.

[211] Le Goff, *Birth of Purgatory*, is the best, if controversial, modern study. See esp. 2–7, 11–13.

high mountain peak, from where he was able to glance up and down, and where men in shining white vestments acted as his guides.[212] Looking up, Geoffrey could see into Paradise, and as he surveyed the scene he was approached by heavenly beings, a column of angels, a choir of virgins, and Our Lady (the patron saint of Le Chalard) in majesty. Below lay a forbidding valley, in the middle of which flared an unspeakably fierce fire. There demons inflicted a number of torments on the pitiful souls (*miserandae animae*) of men described as *miseri*. Evidently the souls were not those of the damned, otherwise the gerundive would have been ill-chosen. Such imagery seems somewhat further removed from the ideas of laymen than the first two visions. The suffering was spatially located, but souls, not bodies, were being tortured, and the agonies would seem to have been understood figuratively for spiritual pains. Furthermore, the vision does not necessarily represent one specific, commonly accepted mental picture of the afterlife, for it was open to visionaries, within the usual conventions, to introduce their own particular details touching on the topography and chronology of life after death.[213] Nevertheless, in broad terms Geoffrey's experience of the valley serves to illustrate further a belief in a middle place where sinners experienced great pain.

Other visions contained in the Life are of interest. One of Geoffrey's loyal disciples had two dreams which he asked his master to interpret.[214] Geoffrey confirmed his follower's worst fears that the appearance of a man in a white habit bearing a pyx, and a vision of a large crowd surrounding Le Chalard, presaged the prior's own death. The crowd, Geoffrey explained, were the saints to whom he had prayed for help when he came to die. Saints' suffrage, therefore, might operate before Judgement. In this context, Geoffrey added that the *labores* he had endured would count in his favour after death. The labours are not clearly defined, but from the Life's preceding description of Geoffrey's final years as a semi-recluse, it appears that they comprised both involuntary afflictions and a rigorous regime of mortification and prayer. Through these Geoffrey hoped that, after his death, God would lift him up from the control of demons,[215] a belief which supports the idea that the demons and torments in the valley of the third vision discussed above were not in Hell but in another place from which release was possible.

Geoffrey's interpretation of the disciple's visions added that he had

[212] 'Vita Beati Gaufredi' (ch. 10), 109–110.
[213] Cf. Gurevich, *Medieval Popular Culture*, 123–4.
[214] 'Vita Beati Gaufredi' (ch. 11), 111–12.
[215] Ibid. 112.

recently begun to recite the Psalter thrice daily in order to achieve three ends, which were defined as the Lord's protection from three fires.[216] These were: the physical fire which might destroy church buildings; the metaphorical fire of sexual desire; and the fire of Hell (*incendium gehennale*) where the souls of the damned (*damnati* as opposed to *miseri/miserandi*) were tortured forever. It is noteworthy that the use of the language of fire did not prompt the author to return to the valley of the third vision, which suggests that his ideas concerning the middle place were somewhat vague. The author's vocabulary, too, reveals the same scope for conflation which is present in the charters: *Gehenna* had a different quality from the valley in terms of its duration and the adjectives used of its inhabitants, but both places were the scenes of sufferings/punishments (*poenae*) inflicted by demons, mainly by means of extreme heat. A layman instructed along the lines of these visions would probably have found it difficult to appreciate the distinction between Hell and the middle place, if not in theory, then at least in terms of what he would have to endure for the foreseeable future after death.

The picture gained from these visions can be complemented by an interesting passage concerning the next world in Orderic Vitalis's *Ecclesiastical History*, which, though written in distant northern France, was nearly contemporaneous (*c.*1133) with Geoffrey's Life. Before turning to this passage, however, some explanation is necessary of the medieval understanding of the origins and significance of visions. Partly but not solely because they usually took place in dreams, visions such as those in Geoffrey's Life would have been understood by an educated, clerical reader in the light of a long-established interpretative scheme first clarified by St Augustine. In his *De cura gerenda pro mortuis* (421 × 423), Augustine had attempted to square Christian dogma with the popular belief of his day that the dead were able to return to this world with their physical substance intact in order to confront the living. The dead who appeared to the living, he argued, had no material existence; as manifestations of God's omnipotence they were no more real than the likeness and speech of a living person which anyone might conjure in his imagination without that person's knowledge.[217] There was more to this argument than an educated ecclesiastic trying to brush aside simple folk-superstition. Augustine was propounding a conception of the interaction

---

[216] Ibid.
[217] Le Goff, *Birth of Purgatory*, 79–81; C. Lecouteux, *Geschichte der Gespenster und Wiedergänger im Mittelalter* (Cologne, 1987), 52–5.

between this world and the afterlife which was to remain very influential. A passage in Robert the Monk's *Historia Iherosolimitana*, for example, demonstrates how Augustinian ideas lay at the heart of the manner in which a vision of the next world would be interpreted.

Robert differed from his fellow historians of the First Crusade in his account of how in 1098 Bohemond of Taranto schemed with a traitor within Antioch—in this version a Turkish Muslim named Pirrus—to gain entry into the city.[218] According to Robert, Pirrus had been impressed by the members of a white host fighting in the Christian army, and during his negotiations with Bohemond he asked who they were. Bohemond was able to explain that what he had seen was the column of martyred crusaders, led by martial saints, which God sent down from Heaven to help the Christians in battle. Pirrus was surprised that the column was celestial in origin, for it had given every impression of being material by inflicting wounds, riding down enemy troops, and killing.[219] Robert did not trust a layman to explain this apparent paradox. In a significant phrase Bohemond conceded that such matters were 'beyond my understanding', and he referred the problem to his chaplain, whose speech became Robert's vehicle for a concise explanation of the meaning of visions. Whenever God wished to send angels or 'just spirits' (that is, the souls of the blessed) to Earth, the chaplain revealed, they assumed a physical appearance in order to be recognizable to man. Normally their 'spiritual essence' made them invisible to mortals. The host appeared armed and mounted, not because there existed in Heaven a ready supply of weapons and horses, but because this was appropriate to the particular earthly conditions in which divine help was provided. In other words, the outward form was a device for the benefit of men's limited understanding; if the divine message should differ, the illusion would vary accordingly.[220] Robert could not rid himself completely of the belief that the host's intervention was physically effective, for he envisaged it coming down from Heaven, finishing its task or business (*negotium*), and returning skywards.[221] But the basic explanation of what Pirrus had seen lay in God's infinite power to mould the spiritual and physical creation (*materia*) in whatever way he chose. The host was thus a sign of God's support for the crusaders.[222]

[218] 'Historia Iherosolimitana' (V. 8–10), 796–8.

[219] Ibid. (V. 8), 796–7.

[220] Ibid. (V. 9), 797.

[221] Cf. ibid. (VII. 10, 13, 18) 830, 832, 836.

[222] Cf. Riley-Smith, *First Crusade*, 99–107, 116–18.

Robert's account establishes that, given God's control over the outward
form of visions, there could be no guarantee that a visionary's picture of
the afterlife corresponded to what actually happened to the dead.[223] In
some respects Orderic Vitalis's account of Herlequin's Hunt is subject to
this caveat. Orderic was a sophisticated Benedictine, raised in the cloister
from childhood. His first instinct, therefore, was to approach the
meaning of visions through Christian doctrine. Orderic was no claustral
introvert, however, and he displayed a particularly keen empathy with
the minds of his lay contemporaries. His account of the Hunt stands at
the blurred border between dogma and popular belief, and was less
subject to the demands of high theology than, for example, Robert the
Monk's efforts to place the First Crusade within the historical sweep of
divine revelation.[224] Orderic's description of the Hunt, moreover, was
not intended as a diversion from the historical narrative of his *Ecclesiastical
History*, and it appears certain that he believed in the literal truth of the
story. The vision is located very precisely in time (1 January 1091) in its
chronologically correct place within a description of the disorder then
prevailing in Normandy. Furthermore, Orderic's informant was the
visionary himself, whose name and place of origin were supplied.[225] The
story of the Hunt, therefore, bears comparison with the same author's
insertion of a Life of St William of Gellone into his account of the
preaching of Gerold of Avranches to the household knights of Earl Hugh
of Chester: a digression which complemented the thrust of the
surrounding narrative.[226] The Hunt story's interest lies in its references to
the theology of penance, intercession, and the afterlife within a semi-
folkloric setting. The account is also particularly important in the light of
Orderic's unusually clear understanding of the First Crusade 'indulgence'
in the context of contemporary penitential discipline.[227]

Orderic's story recounts the appearance to a priest, Walchelin of
Bonneval, of a ghostly procession, the *familia Herlechini*.[228] The vision
took place at night, but Walchelin was awake, and Orderic took care to
stress the priest's youth and mental alertness: there was to be no doubt
in the reader's mind about the informant's trustworthiness.[229] The

---

[223] For a discussion of the allegorical treatment of visions, Dinzelbacher, *Vision*, 171 ff.

[224] For popular legends of ghostly hunts, see Lecouteux, *Geschichte der Gespenster*, 131–2.

[225] OV iv. 236, 248. For the value of eye-witness testimony, see Gurevich, *Medieval
Popular Culture*, 124–6.

[226] OV iii. 216–26.

[227] Ibid. v. 16–18, 26; Riley-Smith, *First Crusade*, 28.

[228] OV iv. 242.

[229] Ibid. 236–8, 242.

procession was made up of distinct groups of sinners which mirrored social divisions in this world. Each group had been prone to particular types of sin, and the torments they now suffered varied accordingly. Plunderers—men on foot, therefore below the status of knight—were burdened by examples of the mundane objects they had habitually stolen.[230] Women, including noblewomen, suffered pains appropriate to their sexual misdoings.[231] Arms-bearers, mounted on black but fiery horses and subjected to extremes of temperature, bore red-hot arms and heavy spurs.[232] One, Landric of Orbec, was treated by his companions to the same contempt which he had himself shown as a judge to poor litigants.[233] As for the sufferings peculiar to the clergy, Orderic chose, unsurprisingly, to be more discreet.[234] In addition to agonies relating to generic sins, pain was inflicted for specific misdeeds. In a passage redolent of Peter Brunus's instructions to his son, a *miles* named William of Glos asked Walchelin to urge his family to restore pledged property to the heir of a poor debtor whom he had mistreated.[235] William had committed many sins typical of his social position—false judgements and rapine, for example—but this one instance of usury, compounded by dynastic greed, had had particularly dire consequences.

In a brief excursus from his description of the Hunt's appearance, Orderic explained that the pains of the dead were purgative (although he also hinted at a probative element). Because Heaven was perfect, he argued, it could not receive any trace of iniquity, disorder, or evil. Souls were consequently subjected to 'purgative fire', from which they emerged fully cleansed of all the stains of sin.[236] The Hunt was not Hell, therefore, nor strictly a prefiguration of eternal damnation, though Orderic did not commit himself to the statement that all the members of the Hunt would eventually be saved.[237] At times, however, Orderic's vocabulary reveals the same scope for the conflation of Hell and the place of purgative/probative pain which is evident in the charters. The sufferings of the dead were described as *poenae* and *supplicia*, and vivid verb-constructions increased the horrific effect (for example, 'intolerabiliter

---

[230] Ibid. 238.
[231] Ibid. 238–40.
[232] Ibid. 240–2, 244, 246–8.
[233] Ibid. 242.
[234] Ibid. 240.
[235] Ibid. 244–6.
[236] Ibid. 240.
[237] Cf. E. Mégier, 'Deux exemples de "prépurgatoire" chez les historiens: à propos de *La Naissance du Purgatoire* de Jacques Le Goff', *CCM* 28 (1985), 56–7.

cruciari' and 'inerrabiliter cruciatus sum').[238] Although the agonies comprised assaults on all the senses,[239] the predominant source of pain was extreme heat.[240] Souls—the 'shades of the dead'[241]—were being tortured, but in bodily form, to the extent that Walchelin could badly burn his hand when trying to mount a ghostly horse, and be knocked over and scarred for life by an aggressive knight.[242]

In a key passage Orderic attempted to explain the vision in terms of the need for penances. Walchelin recognized one of the sufferers as the notorious murderer of a priest, who was now experiencing intolerable torments because he had died within the previous two years having failed to complete the severe penance necessary.[243] The pain, it was stated, represented the balance of penances left unperformed at death. (If the murderer had died unrepentant he would have been damned.) Intercession had an important impact upon the world of the Hunt. The burning clergy, for example, asked Walchelin to pray for them.[244] When Walchelin had celebrated his first Mass, he was told, his own father had escaped further tortures and his brother's burden had lessened.[245] The brother, indeed, anticipated his own complete release in a little more than a year; even so, he asked Walchelin to continue to help him through prayers. The middle place, it follows, obeyed the same chronological rules as this life. Accordingly, precise time-limits could be placed on sufferings. Whereas William of Glos had died 'long ago' (although his wife and son were still alive), Walchelin's brother had fewer than five hundred days more to endure in the Hunt.[246] Among the clerics named in the vision, Bishop Hugh of Lisieux had been dead thirteen years in January 1091, Abbots Mainier of St-Évroul and Gerbert of St-Wandrille less than two years.[247] The inhabitants of the middle place were, then, not without hope. When the pains ended, at least some of them would be 'saved' (*saluari*). Salvation, moreover, might conceivably be attained within the lifetime of a sibling or child.

---

[238] OV iv. 238, 240, 242, 246, 248.

[239] Ibid. 240, 242, 244, 246–8.

[240] Ibid. 238, 238–40, 244 (heat and cold), 246, 246–8. Cf. Orderic's use of 'decoquitur': ibid. 240.

[241] Ibid. 242.

[242] Ibid. 244, 246, 248. Dinzelbacher, *Vision*, 103–4 discusses the broad similarities in medieval visions between Purgatory and its antecedents and Hell.

[243] OV iv. 238.

[244] Ibid. 240.

[245] Ibid. 248.

[246] Ibid. 244, 248.

[247] Ibid. 240.

Unfortunately Orderic did not go on to explain what happened to the soul when it was saved beyond implying that it was released from participation in the Hunt. He came close to saying that each soul entered Heaven as soon as its individual term of purgation was completed, but he avoided trying to reconcile this with belief in the Last Judgement. It is possible that he envisaged some form of restful antechamber to Heaven, a place of *refrigerium*,[248] but this is not made clear. For our present purposes, however, what is significant is Orderic's firm belief in a middle place—of which the Hunt was one possible manifestation—similar in many respects to Hell.

Orderic's account is thus not entirely free from inconsistencies and some muddled thinking. Yet it complements on many points the visions in the Life of Geoffrey of Le Chalard and the language of the charters. Efforts to reconstruct the exact topography of eleventh- and early twelfth-century ideas of the afterlife are bound to be frustrated, for ideas created over centuries by the interaction of church teaching and popular beliefs were inevitably blurred at the edges. On some points, moreover, the uncertainty may not have been entirely accidental. How, for example, could a layman be expected to worry about his dead kinsman's soul if, like Walchelin's brother, it had a finite number of days to suffer, or, like Peter Brunus, was assured of salvation, unless the suffering to which it was being subjected was so appalling that it was virtually indistinguishable from the pains of Hell?

On the other hand, it is hard to conceive how decade after decade nobles and knights could have given valuable property to religious communities and entrusted their own and their kin's souls to the care of monks and canons, unless they had had some clear ideas about death and the afterlife.[249] It is important not to underestimate the capacity of arms-bearers to understand the next world on a level close to that of religious, nor to distinguish rigidly between popular and clerical belief-systems. It is not necessary to posit scholar-knights steeped, as were monks, in the language, metaphors, and imagery of Scripture, especially the Psalms and the Gospels. Laymen's credence in physical rather than spiritual pain and in earthly chronology in the next world points to a tendency to simplify difficult concepts by means of literalization. But, equally, the fact that

---

[248] See Angenendt, 'Theologie und Liturgie', 81–6.

[249] See, *contra*, White, *Custom*, 153: 'there was probably no consensus in this period [1050–1150] about the nature and purpose of gifts to saints.'

efforts were made to simplify in this way points to the regular and intimate interchange of ideas between the laity and the church. Arms-bearers in south-western France (and most probably elsewhere) were not ignorant ruffians wrapped up in some barely Christianized Germanic or Gallo-Roman *Urglaube*, for whom support of the church was only a matter of outward form and the doctrinal content of charters so much incomprehensible verbiage.[250]

It has recently been argued that Purgatory and its antecedent forms were never a central element in medieval popular faith, a simpler belief in ghosts being predominant.[251] While it is incontestable that credence in ghosts was an important part of medieval notions of the afterlife—some of the best ghost stories were written by educated clerics such as Peter the Venerable in the mid-twelfth century and the Cistercian Caesarius of Heisterbach in the early thirteenth[252]—such a belief was, as far as arms-bearers in regular contact with religious communities were concerned, the function of, not the alternative to, conviction in the existence of a middle place. Be it purgatorial, probative, or punitive—from a layman's perspective the distinction would not have mattered greatly—the middle place and the fear it induced dominated all grants to religious com-munities, for the constant intercession of the saints and holy men were held to make good the imbalance of expiation due for sins. When laymen endowed monasteries they were not expressing disagreement with the theory of satisfactory penance. On the contrary, they were affirming a belief in penitential equivalence. It was simply the case that laymen, steeped in a sense of sinfulness which was constantly renewed by seeing the example of religious, and daunted by the number and severity of the penances which they could realistically undertake, needed assistance. Such a desire to make up the difference, fuelled by ideas of sin, pain, and the middle place, links the thought-world of the charters directly to the appeal of the First Crusade.

---

[250] For a recent view that the medieval church was chronically on the defensive against powerful folkloric beliefs deeply rooted in the Germanic or even Indo-European past, see Lecouteux, *Geschichte der Gespenster*, *passim*. It should be stressed that this work draws principally on vernacular literature from the Germanic north, an area converted far later than southern France. See also A. J. Gurevich, 'Popular and Scholarly Medieval Cultural Traditions: Notes in the Margin of Jacques Le Goff's Book', *Journal of Medieval History*, 9 (1983), 78–86.

[251] B. P. McGuire, 'Purgatory, the Communion of Saints, and Medieval Change', *Viator*, 20 (1989), 83.

[252] See J.-C. Schmitt, 'Les Revenants dans la société féodale', *Temps de la réflexion*, 3 (1982), 285–306; Lecouteux, *Geschichte der Gespenster*, 35–6, 57, 84–5, 162, 163–6.

The charters and other types of evidence demonstrate religious ideas—particularly associated with entry into professed religion and benefaction of churches—which throw light on the roots of the ideological response of arms-bearers to the crusade message. There was another form of pious expression, pilgrimage, which had close associations with crusading. It is the subject of the following chapter.

5

Pilgrimage and the Cult of Saints

THE First Crusade was preached and conducted as a species of pilgrimage. The Latin jargon used of the crusade, and the liturgical and devotional practices of the crusaders during the expedition, demonstrate clearly that pilgrimage was not regarded as simply a simile by which to make sense of the campaign's novelty, nor as camouflage for expansionist aggression, but as the core of the crusade's purpose, form, and rituals.[1] It follows that, in order to understand the pious motivations of the first crusaders, it is necessary to examine the appeal and popularity of pilgrimage among arms-bearers in the years before 1095. Pilgrimage has been a popular area of study among scholars in recent times,[2] and as a subject it does not lend itself fully to a narrowly regional focus.[3] Consequently, there is no need to attempt an exhaustive survey of the growth of pilgrimages in the eleventh century and the various political, religious, and economic conditions which stimulated it. It is, rather, useful to concentrate on establishing two fundamental points: that one reason why pilgrimage was very popular in the eleventh century was the variety of experiences it

---

[1] J. S. C. Riley-Smith, *The First Crusade and the Idea of Crusading* (London, 1986), 22–4, 35, 84–5, 108, 113, 126–9; H. E. Mayer, *The Crusades*, trans. J. Gillingham, 2nd edn. (Oxford, 1988), 14, 26–30. For a different view see A. Becker, *Papst Urban II. (1088–1099)* (MGH Schriften, 19; Stuttgart, 1964–88), ii. 396–8. See the well-known passage in *Gesta Francorum et aliorum Hierosolimitanorum*, ed. and trans. R. Hill (London, 1962), 8, in which the author states that natives of the Balkans could not accept the crusaders as pilgrims.

[2] P.-A. Sigal, *Les Marcheurs de Dieu: pèlerinages et pèlerins au Moyen Âge* (Paris, 1974) and J. Sumption, *Pilgrimage: An Image of Mediaeval Religion* (London, 1975) are good introductions. See also B. Töpfer, 'Reliquienkult und Pilgerbewegung zur Zeit der Klosterreform im burgundisch-aquitanischen Gebiet', in H. Kretzschmar (ed.), *Vom Mittelalter zur Neuzeit: Zum 65. Geburtstag von Heinrich Sproemberg* (Forschungen zur Mittelalterlichen Geschichte, 1; Berlin, 1956), 420–39; E.-R. Labande, 'Recherches sur les pèlerins dans l'Europe des XIᵉ et XIIᵉ siècles', *CCM* 1 (1958), 159–69, 339–47; R. A. Fletcher, *Saint James's Catapult: The Life and Times of Diego Gelmírez of Santiago de Compostela* (Oxford, 1984), 83–101.

[3] But see L. Musset, 'Recherches sur les pèlerins et les pèlerinages en Normandie jusqu'à la Première Croisade', *Annales de Normandie*, 12 (1962), 127–50. See also P. Bonnassie, *La Catalogne du milieu du Xᵉ à la fin du XIᵉ siècle: croissance et mutations d'une société* (Publications de l'Université de Toulouse-Le Mirailᴬ, 23 and 29; Toulouse, 1975–6), ii. 938–42.

embraced; and that pilgrimage was not an exotic exercise distinct from the faithful's more mundane devotional routines, but rather one expression of those same religious beliefs which are also demonstrated by laymen's relations with local ecclesiastical institutions.

The Methods and Aims of Pilgrims

Two basic points need to be made immediately. First, the mounted classes, although pilgrimage did not appeal exclusively to them, enjoyed important advantages in pilgrimaging. Writing of the 'innumerable multitude from the entire world' who travelled to Jerusalem in the years around 1030, Ralph Glaber stressed the social eclecticism of the pilgrims. In fact he believed that there had been a process of social progression, with the 'lesser, humble folk' first exhibiting enthusiasm for pilgrimage, followed by the 'middling sort', the high nobility, and finally noblewomen and, out of sequence, the very poor.[4] Glaber's scheme is rather too neat to be trusted fully, but his general emphasis upon pilgrimage's broad social appeal is confirmed by similar observations made by Adhemar of Chabannes.[5] It is important to note also that the sources are much more likely to record socially important pilgrims, and that nobles and knights did not have a monopoly of enthusiasm for cult centres.[6] Nevertheless, it is reasonable to assume that arms-bearers enjoyed certain advantages over poorer laymen by reason of their greater wealth and freedom from the constraints of the agricultural seasons. In particular, the ability to use horses and pack-animals enhanced their mobility. In *c.*1140 the compiler of the *Pilgrim Guide* assumed that those who could so afford rode to Compostela or at least walked with animals taking the weight of their baggage.[7] A miracle story from the same period suggested that a knight who let a destitute fellow pilgrim ride on his mule, and carried a poor woman's sack, exhibited particular virtue.[8] Other miracle stories worked from the presumption that mounted pilgrims were an unremark-

---

[4] Glaber (IV. 6), 198–200.
[5] Lair, 244 [C]; cf. Adhemar, *Chronique* (ch. 68), 194.
[6] For peasants as adherents of cults, see P. R. Morison, 'The Miraculous and French Society, circa 950–1100', D. Phil. thesis (Oxford, 1984), 110–15.
[7] *Guide* (chs. 2, 6, 7), 4, 12–14, 16, 20.
[8] *Memorials of St. Anselm*, ed. R. W. Southern and F. S. Schmitt (Auctores Britannici Medii Aevi, 1; London, 1969), 196–200; *Liber Sancti Jacobi: Codex Calixtinus*, i. *Texto* (II. 16), ed. W. M. Whitehill (Santiago de Compostela, 1944), 276–8.

able sight.[9] The use of horses and pack-animals (and litters, when necessary) had the important practical result of increasing better-off pilgrims' opportunities to travel in groups.[10] We must therefore rid ourselves of any mental picture of lonely and simply clad men (and women) trudging long distances. The fact that the idea of pilgrimage was not invariably, or even usually, associated in people's minds with a solitary and vulnerable ordeal must have been one of the practical reasons why arms-bearers were able to be sympathetic to the novel suggestion of the armed pilgrimage in 1095–6 (although, of course, the First Crusade 'indulgence's' attraction depended on the belief that the *iter* would be physically and emotionally strenuous).

The second basic point is that religious impulses, although they alone cannot explain all pilgrims' motives, were the most important element of pilgrimage's appeal. It would be unwise, of course, to underestimate the significance of mundane factors such as some form of proto-touristic curiosity and the desire for a change from routine. Innumerable miracle stories from this period involving cures and remedies demonstrate the importance, too, of illness, injury, or domestic misfortune as incentives to go on pilgrimage.[11] In one remarkable case, for example, a *miles* called William Gerald of Arsac gave a vineyard to St-Seurin, Bordeaux before setting out for the Saragossa crusade of 1118 in the hope that the expedition (*iter*) would improve his very poor health.[12] Nevertheless, there is a danger of stripping away pilgrimage's religious character too far to leave only unsatisfactorily vague socio-economic and socio-medical generalizations as explanations.[13] Pilgrimage makes sense when it is treated first and foremost as an expression of religious enthusiasm.

In strict theory pilgrimages were pious exercises in two general ways. They might be voluntary expressions of devotion towards a particular

---

[9] e.g. Hugh of Fleury, 'Vita S. Sacerdotis Episcopi Lemovicensis' (ch. 4), *AASS* (May) 2. 19.

[10] e.g. the sworn association of 30 Lorrainers believed to have travelled to Compostela in 1080, and the journey there of Count Pons of Toulouse (1037–61) and nearly 200 pilgrims 'sue societatis': *Liber Sancti Jacobi* (II. 4, 18), 265–6, 282–3. See P.-A. Sigal, *L'Homme et le miracle dans la France médiévale (XIᵉ–XIIᵉ siècle)* (Paris, 1985), 118–21; Sumption, *Pilgrimage*, 123–5, 127–8.

[11] See Sigal, *Homme*, 227–64.

[12] St-Seurin, 63. The editor dates the charter to c.1110 × 1143, but the Saragossa campaign must be what is referred to. The charter's vocabulary makes it highly unlikely that William Gerald's 'illness' was simply a metaphor for his spiritual state.

[13] See the judicious remarks of Fletcher, *Saint James's Catapult*, 83.

saint or shrine; or they might be enjoined as penances.[14] Pope Urban II implicitly recognized this distinction in 1089, for example, when he declared that support of the frontier church of Tarragona in Spain could be reckoned the equivalent of a pilgrimage to Jerusalem.[15] Writing in or before *c.*1124, Orderic Vitalis described Duke Robert I of Normandy's pilgrimage to Jerusalem in 1035 as a *spontanea peregrinatio* inspired by fear of God, meaning that Robert's decision to leave for the East was his own rather than his confessor's.[16] It has been seen, however, that charters often blurred the distinction between spontaneous and penitential acts of benefaction towards churches, in ways that suggest that it was considered more appropriate to record a generally reverential and contrite state of mind on the donor's part than the technical details of a particular penance. The same blurring process operated with regard to the motivations behind pilgrimages. For example, Ralph Glaber records that Count Fulk Nerra of Anjou went to Jerusalem in 1003 because he was troubled by responsibility for the deaths of his enemies—Glaber does not say so expressly, but he implies that the bloody Conquereuil campaign of 992 against the Bretons was foremost in Fulk's mind. It is quite likely that Fulk's pilgrimage was a formal penance; the point is that Glaber did not feel it necessary to say so, but rather concentrated on Fulk's personal contrition and religious fears.[17] Similarly, some charters recording pilgrimages tended to be vaguely expressed. When Viscount Adhemar II of Limoges went to Compostela in 1068 × 1090, he gave allodial property to Uzerche 'admonished by Abbot Gerald'; we cannot tell whether the pilgrimage or the donation were formal penitential acts required by Gerald in his capacity as Adhemar's confessor, or whether Gerald was simply acting more informally as a spiritual adviser.[18] Similarly, Viscount Bernard of Comborn went to Rome in 1097 × 1108 'admonished by Abbot Gauzbert [of Uzerche]'; it is possible, but there is no direct proof, that Bernard's journey was linked to a local *cause célèbre*

---

[14] C. Vogel, 'Le Pèlerinage pénitentiel', in *Pellegrinaggi e culto dei Santi in Europa fino alla IᴬCrociata* (Convegni del Centro di Studi sulla Spiritualità Medievale, 4; Todi, 1963), 37–94, esp. 39–40, 52–68, 90.

[15] *La documentación pontificia hasta Inocencio III (965–1216)*, ed. D. Mansilla (Monumenta Hispaniae Vaticana, Sección Registros, 1; Rome, 1955), no. 29, pp. 46–7.

[16] OV ii. 10; see also William of Jumièges, *Gesta Normannorum Ducum* (VI. 11[12]), ed. J. Marx (Rouen, 1914), 111.

[17] Glaber (II. 3–4), 58–60. The evidence does not warrant the assertion by Vogel, 'Pèlerinage pénitentiel', 60 that Fulk's pilgrimage was definitely penitential.

[18] Uz. 543.

from about that time, his vicious mutilation and murder of his nephew Ebles.[19] Given the overlapping between penitential and devotional acts, it follows that when Urban II preached the crusade pilgrimage as a satisfactory penance[20] he was not proposing some unattractive chore alien to laymen's own pious instincts and practices.

Pilgrimage in Narrative Sources

The history of pilgrimage in south-western France in the eleventh century is the story of two types of sources, the narrative and the documentary. In his chronicle Adhemar of Chabannes noted a number of pilgrimages to Jerusalem and Rome which had been undertaken during his lifetime by nobles from the Limousin and adjoining regions. Around the time that Adalbald succeeded Geoffrey as abbot of St-Martial (998), for example, Viscount Guy I of Limoges and his brother Bishop Hilduin returned from a pilgrimage to Jerusalem; Guy went to Rome a little later.[21] Count Boso II of La Marche travelled to Rome at about the same time.[22] Adhemar of Chabannes's maternal uncle Ainard, who was Abbot Peter of Le Dorat's provost when the latter acted as regent for Count Bernard of La Marche, died in Rome; another uncle died in Jerusalem; and when Peter of Le Dorat's power in La Marche was broken by William V of Aquitaine, he went on what may have been a penitential pilgrimage to Jerusalem.[23] Bishops Hilduin and Gerald of Limoges journeyed to Rome in the 1010s.[24] In c.1010 Bishop Ralph of Périgueux expired after returning from Jerusalem with the news of the destruction of the Holy Sepulchre by the caliph Al-Hakim.[25] It is clear that Adhemar kept himself in touch with developments effecting pilgrimage to Jerusalem. He reported, for example, that the 'via Hierosolimae' in southern Italy was disrupted for three years from c.1016 because of warfare between the Byzantines and Normans in Apulia.[26] Adhemar's intense interest in pilgrimage to

---

[19] Uz. 463; Geoffrey of Vigeois, 'Chronica' (ch. 25), ed. P. Labbe, *Novae Bibliothecae Manuscriptorum Librorum* (Paris, 1657), ii. 291. See also Uz. 478 (1097 × 1108).

[20] Riley-Smith, *First Crusade*, 27–9.

[21] Adhemar, *Chronique* (ch. 40), 162; *Les Miracles de saint Benoît* (III. 5), ed. E. de Certain (Paris, 1858), 141–2.

[22] Adhemar, *Chronique* (ch. 35), 159.

[23] Ibid. (ch. 45), 168.

[24] Ibid. (ch. 49), 171, 173.

[25] Ibid. (ch. 48), 171; Lair, 194 [C].

[26] Adhemar, *Chronique* (ch. 55), 178.

Jerusalem is also clearly revealed in his detailed description of a large overland pilgrimage between October 1026 and June 1027 which was led by Count William IV of Angoulême and Abbot Richard of St-Cybard (who died *en route*). On their return journey the pilgrims were solemnly received outside Limoges by the monks of St-Martial, and as they approached Angoulême they were escorted by a procession of monks and clergy.[27] Adhemar believed that Count William's example directly inspired many others to go to the Holy Land, including Bishops Isembert of Poitiers and Jordan of Limoges (whom we know left for the East in or shortly after 1028)[28] and Count Fulk Nerra of Anjou.[29] In this instance Adhemar undoubtedly presents us with an over-neat picture of the connections between different pilgrimages to Jerusalem, but he still provides solid evidence for a fashion for long-distance travel, stimulated by the example of peers, among ecclesiastical and secular princes in the first third of the eleventh century.[30]

The chronicles of Adhemar and Ralph Glaber might seem to give the impression that the period *c.*1000–*c.*1035 was a unique, heroic age of pilgrimage, especially to Jerusalem. Certain statements by Glaber have been used to advance the view that there was a peak of interest in Jerusalem and other pilgrimage centres in these years, coinciding with outbursts of religious fervour associated with the millenia of the Incarnation and Resurrection.[31] It is worth noting, however, that Glaber in fact displayed some reserve about the belief that the pilgrimages of *c.*1033 heralded the imminent advent of Antichrist; and in another part of his

---

[27] Ibid. (ch. 65), 189–90.

[28] Geoffrey of Vigeois, 'Chronica' (ch. 9), 283.

[29] Adhemar, *Chronique* (ch. 68), 194. Adhemar seems to have exaggerated the links between the various pilgrimages. Count Fulk already had a personal tradition of pilgrimage to Jerusalem dating back to 1002–3. If Adhemar was correct, Fulk's pilgrimage must be dated to 1027 X 1029, but there is no other evidence for it: L. Halphen, *Le Comté d'Anjou au XIe siècle* (Paris, 1906), 213–18. It is possible, however, that Adhemar's statement may be corroborated by Glaber (IV. 9), 212–14, who reports that Fulk had already been to Jerusalem thrice when he went there a final time in 1039–40; only two earlier journeys are documented. See, *contra*, Halphen, *Comté*, 215–16. Adhemar's belief that Count William's expedition was accompanied by Abbot Richard of St-Vanne, Verdun probably does not refer to its early stages. Hugh of Flavigny's long account of the large pilgrimage funded by the duke of Normandy and led by Abbot Richard makes no mention of companions from Aquitaine: 'Chronicon' (II. 18–23), *MGH SS* 8. 393–7. See, generally, H. Dauphin, *Le Bienheureux Richard abbé de Saint-Vanne de Verdun* (Bibliothèque de la Revue d'Histoire Ecclésiastique, 24; Louvain, 1946), 281–96.

[30] Cf. Hugh of Flavigny's statement that Richard of St-Vanne was inspired by the recent pilgrimage of an inhabitant of Autun: 'Chronicon' (II. 18), 393.

[31] Glaber (III. 4, 6; IV. 5, 6), 114–16, 126–8, 194, 198–204; cf. the comments of J. France, ibid., p. xxiii.

Histories he showed himself perfectly capable of recognizing an important practical cause of the growth in pilgrimage to Jerusalem, namely the conversion of Hungary which opened up a land route safer than the journey by sea.[32] The apparent uniqueness of the early eleventh century is simply a reflection of the unusually abundant narrative evidence available for central and south-western France. The charter evidence, though more tersely worded and chronologically diffuse, reveals that, far from diminishing with the easing of supposed millenarian fears, enthusiasm for pilgrimage continued throughout the eleventh century and beyond.

Pilgrimage and Ties to Local Religious Communities

The charters record pilgrimages to shrines several hundreds of miles distant for two reasons. First, because journeys over great distances were naturally reckoned to be the most hazardous, potential pilgrims were often particularly anxious to create or confirm ties of intercession with local religious communities. The second reason, which is discussed later, was the considerable expense which pilgrimage usually entailed.

It was common for pilgrims to reinforce their existing bonds with a community. For example, Viscount Bernard of Comborn, whose family had a long and impressive tradition of support of Tulle, gave that abbey five *mansi* before he started out for Jerusalem in 1119.[33] In 1112 × 1150 Adhemar of Rouffignac confirmed all the donations made by his father, uncle, and sister in the same abbey's chapterhouse before leaving for the Holy Land.[34] A note of a charter (now lost) from Uzerche records that Roger of Courson went to Jerusalem 'for reasons of prayer' (the phrase does not preclude his travelling under arms) in 1103.[35] Roger belonged to a well-connected family whose members appear as close associates, probably as *fideles*, of the viscounts of Comborn from the second quarter of the eleventh century.[36] The family were benefactors of Uzerche in their own right, and were further drawn into association with the monastery by

---

[32] Ibid. (III, 1; IV. 6), 96, 204.

[33] T. 140. See also ibid. 84, 90, 103, 128, 139, 152, 177, 190.

[34] Ibid. 175.

[35] Uz. 493. See G. Constable, 'Medieval Charters as a Source for the History of the Crusades', in P. W. Edbury (ed.), *Crusade and Settlement: Papers read at the First Conference of the Society for the Study of the Crusades and the Latin East and presented to R. C. Smail* (Cardiff, 1985), 75.

[36] Uz. 319 (1059 × 1086), 441 (1030), 460 (1048), 464 (1068), 478 (1097 × 1108), 481 (1097 × 1108).

reason of their ties to the Comborns. Roger's father and uncle gave property to the abbey in 1055, and in 1086 × 1096 Roger gave a quarter of a tithe with the consent of Viscount Bernard.[37] Although Uzerche appears to have been its favoured community, the family also had links to Tulle which were similarly reinforced by the Comborn connection. Roger's ex-wife Emma made a death gift to the monks there in 1119, as had his aunt earlier.[38] The daughter of Roger's cousin Hugh of Courson married Viscount Bernard *c*.1086 × 1096. Roger's mother Aina was the sister of Guy of Laron, bishop of Limoges (1073–86), which makes it likely that the Rigald of Courson attested as a canon of St-Étienne during Guy's episcopate was Roger's brother.[39] Aimo Bernard of Tulle, who endowed his local abbey and went on pilgrimage to Jerusalem in 1087 or 1088, was Roger's father-in-law.[40] Thus Roger belonged to a kindred with an impressive record of support for the church.

Further examples reveal that pilgrimage could form part of a broader pattern of contacts with religious communities. Before Peter of Noailles, near Brive, left for Jerusalem in 1111 × 1124, he gave *post obitum* to the monks of Vigeois a *mansus* and in addition surrendered properties he held from the abbey. The scribe who noted the transaction thought it relevant that the abbot of Vigeois, Rainald of Rouffignac, was Peter's (probably maternal) uncle and that a monk named Gerald who witnessed the gift was his brother. Such were the close family ties with the abbey that, after Peter's death, Abbot Rainald's paternal kinsmen tried to seize the *mansus*, but were persuaded to surrender their claim, again with the monk Gerald in attendance.[41] Two brothers named Gerard Cabrols and Ranulf of Chazarein gave two *mansi* to Tulle when they went to the Holy Land in 1053 × 1084. Their father Robert and brother Archambald had been benefactors of the abbey; a brother and two nephews were monks there. Shortly before he died, probably childless, Archambald's son Peter gave all his allodial property to Tulle. Peter's niece took the veil at Aureil, but in 1118 was given leave to travel to Tulle to give further allods to the monks.[42] In 1068 × 1096 Peter of Vic, in Fezensac, recognized that he unjustly held the church and tithes of Vic contrary to the rights of the cathedral of Auch, and he surrendered his claims to them. His lord,

[37] Ibid. 495, 1195.

[38] T. 9 (*c*.1040), 308 (mid-11th C.), 309 (1059 × 1084), 310 (? 3rd qr. 11th C.), 337 (1119), 356 (mid-11th C.).

[39] Uz. 67², 399, 627–8; cf. A. 161¹¹ (late 11th C.).   [40] T. 337 (1119).   [41] V. 211.

[42] T. 114 (later 11th C.), 115 (1118), 116 (1073 × 1084), 121 (1053 × 1084), 209 (*c*.3rd qr. 11th C.).

the count of Fezensac, similarly made over all his rights of lordship in the settlement. When Peter later went to Jerusalem, he surrendered those jurisdictional rights over Vic which he had managed to retain, evidently with the count's connivance; and he offered a son to be a canon of Auch, reserving part of his *honor* for his other sons to ensure that they could continue to serve their lord. The son's entry gift comprised four churches and their tithes and valuable milling- and pasturage-rights; his heirs were allowed the right to recover one of the churches from the cathedral on condition that the services of one peasant be reserved for the cathedral and the son-canon. The whole grant was made so that the clerics installed in the church at Vic under the supervision of the canons should pray 'all the time' for the donor, his parents, and kinsmen.[43]

Enthusiasm for pilgrimage could stimulate future links with local religious communities. In 986 Gerald of La Valène, about to leave for Jerusalem, and his wife Garsendis of Malemort gave widespread properties to Tulle and offered their son as an oblate.[44] Between 1073 and 1084 Bernard of Chanac, three miles east of Tulle, gave that abbey half an allodial property if he were to die on his intended journey to Jerusalem. Bernard almost certainly did die on pilgrimage, for his brother Peter gave the whole property to Tulle a little later.[45] In 1084 X 1091 five brothers contested a gift of rights in one *mansus* which had been made to Tulle by their father. When one of the brothers, William of Bort, went to Jerusalem, he surrendered his putative share in the disputed property. Doubtless by breaking ranks in this way, William persuaded the other brothers to release their claims to Tulle for 30*s*. (Whether any part of this money went to finance William's journey is unrecorded.)[46] Around the middle of the eleventh century Arbert of La Vallette, wishing to go on pilgrimage to Compostela, gave to Uzerche measures of corn and grain which he took from a tithe. The almoner of Uzerche witnessed the gift. When later Arbert found himself dying he confirmed the grant. It was probably this same man who in 1071 consented to the oblation at Uzerche of his son Adhemar by Viscount Adhemar II of Limoges, and himself gave one *mansus*. In 1097 X 1108 a Gerald of La Vallette

---

[43] Auch, 6. The charter is dated by the archiepiscopate of Archbishop William of Auch (1068–96), so it is possible that Peter was a first crusader. Peter was alive in 1094; his lord Count Astanove went on the First Crusade: ibid. 39, 57.

[44] T. 46.

[45] Ibid. 32–3.

[46] Ibid. 119.

(probably Arbert's son, for he had a kinsman named Arbert) gave Uzerche half of a *villa* which he held of the lords of Pierrebuffière; and the younger Arbert's wife Audenos was a generous benefactress of Uzerche in the first third of the twelfth century.[47] In *c.* 1060 X 1080 Sancho Auriol granted to the monks of St-Savin, Lavedan, before he set out for Jerusalem, property which had been pledged to him for 100*s.* by his kinsman William Fort of Ayzac. William Fort's exact reaction is not clear, but the fact that he surrendered his claim on the property to the abbey for only 15*s.* suggests that respect for Sancho Auriol's pious reasons for the gift (or, quite possibly, sale) placed him in a weak bargaining position against the monks.[48]

As well as being occasions on which individuals and families wished to strengthen their ties to a religious body, pilgrimages also feature in the charters because they involved the sort of expenses which might oblige a layman to sell or pledge some of his property. Religious communities were very important sources of specie or ingots. To say that monks and canons acted as bankers for their lay neighbours is anachronistic, but they often displayed a willingness to finance enterprises of which they approved, or at least to make some contribution towards their funding. Pilgrimage was one such approved enterprise. For example, sometime in the first half of the twelfth century Bernard of Meiras decided to go to Compostela and asked the canons of Aureil 'that they might render him assistance in undertaking such a great journey'. He felt able to do so because his father Bernard had granted Aureil the church and priest's fief of Eyjeaux, near Pierrebuffière, with the consent of his sons and other relatives. In return for a confirmation of the gift, the canons provided the younger Bernard with a mule worth 4*l.* Sometime after he returned from Spain, Bernard resumed possession of the priest's fief in protest at a chapel which the canons had erected within the parish of Eyjeaux; but he subsequently struck a second deal with the canons in return for 230*s.* when he decided to go to Jerusalem (possibly on the Second Crusade but more probably earlier). Bernard was prepared to exploit his potential claims to rights held by Aureil, but he cannot be seen as a robber-lord preying upon the church. Before his pilgrimage to Compostela he and his brothers had petitioned Aureil to receive three canons so that they might be assured regular intercessory prayer; the only condition was that the

---

[47] Uz. 229, 325, 532, 1148; cf. ibid. 539, which illustrates the family's important connections.

[48] *Cartulaire de l'abbaye des bénédictins de Saint-Savin en Lavedan (945–1175),* ed. C. Durier (Cartulaire des Hautes-Pyrénées, 1; Paris, 1880), 2 at p. 6.

brothers could retain two horses worth 500*s*. which their father had received from the canons at the time of or shortly after the original grant. Bernard's pilgrimages, therefore, fitted neatly into a tradition of securing Aureil's support for both spiritual and practical purposes.[49]

The financing of pilgrimages by religious communities was not new to the twelfth century. Around the middle of the eleventh century William of Sadran pledged a *mansus* to Vigeois for 5*s*. and the promise of the usufruct of the property should he return from Jerusalem.[50] Towards the end of the century Fort Sancho of Godz, a member of a family with established links to St-Mont, pledged his share of the family church in return for 5*s*. before he set out for Rome.[51] In 1087 or 1088 Aimo Bernard of Tulle gave a *mansus* to the community at Tulle and confirmed the donations made by his father, 'for the sake of money', when he departed for Jerusalem.[52]

A remarkable series of events in the Bas-Limousin in 1091–2 demonstrates that the practical help which a religious community might offer pilgrims was not considered to be formally distinct from its spiritual assistance. In 1091 Viscount Boso I of Turenne granted half of a wood to Tulle when he offered his son (the future Abbot Ebles) as an oblate, allowing the monks the option of buying the other half at their discretion. The woodland, at Auriol on high ground seven miles north-east of Turenne, was specifically donated so that the monks might build there a church dedicated to St Nicholas. Within a year Boso fell mortally ill at Jerusalem, and as he expired he asked two companions to take back to the Limousin his instructions that two *mansi* close to the new church at Auriol be granted to Tulle. In 1092 Boso's son and successor Raymond (a future first crusader) persuaded the monks to exercise their rights of purchase in the remaining woodland—probably sooner than they would have wished. The price, which was raised by Abbot William, was 200*s*. and a mule 'quam praestavit ei apud Romam et Sanctum Nicolaum'.[53] The meaning of this phrase is ambiguous, since the whole notice from

---

[49] A. 207, 209–10. The *placitum* negotiated before the journey to Jerusalem was witnessed by a priest who probably appears in ibid. 39 (1125). The church of Eyjeaux was in Aureil's possession by 1140 × 1147, when it was contested by St-Étienne, Limoges: ibid. 208. See also 'Chartes', ed. A. Leroux *et al.*, *Documents historiques bas-latins, provençaux et français concernant principalement la Marche et le Limousin* (Limoges, 1883–5), i. 14 at p. 134 (1141), which may represent the resolution of the dispute.

[50] V. 5. For the dating cf. ibid. 19 (1031 × 1060).

[51] St-M. 26.

[52] T. 105, 474, which, though dated by different years, must refer to the same event.

[53] Ibid. 498–501. The quotation is from 499.

which it comes is in the third person and perfect tense. It may signify that Abbot William or Raymond had recently been to Italy, possibly accompanying Boso at least part of the way to Jerusalem, or that Raymond was planning a pilgrimage of his own. In either event, this episode is good evidence for the exceptionally close links between the devotional practices of the viscounts of Turenne, the growth of a local religious community's landed power, and the appeal of pilgrimage. It seems probable that by 1091 Boso or someone very close to him had already been on pilgrimage to the shrine of St Nicholas at Bari in Apulia, probably *en route* to the Holy Land. The relics of St Nicholas had been translated from Asia Minor to Bari as recently as 1087.[54] The fact that Boso decided to dedicate his foundation (which he probably intended to be his last) to this saint, who was little known in the Limousin before then, is interesting evidence for the way in which the enthusiasm for distant cults generated by pilgrimage could be accommodated in the network of devotion around a local religious community.[55] In fact Boso's actions clearly anticipate the endowment of communities with relics brought back from the East by returning first crusaders.[56] Tulle's rights in Auriol were contested by the abbey of Cluse and its priory at Albignac, but it managed to retain control of its new church, which by 1096 was served by one monk. Raymond of Turenne continued to support his father's creation, extending Tulle's rights in the area when his mother died in 1103, for a brother's soul in 1105, in 1116, and when his brother Archambald died in 1117.[57] A religious foundation made possible by pilgrimage contacts thus became an important focal point of a major family's spiritual provision.

The combination of devotional and practical reasons for associating a religious community with an intended pilgrimage can also be seen in

[54] Sumption, *Pilgrimage*, 33. See P. J. Geary, *Furta Sacra: Thefts of Relics in the Central Middle Ages*, rev. edn. (Princeton, NJ, 1990), 94–103. It is noteworthy that a century later Geoffrey of Vigeois thought this episode of interest: 'Chronica' (ch. 19), 289.

[55] There is a faint trace of a connection between the viscounts of Turenne and the Norman world in Bernard of Angers's book II of the *Miracles of Saint Faith*, which records that in *c.*1010 × *c.*1018 Ebles I of Comborn, then lord of Turenne, was married to Beatrice, the sister of a Count Richard of Rouen (i.e. duke of Normandy). There is no reason to doubt Bernard's testimony, for he claimed to have met Beatrice at Duke William V of Aquitaine's court. Bernard's chronology is imprecise, but it seems probable that the brother was Richard II (996–1026): *Liber Miraculorum Sancte Fidis* (II. 6), ed. A. Bouillet (Paris, 1897), 109–11. But see Geoffrey of Vigeois, 'Chronica' (ch. 23), 290, according to which Ebles was the son of Richard of Normandy's sister.

[56] Riley-Smith, *First Crusade*, 122–3.

[57] T. 502–7, 510.

instances of would-be pilgrims settling a dispute with local religious or trying to anticipate potential disagreements in later generations. This was the case with Bernard of Meiras's dealings with Aureil noted earlier. Similarly, in about 1080 Count Aymeric II Fort of Fezensac recognized that he had built mills in Auch contrary to the rights of the city's archbishop and cathedral canons. The terms of the deal (*negotium, placitum*) which Aymeric arranged with Archbishop William I required that he give the mills to Ste-Marie *post obitum*, but that if in the mean-time he went to Jerusalem the cathedral community might enjoy the use of them during his absence.[58] A remarkable case from 1119 demonstrates how pressure could be placed on an intending pilgrim to make his peace with a local monastery. In 1064 Peter of Rouffignac had pledged to Tulle for 100s. a *mansus* which had (as far as the evidence can reveal) first been given to the monks more than thirty years before, probably by a vassal of his family.[59] Peter's son Robert subsequently contested the grant. In 1119 Robert decided to accompany Viscount Bernard of Comborn on pilgrimage to Jerusalem, and therefore attended a large gathering at Tulle at which Bernard made his preparations for the journey. The monks used the opportunity of Robert's presence to raise the matter of the disputed property and of a second *mansus* which Robert's sister Petronilla had given to the abbey. Robert, however, refused point-blank to surrender the properties. Events then took a strange turn, for later that day, as the monks were sitting in refectory, Robert, accompanied by his son and grandson, burst into the room in a state of emotional excitement and surrendered his claims at the abbot's table.[60] Before this date Robert's relations with local monasteries had not always been easy. He had contested a gift of five *mansi* made to Tulle by his brother Stephen, and then surrendered his claims in return for very extensive spiritual benefits; and in 1116 he had released for 25s. all the rights he and his kinsmen had ever granted to Vigeois.[61] But his kindred had very close ties to these communities. The genealogy of the Rouffignacs is confused; it seems that there were at least two, and probably three, collateral lines by the second

[58] Auch, 46. The dating of the charter presents problems. It is dated 1088 but 'vigente Gregorio papa VII [1073–85]'. It was witnessed by Bishop Pons of Bigorre, who was elected in 1079 or 1080 and was dead or retired by 1087. The synchronisms mean that there is no reason to suppose that Aymeric was a *fidelis Sancti Petri* involved with Pope Gregory's plans for an expedition to Jerusalem in 1074: Riley-Smith, *First Crusade*, 8 and n. 16.

[59] T. 268–9; cf. ibid. 205.

[60] Ibid. 269[2–3]. See also ibid. 270 (1121).

[61] Ibid. 180; V. 210.

half of the eleventh century descended from a Peter of Rouffignac and/or a Gerard who are attested in *c.* 1000–30.[62] It is reasonably clear, however, that Robert belonged to a pious kindred with a solid record of support for the church. His cousin Rainald, for example, was abbot of Vigeois in 1111–24. Another cousin, Bernard, is attested as prior of Tulle in the years around 1100. Yet another cousin had been given to Tulle as an oblate in 1109. Robert himself appears second, after Bernard of Comborn, in a list of Tulle's *confratres* drawn up in *c.* 1103.[63] In the dramatic events of 1119 it seems that Robert's son and grandson influenced his change of heart. Clearly they valued their kindred's links with Tulle even if Robert at first did not.

Short- and Middle-Range Pilgrimage

The foregoing examples suggest that the charters give a slightly distorted picture of pilgrimage because longer journeys were much more likely to create situations suitable to be put in writing. It is important to emphasize that pilgrimage was a variegated practice, its appeal not being confined to a few famous shrines. If pilgrimage had been associated solely with a handful of distant cult centres, which even the most pious layman could have visited only occasionally, it would be hard to explain the mass appeal of the crusade as an armed pilgrimage in 1095–6. In fact one of the most significant features of pilgrimage in the eleventh century was its variety of forms and destinations.

When a layman visited a local religious community in order to make a donation, this very simple act could be construed as a type of pilgrimage. The charters recording such gifts often specified that the donation was made, in the first instance, to the community's patron saint.[64] This was not a matter of form. The laity must have shared the brethren's belief that churches belonged to a particular saint or saints (though this notion was revised by greater legalistic precision from the twelfth century), for this was the basis of one of the links of the intercessory chain which connected the faithful to, ultimately, God. Local shrines and cults could have a very strong appeal. One important Limousin cult centre was St-Léonard, Noblat, which is discussed below. Another example of a famous saint

---

[62] T. 268; V. 24.

[63] V. 25, 210, and p. 286; T. 97, 144, 247², 255, 259, 499. See also ibid. 207, 252–3, 256.

[64] e.g. T. 164 (1084 × 1091).

intimately connected with the Limousin is St Martial, the patron of the most important abbey in Limoges. Unfortunately no collection of Martial's miracles from this period survives, and relatively few of St-Martial's documents are extant.[65] The best evidence for the popularity of Martial's cult, though very *parti pris*, is the work of Adhemar of Chabannes, who recorded, for example, that under Abbot Guy (974–91) a monk named Gauzbert created a 'golden icon' of the saint, the right hand of which extended outwards in a gesture of blessing from its position on the high altar. This was most probably a three-dimensional reliquary, or *majestas*, of the sort common in south-central France at that time.[66] Adhemar also stated that Martial corruscated with miracles in *c.*1010, and he noted the presence of noble pilgrims at St-Martial during the abbacy of Geoffrey II (1007–18).[67] Some measure of St Martial's appeal is provided by a disaster which took place during Lent 1018, when fifty-two pilgrims pressing to enter the west door of the basilica to hear Matins were crushed to death; a new and much larger basilica of St-Sauveur was begun around this time with at least the partial aim of accommodating more pilgrims.[68] It was noted above that the hagiographical material and sermons associated with the Limousin Peace of God celebrated Martial as the foremost patron of the region. There is no reason to believe, however, that his cult became so closely bound up with the Peace that it declined when the Peace movement lost its impetus in

[65] Numerous references to Martial's miracle-working powers are scattered, however, throughout the sermons of Adhemar of Chabannes: D. F. Callahan, 'The Sermons of Adémar of Chabannes and the Cult of St. Martial of Limoges', *RB* 86 (1976), 286–8 and 289 n. 7.

[66] Adhemar of Chabannes *et al.*, 'Commemoratio abbatum Lemovicensium basilice S. Marcialis, apostoli', ed. H. Duplès-Agier, *Chroniques de Saint-Martial de Limoges* (Paris, 1874), 5. For similar reliquaries from the later 10th and 11th C., see *Liber Miraculorum Sancte Fidis* (I. 13–15), 46–51; *Les Miracles de saint Privat suivis des opuscules d'Aldebert III, évêque de Mende* (chs. 5, 7, 9), ed. C. Brunel (Collection de textes pour servir à l'étude et à l'enseignement de l'histoire, 46; Paris, 1912), 9, 10, 14–15, 16–17. The best surviving example of a *majestas*, somewhat altered since the 11th C., is that of St Faith from Conques, for which see A. Bouillet and L. Servières, *Sainte Foy, vierge et martyre* (Rodez, 1900), i 167–83; I. H. Forsyth, *The Throne of Wisdom: Wood Sculptures of the Madonna in Romanesque France* (Princeton, NJ, 1972), 12, 40–1, 42, 67–9, 78–9, 85–6.

[67] Adhemar, *Chronique* (chs. 49, 52), 171–2, 175.

[68] Adhemar of Chabannes *et al.*, 'Commemoratio abbatum', 7; Bernard Itier, 'Chronicon', ed. Duplès-Agier, *Chroniques de Saint-Martial de Limoges*, 46; C. de Lasteyrie, *L'Abbaye de Saint-Martial de Limoges: étude historique, économique et archéologique, précédée de recherches nouvelles sur la vie du saint* (Paris, 1901), 294. The attempt by R. Landes ('The Dynamics of Heresy and Reform: A Study of Popular Participation in the "Peace of God" (994–1033)', *HR* 14 (1987), 501–6, 510) to link the disaster of 1018 to a putative heretical movement in the Limousin pushes the evidence much further than it warrants.

the 1030s. When St-Martial was granted to Cluny in 1062, some of its monks fiercely resisted their incorporation.[69] An unstated but surely very potent cause of their dissatisfaction was the fear that St Peter might eclipse St Martial: it cannot be insignificant that the Feast of the Apostles Peter and Paul (29 June) clashed with the Vigil of Martial's principal Feast in the pilgrimage season. The Cluniac connection, however, soon enhanced St-Martial's prestige. In the second half of the twelfth century the abbacy of Adhemar (1063–1114) was remembered as a golden age in the monastery's history when its material resources had increased greatly.[70] This would have been impossible without the continuing popularity of St Martial.

The importance of major regional cults such as that of St Martial can be gauged by the competitive spirit which sometimes informed relations between their supporters. Between the second half of the tenth century and the early years of the twelfth, for example, the cult of St Front, the patron saint of the cathedral of Périgueux, was developed to rival that of St Martial. According to a primitive Life of St Front which was in existence by the first half of the ninth century, the saint had been a native of Périgord, born (it is implied) of Christian parents in the first century. He visited Rome by way of Egypt, and was made the first bishop of Périgueux by St Peter.[71] In his account of the Council of Limoges in 1031, Adhemar of Chabannes provides important evidence that the canons of St-Frond had tried to enhance their patron's status in the tenth century. He wrote that during the council Abbot Gerald of Solignac delivered a passionate speech in favour of St Martial's apostolicity, only to be interrupted by a Périgourdin cleric claiming the same status for Front. In a venomous reply the abbot poured scorn on a *scriptura nova* concerning Front which, he claimed, had been produced by Gauzbert, choir-bishop of Limoges under Bishop Hildegar (969–90). The exchange formed part of Adhemar's elaborate fraud to publicize St Martial the Apostle, so it cannot be treated as a verbatim account of a real argument. But it is clear that Adhemar was writing on the basis of genuine and deep hostility

---

[69] Geoffrey of Vigeois, 'Chronica' (ch. 13), 284; Lasteyrie, *Abbaye de Saint-Martial*, pièce just. no. 7, pp. 426–9; H. E. J. Cowdrey, *The Cluniacs and the Gregorian Reform* (Oxford, 1970), 90–3. A. Sohn, *Der Abbatiat Ademars von Saint-Martial de Limoges (1063–1114): Ein Beitrag zur Geschichte des cluniacensischen Klösterverbandes* (Beiträge zur Geschichte des alten Mönchtums und des Benediktinertums, 37; Münster, 1989), 46–78 provides a thorough examination of the evidence.

[70] Adhemar of Chabannes *et al.*, 'Commemoratio abbatum', 9–10.

[71] 'La Vie ancienne de S. Front' (chs. 1, 4–6), ed. M. Coens, *AB* 48 (1930), 343, 345–8.

between the religious of St-Martial and other Limousin communities on the one hand, and the canons of St-Frond on the other.[72] The Périgourdin cleric, for example, was accused of being unlearned, and Gauzbert was reviled for having charged a fee for his work.[73]

It has been convincingly argued that Adhemar's 'new writing' was in fact a revised Life of St Front.[74] According to this version Front continued to be a Périgourdin, living at a time when Gaul was still substantially unconverted. Baptized by St Peter at Rome, he was sent to evangelize in Gaul as the *primus inter pares* of such early missionaries as SS Saturninus (Toulouse), Ursinus (Bourges), and Austremonius (Auvergne).[75] As in the original Life, Front was persecuted in Périgueux by a man named Squirius, the city's *preses*, who was eventually converted; but in the later version Squirius's example of receiving baptism led directly to the conversion of all the remaining inhabitants of Périgord.[76] A greater emphasis on Saint Front's mass appeal is further evident in a miracle story appended to the new Life which tells of how in recent times his shrine had attracted many pilgrims.[77]

By about the beginning of the twelfth century St Front's cult had so developed that much larger claims could be made for him. A third Life (the *Pseudo-Sebald*) was produced which stated that Front had been a Jew of the Tribe of Judah who was baptized by St Peter on Christ's express command, became one of the seventy-two disciples, attended the Last Supper, received the Holy Ghost at Pentecost, and accompanied St Peter to Antioch and Rome, whence he was sent to Périgord to preach.[78] At Périgueux he cured the husband of an early convert, the 'venerable matron' Maximilla, prompting the couple's large *familia* to seek baptism;

---

[72] See Adhemar's comments in 'Epistola de apostolatu Sancti Martialis', *PL* 141. 99–100. Relations improved by the late 11th C.: see Sohn, *Abbatiat Ademars*, 198–200.

[73] Mansi, 19. 513–15.

[74] M. Coens, 'La *Scriptura de sancto Fronto nova*, attribuée au chorévêque Gauzbert', *AB* 75 (1957), 345–7.

[75] 'Vita S. Fronti ep. Petragoricensis' (chs. 2–4, 6), ed. M. Coens, '*Scriptura*', 352, 352–4, 354–5. Note the absence of Martial from the list of Front's peers: see ibid. 346, 352 n. 1.

[76] 'Vie ancienne de S. Front' (chs. 3, 8, 13–17), 344–5, 350–1, 355–60, esp. (ch. 17) 360 (where the link between Squirius's conversion and that of 'omnis plebs' is less forcibly expressed than in the later version); 'Vita S. Fronti' (chs. 8, 14–19), 356, 359–63.

[77] 'Vita S. Fronti' (ch. 21), 363.

[78] 'Vita S. Frontonis, Episcopi Petragoricensis' (ch. 1), *AASS* (Oct.) 11. 407–8. This reproduces a 14th-C. abridgement (*BHL* 3185) of the full *Pseudo-Sebald* (*BHL* 3183) inserted by Bernard Gui into his sanctorale: see *Catologus codicum hagiographicorum latinorum antiquiorum saeculo XVI qui asservantur in Bibliotheca Nationali Parisiensi*, ed. Société des Bollandistes (Paris, 1889–93), iii. 560.

the city's count converted and lavishly supported Front's new church; when the son of a noble was cured, his parents and all their *familia* sought baptism; the cure of the son of a 'distinguished man' led seven thousand to convert.[79] All these stories directly recalled elements of St Martial's *Vita Prolixior*. Squirius, moreover, was promoted by obvious analogy to Duke Stephen; he became Emperor Claudius's kinsman, sent to subject Aquitaine.[80] The account of Front's mission in Bordeaux refers to a Count Sigebert, a direct borrowing from the *Prolixior*.[81] The author of the *Pseudo-Sebald* even took the fight right into the enemy's camp, for it was now claimed that Front evangelized the Limousin, performing miracles and erecting a church there.[82] In *c.*1140 the compiler of the *Pilgrim Guide* gave a synopsis of the *Pseudo-Sebald* tradition and noted (with a hint of reserve) that some people believed that Front had been a disciple of Christ, an indication that the developed cult of St Front the Apostle had gained currency.[83] The precise details of the competition between the cults of Martial and Front cannot now be reconstructed; but if it is accepted that the *Prolixior* reached its definitive form after 994 (that is, after the 'new writing'), then it would follow that St Front's cult stole a lead on Martial's in the later tenth century. The borrowings of the *Pseudo-Sebald*, on the other hand, reveal that by the later eleventh or early twelfth century it was the turn of Front's cult to do the catching up. The *Pseudo-Sebald* was a backhanded compliment to the popularity of St Martial and his close identification with Aquitaine in general and his region in particular.[84]

Two other Limousin saints, though never approaching St Martial's popularity, were also important. St Aredius (Fr. Yrieix) was a sixth-century Limousin abbot known and respected by Gregory of Tours.[85] By the eleventh century the monastery he had founded, St-Yrieix-la-Perche, was staffed by a community of canons. Few documents from St-Yrieix survive by which to judge its reputation in our period, but it is note-

---

[79] 'Vita S. Frontonis, Episcopi Petragoricensis' (ch. 1), 408.
[80] Ibid. 409.
[81] Ibid. (ch. 2), 410–11.
[82] Ibid. 411.
[83] *Guide* (ch. 8), 56–8.
[84] For further interrelationships between hagiographical legends in south-western France, see L. Duchesne, *Fastes épiscopaux de l'ancienne Gaule*, 2nd edn. (Paris, 1907–15), ii. 117–24, 135–7.
[85] Gregory of Tours, *Libri Historiarum X* (x. 29), *MGH Scriptores rerum Merovingicarum*, rev. edn., 1^1. 522–5; id., 'Liber in gloria martyrum' (chs. 36, 41), *MGH Scriptores rerum Merovingicarum*, rev. edn., 1^2. 61, 66. See also 'Vita Aridii abbatis Lemovicini', *MGH Scriptores rerum Merovingicarum*, 3. 576–609, esp. (ch. 3) 582.

worthy that it was very close to the castle of the enthusiastic crusading lords of Bré and could not have flourished without at least some support from them.[86] A second popular saint was St Pardulphus (d. c.737), another Limousin, who founded a monastery at Guéret. By the tenth and eleventh centuries Pardulphus was the patron saint of a large number of churches in the diocese of Limoges.[87] Some measure of his appeal comes from Geoffrey of Vigeois's story of how, in c.1015, Guy Niger of Lastours arranged to steal the saint's relics from a church close to St-Sacerdos, Sarlat, and then transfer them to his new foundation at Arnac.[88]

As well as long- and short-range pilgrimages there was a third and intermediate type of journey which must have been much more common than the charter evidence reveals: to shrines between approximately twenty and two hundred miles distant with which the pilgrim need not have had family ties based on the giving of land, but which could be reached in a matter of days or weeks without prohibitive expense. Narratives furnish a few examples. Emma, wife of Viscount Guy I of Limoges, was captured by Vikings while on pilgrimage to St-Michel-en-l'Herm in the Vendée; according to Adhemar of Chabannes, she was held to ransom for three years.[89] A bizarre story recounted by Adhemar concerning the persecution of a Jew involved the presence in Toulouse over Easter of Viscount Aymeric I Ostafrancs and his chaplain Hugh; the circumstances of the story suggest that they were on pilgrimage to St-Seurin.[90] According to Geoffrey of Vigeois, Peter II of Pierrebuffière (a former first crusader) was waylaid and mortally wounded while returning from a pilgrimage to Charroux in about 1114.[91] It is noteworthy that in the above cases pilgrimage merely formed an incidental background detail in the story. The authors treated the act of pilgrimage as commonplace.

Evidence that interest in pilgrimage centres was not restricted by traditions of endowment and benefaction is provided by Helgaud of Fleury's Life of King Robert the Pious (d. 1031). Helgaud reports that in the latter part of his reign, and motivated by intense fear of the Day of Judgement, Robert embarked on a grand tour of shrines in central and southern Gaul. Robert did not, as far as we know, pass through the Limousin, and Gascony was too distant to interest him, but he visited

[86] For the Brés' enthusiasm for crusading, see below, Ch. 6.
[87] M. Aubrun, L'Ancien Diocèse de Limoges des origines au milieu du XIᵉ siècle (Publication de l'Institut d'Études du Massif Central, 21; Clermont-Ferrand, 1981), 321–4.
[88] 'Chronica' (chs. 3–4), 280–1.
[89] Adhemar, Chronique (ch. 44), 166–7.
[90] Ibid. (ch. 52), 175.
[91] 'Chronica' (ch. 18 [recte 38]), 299.

famous cult centres in, *inter alia*, Berry, the Auvergne, the Rouergue, and the Toulousain, dispensing gifts and alms as he went. The wide sweep of Robert's itinerary, which passed through areas where royal authority was minimal or non-existent and where the king had no land to give, contrasts strongly with Helgaud's account of the religious communities which Robert founded and endowed. With the exception of a church in Autun and of St-Paul, Chanteuges, near St-Julien, Brioude in the Auvergne, these communities lay in an area of central northern France based on the royal domain.[92]

Pilgrims' mobility meant that the guardians of a shrine might worry that the appeal of a more distant cult centre would eclipse their own. In book III of the *Miracles of St Benedict*, Aimo of Fleury told the story of a crippled girl whose noble parents led her round a number of churches in the vain search for a cure. The shrines they visited included St-Denis and St-Martial, which, Aimo explained, 'were both corruscating with miracles at that time', a phrase suggesting that the parents had been beguiled by the transitory celebrity of distant cults. The child's mother was warned by a heavenly voice that travelling at great expense to remote shrines was futile, and that the family should visit Fleury, only some eighteen miles from their home. Predictably, a visit to St Benedict's shrine cured the girl, and the parents cemented their links with the local monastery by offering her to the saint.[93]

The best evidence for the provenance of pilgrims to a given shrine usually comes from collections of miracle stories. A late but very important example from south-western France is the miracle collection from Our Lady of Rocamadour, a cult centre in Quercy about twenty miles south of the Limousin. The collection comprises 126 miracle stories, arranged into three books, which were written up in or shortly after 1172 and mostly relate events which took place after 1166.[94] In almost every case the scribe employed at Rocamadour to record the miracles noted the place of origin of his informant. In the collection nearly all the Romance-

[92] Helgaud of Fleury, *Vie de Robert le Pieux* (chs. 27–8), ed. and trans. R.-H. Bautier and G. Labory (Sources d'histoire médiévale, 1; Paris, 1965), 124–32. See C. Lauranson-Rosaz, *L'Auvergne et ses marges (Velay, Gévaudan) du VIIIe au XIe siècle* (Le Puy, 1987), 442–51.

[93] *Miracles de saint Benoît* (III. 20), 169–70.

[94] *Les Miracles de Notre-Dame de Roc-Amadour au XIIe siècle*, ed. and trans. E. Albe (Paris, 1907); R. Pernoud, 'Le Livre des miracles de Notre-Dame de Rocamadour: étude des manuscrits de 1172', in M. François (ed.), *Le Livre des miracles de Notre-Dame de Rocamadour* (2e Colloque de Rocamadour; Rocamadour, 1973), 9–23. See also B. Ward, *Miracles and the Medieval Mind: Theory, Record and Event 1000–1215*, rev. edn. (Aldershot, 1987), 145–50.

speaking world is represented, including Italy, Spain, the Latin settlements in the Levant, and the Angevin empire; there are also some
Germanic-speaking pilgrims recorded.[95] Rocamadour seems to have
been particularly popular in Gascony. One miracle story noted that
many Gascons came there to pray, and other miracles involved pilgrims
described variously as Gascons or from Lectoure, the Bazadais, Couserans,
or St-Sever.[96] Quercy and such neighbouring regions as Périgord, the
Toulousain, the Rouergue, and the Auvergne also provided pilgrims.[97] It
is therefore odd that Limousins do not feature prominently in the collection. One rather unusual story involves Almodis of Pierrebuffière, a
noblewoman, whose pet starling escaped from its cage but was retrieved
through Our Lady's intervention.[98] Another account records that pilgrims
bound for Rocamadour from northern Burgundy stayed at Uzerche.[99]

The collection's relative silence concerning the Limousin should not be
taken to mean that Rocamadour was not popular there, for the cult centre
in fact had very close connections with the area. For example, an account
of the life of St Amadour, surviving in a late medieval version from Italy
which was very probably based on a lost Life contemporaneous with the
Miracles, displays close borrowings from the Life and Miracles of St
Leonard.[100] Rocamadour also had intimate ties with Tulle. By 1113 the
monks of Tulle were claiming that the church at Rocamadour had been
built on allodial land given to them in *c.*930, and that Bishop Frotarius
of Cahors had confirmed them in possession of it in 968.[101] In 1113
Bishop William of Cahors confirmed their rights in the church.[102] The
monks' interest in their possession seems to have developed only recently,
for Rocamadour appears in a papal confirmation of their rights and
properties in 1105 and again in a bull of 1114.[103] By the second half of

---

[95] *Miracles de Notre-Dame* (I. 4, 9, 10, 25, 45; II. 6, 20, 34, 42; III. 10, 13), 77–9,
90–4, 116, 147–9, 182–4, 213–15, 240–2, 254–6, 285–8, 293–4. Cf. Sigal, *Homme*,
204–5.

[96] *Miracles de Notre-Dame* (I. 1, 5, 11, 36; II. 16, 24, 46–8; III. 24), 70–1, 79–82,
94–6, 132–5, 207, 221–9, 260–2, 313–19.

[97] Ibid. (I. 6, 14, 21, 31, 33, 38–9, 41, 43, 51; II. 21, 25, 36, 40, 49; III. 11, 18)
82–3, 100–1, 109–10, 123–5, 127–8, 137–9, 141–2, 145–6, 158–60, 215–18,
230, 245–6, 252, 262–4, 288–92, 301–2.

[98] Ibid. (II. 14), 199–200.

[99] Ibid. (II. 11), 193–5.

[100] E. Albe, 'La Vie et les miracles de S. Amator', *AB* 28 (1909), 58–65, 71, 72–3 and
73 n. 1, 76 and n. 2, 77 and n. 2, 81–2 and 81 n. 1.

[101] T. 301–2; see ibid. 14.

[102] Ibid. 302.

[103] Ibid. 3, 601.

the twelfth century, the abbey's claim to Rocamadour was well established. In 1154 Pope Adrian IV listed Rocamadour among the abbey's possessions; and in 1181 King Alfonso VIII of Castile granted two properties to Our Lady of Rocamadour at the petition of Abbot Gerald of Tulle.[104] By mid-century Rocamadour had obviously become a valuable prize, for Tulle's rights were hotly contested by the Cahorsin abbey of Marcillac, only to be vindicated in 1193 after an appeal to Pope Celestine III.[105] The early history of the cult of Our Lady at Rocamadour is very obscure, but the development of Tulle's interest in its dependency suggests that pilgrimage there grew in popularity from *c*.1100–10, and was firmly established by the mid-twelfth century.[106] When the body of St Amadour was miraculously 'discovered' in 1166, this only served to heighten interest in an established cult.[107] The apogee of Rocamadour's popularity thus falls just outside our period. The shrine's appeal was also, it is worth noting, an expression of the Marian devotion more characteristic of the twelfth century than the eleventh. Nevertheless, Rocamadour is an excellent example of how a local cult in south-western France might develop to attract pilgrims from both middle-range and long distances; and as a model it applies equally well to the eleventh century as to the twelfth.

Because of the Limousin's geographical position and the fairly easy communications into it, a list of the middle-range shrines which might have attracted Limousin pilgrims would necessarily include every cult centre in central and southern Gaul. It is unnecessary to list them all, although it is worth emphasizing the sheer variety of shrines on offer. What follows is only a selective sample of some of the pilgrimage centres with recorded links with the Limousin.

Aimo of Fleury knew of a monk from Beaulieu in the Limousin who was a habitual pilgrim to the shrine of St Benedict at Fleury in the later tenth century. The monk generated interest in the abbey and its school among nobles as well as monks in Aquitaine.[108] The viscounts of Limoges had connections with Fleury in the years either side of 1000 by reason of

---

[104] Ibid. 602, 605; cf. ibid. 606.

[105] Ibid. 623–4; E. Rupin, *Roc-Amadour: étude historique et archéologique* (Paris, 1904), 87–93, 96–99 and doc. at 88 n. 2.

[106] For a slightly different view, see J. Juillet, 'Lieux et chemins', in François (ed.), *Le Livre des miracles de Notre-Dame de Rocamadour*, 26–9.

[107] Robert of Torigny, *Chronica*, ed. R. Howlett (RS 82[4]; London, 1889), 248; cf. ibid. 292–4.

[108] A. Vidier, *L'Historiographie à Saint-Benoît-sur-Loire et les Miracles de saint Benoît* (Paris, 1965), 228; cf. ibid. 189–90.

their close links to the abbey's dependency at Sault in southern Berry. Viscount Guy I's sons Adhemar and Peter gave Fleury measures of wine which they received annually from the monks of Sault.[109] According to a legend recorded at Fleury, another of Guy's sons, Gerald, was punished for his support of men in dispute with Sault by being afflicted in the throat. Warned in a vision by Our Lady, Gerald went as a penitent to Fleury and repented of his error, only to find on his return home that, as the vision had warned, his wife and son had died.[110] Other shrines of pan-regional importance which attracted Limousin and other Aquitanian support included Ste-Foi, Conques, and Charroux, which housed the principal relic of the True Cross in southern Gaul.[111]

A further cult centre with close links to the Limousin was St-Géraud, Aurillac, in the Auvergne. Odo of Cluny's well-known Life of St Gerald (d. 909) was written in *c*.925 × 930 at the request of Bishop Turpio of Limoges (908–44) and Abbot Aimo of Tulle (later abbot of St-Martial, d. 943). In his dedicatory letter prefacing the Life, Odo implied that these two had been among those who had silenced his original doubts about Gerald's sanctity.[112] The ties binding the three men were very close. Turpio and Aimo were brothers, and it was Turpio who engaged Odo to write another of his works, *The Collations*. Furthermore, Odo, who is known to have had a pronounced personal devotion to St Martin, Tulle's patron, held the *abbatia* of Tulle immediately after Aimo.[113]

Odo's Life of St Gerald is often treated as the *locus classicus* of the central medieval church's acceptance that laymen, especially those in positions of authority, could demonstrate genuine and deep religious devotion of the sort heavily influenced by monastic role models.[114] As

---

[109] Andrew of Fleury, *Vie de Gauzlin, abbé de Fleury* (ch. 32), ed. and trans. R.-H. Bautier and G. Labory (Sources d'histoire médiévale, 2; Paris, 1969), 72.

[110] *Miracles de saint Benoît* (IV. 6), 181; Andrew of Fleury, *Vie de Gauzlin* (ch. 25), 66–8.

[111] *Cartulaire de l'abbaye de Conques en Rouergue*, ed. G. Desjardins (Paris, 1879), 79, 495, 524; *Liber Miraculorum Sancte Fidis* (I. 10, 30; II. 6), 37–8, 74–6, 109–11; A. Frolow, *La Relique de la Vraie Croix: recherches sur le développement d'un culte* (Archives de l'Orient Chrétien, 7; Paris, 1961), pièces just. nos. 75⁶⁻⁸, 130, 184, 202, 272, 283⁵, 295, pp. 206–8, 230, 259, 265, 295–6, 303, 309.

[112] 'Vita sancti Geraldi Auriliacensis comitis' (pref.), *PL* 133. 639–42. For further Limousin influence on Odo, see ibid. (I. 39), 666.

[113] Id., 'Collationum Libri III', ed. M. Marrier and A. Du Chesne, *Bibliotheca Cluniacensis* (Paris, 1614), 159–61; John of Salerno, 'Vita sancti Odonis' (I. 10), *PL* 133. 48; T. 15 (933), 216 [*recte* 932], 229 (932), 297 (935); A. Poncelet, 'La Plus Ancienne Vie de S. Géraud d'Aurillac', *AB* 14 (1895), 104–6.

[114] D. Baker, '*Vir Dei*: Secular Sanctity in the Early Tenth Century', in G. J. Cuming and D. Baker (edd.), *Popular Belief and Practice* (Studies in Church History, 8; Cambridge, 1972), 41–53; J.-C. Poulin, *L'Idéal de sainteté dans l'Aquitaine carolingienne d'après les sources*

such, the work has attracted the attention of historians of the origins of the Peace of God, the crusades, and the chivalric ethos.[115] It is very difficult, however, to establish clearly what influence Gerald's Life may have had on the religious ideas of arms-bearers in the tenth and eleventh centuries. It has been argued to good effect that Odo directed the Life's message principally at a professed audience as a polemic in favour of monastic reform.[116] But what is true of the work as pure text need not have applied to popular legends and oral traditions which contributed to or developed from it.

There is some evidence that Count Gerald of Aurillac's memory survived in the Limousin. The Bas-Limousin and northern Quercy were on the western fringes of Gerald's area of authority, and much of his military activity was directed against lords from that area.[117] Odo of Cluny believed that Gerald had contacts with Limousin churches, including the Benedictine abbey of Solignac.[118] By the second half of the twelfth century Gerald's connections with the Limousin had become embroidered by legends. Stating that it had been the saint's practice to visit the shrine of St Martial regularly, Geoffrey of Vigeois reported that the canons of St-Étienne, Limoges claimed they owned certain of Gerald's gifts, including a ring. In addition, Gerald had once been waylaid by local *milites* at Puy-de-Grâce (three miles east of Vigeois on the road from the Dordogne to Limoges), and his curse had brought about their ruin; this story (described as a *fama* and not found in the Life) had been handed down over the generations at Vigeois.[119]

---

hagiographiques (750–950) (Travaux du laboratoire d'histoire religieuse de l'université Laval, 1; Quebec, 1975), 81–98, 126–31; F. Lotter, 'Das Idealbild adliger Laienfrömmigkeit in den Anfängen Clunys: Odos Vita des Grafen Gerald von Aurillac', in W. Lourdaux and D. Verhelst (edd.), *Benedictine Culture 750–1050* (Mediaevalia Lovaniensia[1], 11; Louvain, 1983), 76–95.

[115] C. Erdmann, *The Origin of the Idea of Crusade*, trans. M. W. Baldwin and W. Goffart (Princeton, NJ, 1977), 87–9; Becker, *Papst Urban II.*, ii. 282–3; G. Duby, 'The Origins of Knighthood', *The Chivalrous Society*, trans. C. Postan (London, 1977), 166–7; id., *The Three Orders: Feudal Society Imagined*, trans. A. Goldhammer (Chicago, 1980), 97–9; M. H. Keen, *Chivalry* (New Haven, Conn., 1984), 52. For a salutary revision of the Life's importance, see J. Flori, *L'Idéologie du glaive: préhistoire de la chevalerie* (Travaux d'histoire éthico-politique, 43; Geneva, 1983), 109–12.

[116] P. Rousset, 'L'Idéal chevaleresque dans deux *Vitae* clunisiennes', in *Études de civilisation médiévale (IXᵉ-XIIᵉ siècles): mélanges offerts à Edmond-René Labande* (Poitiers, 1974), 628, 631.

[117] Odo of Cluny, 'Vita sancti Geraldi' (I. 35–9), 663–6; A. R. Lewis, 'Count Gerald of Aurillac and Feudalism in South Central France in the Early Tenth Century', *Traditio*, 20 (1964), 49–50, 54.

[118] 'Vita sancti Geraldi' (II. 10, 32), 676–7, 687–8.

[119] 'Chronica' (ch. 20), 289; Aubrun, *Ancien Diocèse*, 144 n. 30. According to a later

Geoffrey of Vigeois also reported that the lords of Malemort in the southern Limousin claimed descent from St Gerald through the female line.[120] It is tempting to link this belief to Adhemar of Chabannes's favourable account of his contemporary, Gauzbert of Malemort. Gauzbert, described as very *ecclesiasticus*, died on pilgrimage to Jerusalem; a cult then developed around him stimulated by miracles (unfortunately Adhemar does not specify where this cult was centred).[121] It is no less tempting to believe that the active religious career of the devout matriarch Engalsias of Malemort in the eleventh century reflected a family tradition of pronounced piety predicated on Gerald's memory and communicated by her marriages to the important Limousin families of Laron and Lastours.[122] Unfortunately, however, there can be no sure means of establishing whether the Aurillac connection influenced the religious careers of the Malemorts in the eleventh century. The documents of two important religious communities near their *caput*, St-Sour, Terrasson and St-Martin, Brive are virtually all lost, making it unlikely that a complete picture of the family's religious benefactions can ever be reconstructed. The family's appearances in the surviving cartularies of the Benedictine communities of the Bas-Limousin do suggest, however, that it was often an active supporter of the church. For example, Peter of Malemort (also known as Peter of Beaufort), who was probably Gauzbert's son, was a benefactor of Uzerche; so too his wife Aimilina, the sister of Viscount Adhemar II of Limoges. One of their sons, Hugh, became an oblate of that abbey, probably at a fairly advanced age, in 1072, and another son, Guy, was a monk there; a daughter, Petronilla, married into the family of the lords of Nontron, the kinsmen of Abbot Gerald of Uzerche (1068–97), and she took the veil at the abbey late in life.[123] Peter's brother

document, Gerald held property near Vigeois, which he gave to St-Étienne, Limoges: SEL 178[156].

[120] 'Chronica' (ch. 3), 280.

[121] Adhemar, *Chronique* (ch. 48), 171. The suggestion of Aubrun, *Ancien Diocèse*, 199, that Adhemar's account of Gauzbert's release from Ebles of Comborn's gaol by a band of peasants reflected Gauzbert's reputation among the people as a holy man, pushes too far the sense of a cryptic passage. But for another popular, albeit transitory, cult centred on a Limousin noble, see Geoffrey of Vigeois, 'Chronica' (ch. 25), 291.

[122] Geoffrey of Vigeois, 'Chronica' (chs. 3, 4, 6), 280, 281; B. 14 (1062 × 1072). Engalsias was the daughter of a man named Hugh. A Hugh appears as the father of a Gauzbert of Malemort in a fragmentary document which is dated by its editor to the mid-11th-C. but which more likely dates from the late 10th: Uz. 281. In 1062 × 1072 Engalsias was old enough to have three great-grandsons (B. 14), so it is quite possible that she was Adhemar of Chabannes's Gauzbert's sister.

[123] Uz. 87 (1073 × 1085), 111 (1072), 293 (1113 × 1133), 473 (1068 × 96: probably post-1084: see T. 196), 790 (1053 × 1067: this notice establishes that Peter's father was named Gauzbert).

Girbert gave two *mansi* to Vigeois around the middle of the eleventh century, and junior members of the kindred gave properties to Tulle, where Peter's daughter Emma took the veil.[124] Helias of Malemort, the nephew of Peter's son Gauzbert, is recorded going to Jerusalem, though it is not clear whether this was on the First Crusade.[125]

In short, the Malemorts show themselves to have been a very pious kindred whose traditions of ties with religious communities were cemented and extended by marriage alliances. Geoffrey of Vigeois wrote in 1183, at a time, that is, when many important families were recording their family trees and pushing them back to a legendary or prestigious ancestor—typically from the later ninth or early tenth centuries.[126] It is therefore impossible to be certain that the Malemorts claimed descent from Gerald of Aurillac as early as the eleventh century, or, if they did, whether they felt it appropriate to mould their pious behaviour accordingly. Adhemar of Chabannes's account of Gauzbert of Malemort may suggest that St Gerald's Limousin connections were remembered in the early eleventh century, but Adhemar was a monk with access to the Life, and he had very strong personal reasons for perpetuating the link because he claimed descent from the kindred of Bishop Turpio and Abbot Aimo.[127] Disappointing as it may be for students of crusade origins, it is, on balance, prudent to treat St Gerald as simply one popular saint amongst many in the eleventh century, not as the focus of pronounced aristocratic piety communicated to particular family groups by memories of kinship.

Shrines did not need to have great regional or international reputations to flourish. St-Sacerdos, Sarlat, fifteen miles south of the Limousin on the borders of Périgord and Quercy, is a good example of a modest cult centre developing around the time of the First Crusade. The Benedictines of Sarlat made a firm decision to enhance the appeal of their patron's cult when, in the first decade of the twelfth century, they invited Hugh of Fleury to revise their unsophisticated Life and to set down some of the miracles associated with St Sacerdos.[128] Sacerdos was believed to have

---

[124] V. 15 (1031 × 1060); T. 2 and 640 (1060 × 1084: Aimeric, possibly the son of Peter's brother Girbert: see Uz. 109), 196 (1084 × 1091).

[125] Uz. 552 (undated). Helias is attested in the early 12th C.: V. 133 (1106 × 1108), 222 (1111 × 1124); T. 252 (1112).

[126] L. Genicot, *Les Généalogies* (Typologie des sources du Moyen Âge occidental, 15; Turnhout, 1975), 19–21.

[127] Adhemar, *Chronique* (chs. 21, 45), 140–1, 167–8; Adhemar of Chabannes et al., 'Commemoratio abbatum', 4.

[128] Hugh of Fleury, *Chronicon, quingentis ab hinc annis et quod excurrit, conscriptum*, ed. B. Rottendorff (Münster, 1638), 127–8; T. Head, *Hagiography and the Cult of Saints: The Diocese of Orléans 800–1200* (Cambridge, 1990), 91–3.

been a contemporary of King Clovis (d. 511) born into a wealthy and noble family from southern Gaul. He became a monk at Calabrum in Quercy and rose to be bishop of Limoges. His relics were translated to Sarlat during the Carolingian period.[129] The miracle stories which were associated with St Sacerdos addressed nobles and knights in two ways: retributive and hortative. For example, a man named Hubert was granted the abbey by a Count William (probably William of Périgueux, 886–918) and exploited its property until he was driven mad by violent visions of Sacerdos. A 'parasite' in the service of a *miles* was possessed by a demon and died after he had robbed a poor couple travelling to Sarlat. A *miles* met a pilgrim returning from Sarlat with whom he had a vendetta, but let him go in peace for fear and reverence of St Sacerdos 'whose awe-inspiring miracles he had often seen and more often heard about', whereas a companion who was prepared to attack the pilgrim was thrown from his horse and died before he could be carried to Sarlat to benefit from the saint's intercession.[130] Instances of Sacerdos helping the devout included the cure of a deaf, dumb, and demented *pauperculus* raised as an act of charity by a nobleman named Ebles. A local castellan, Adhemar of Montigniac, was overwhelmed by a vision of St Sacerdos during Mass in the abbey church, repented of his life spent burdening peasants and waylaying merchants and pilgrims, took the habit, and promptly died; his pious end, it was claimed, moved many to do penance. The lord and inhabitants of a castle near Sarlat were miraculously encouraged to attend the night Office on Christmas Eve at St-Sacerdos, not in the castle's chapel.[131] The miracle stories suggest that Sacerdos's cult was not directed totally at arms-bearers, for there are a number of miracle cures whose subjects are described simply as *puella*, *mulier*, or *pagensis*.[132] But, equally, the hagiographical material surrounding St Sacerdos is good evidence for the ways in which the promoters of a cult might attempt to establish links with the local aristocracy.

Gascony and Compostela

Unlike the Limousin and its neighbouring regions, Gascony had no major cult centre of its own. Although the *Pilgrim Guide* (*c.*1140) recommended that a traveller to Spain on the route from Tours should visit St-Romain,

---

[129] Hugh of Fleury, 'Vita S. Sacerdotis' (chs. 1, 2, 3), 14–15, 16, 17.
[130] Ibid. (ch. 4), 18–19, 19.   [131] Ibid. (ch. 5), 21.   [132] Ibid. 20–1.

Blaye and St-Seurin, Bordeaux, it is significant that, as far as the *Guide's* compiler was concerned, Gascony's chief interest lay in its collection of relics and shrines associated with the legend of Roland: Roland's body itself in the basilica of St-Romain; his horn at St-Seurin; a chapel built at Roncesvalles, scene of his death; the bodies of Oliver, Gandelbald of Frisia, Ogier the Dane, and other of Charlemagne's companions at Belin in the Landes.[133] These sites were suggested by a corpus of legends contained principally in the *Chanson de Roland* and a Latin prose version of the Roland story, the *Pseudo-Turpin*. The *Pseudo-Turpin* comprises the bulk of book IV of the curious hybrid collection of texts, the *Codex Calixtinus*, of which the *Guide* forms book V.[134] The *Pseudo-Turpin* post-dates the First Crusade by two or three decades, and there is no firm evidence that Roland and his companions were venerated as saints in Gascony before 1095.[135] Gascony was not a relic desert of course, and doubtless local cults attracted loyalties as they did throughout Latin Christendom. St Romanus, for example, was an associate of St Martin of Tours known to Gregory of Tours as a patron of sailors in peril; this was still a feature of his cult in the tenth century.[136] St Severinus (Fr. Seurin) was a late antique bishop of Bordeaux also venerated in Gregory of Tours's day.[137] A third Gascon saint of some significance was St Orientius, the patron of the priory of St-Orens, Auch, and the abbey of Larreule near Tarbes in Bigorre (not to be confused with the Béarnais Larreule de Sauvestre, a Benedictine abbey near Orthez dedicated to St Peter). Orientius's cult extended over the Pyrenees.[138] But compared to other

---

[133] *Guide* (ch. 8), 78–80.

[134] The *Codex Calixtinus* is a highly complex work which has generated a great deal of debate and confusion. See P. David, 'Études sur le livre de Saint-Jacques attribué au Pape Calixte II', *Bulletin des études portugaises et de l'Institut Français en Portugal*, 10 (1945), 1–41; 11 (1947), 113–85; 12 (1948), 70–223; 13 (1949), 52–104; C. Hohler, 'A Note on *Jacobus*', *Journal of the Warburg and Courtauld Institutes*, 35 (1972), 31–80; K. Herbers, *Der Jakobuskult des 12. Jahrhunderts und der 'Liber Sancti Jacobi': Studien über das Verhältnis zwischen Religion und Gesellschaft im hohen Mittelalter* (Historische Forschungen, 7; Wiesbaden, 1984), esp. 13–47; A. Moisan, 'Aimeri Picaud de Parthenay et le "Liber Sancti Jacobi"', *Bibliothèque de l'École des Chartes*, 143 (1985), 5–52. See also Fletcher, *Saint James's Catapult*, 304–5.

[135] Hohler, who doubts whether the *Guide* was written with an entirely straight face, argues that its compiler concocted the Roland 'relics', which then appeared at the pertinent places when the work gained currency: 'A Note on *Jacobus*', 33, 50–1.

[136] Gregory of Tours, 'Liber in gloria confessorum' (ch. 45), *MGH Scriptores rerum Merovingicarum*, rev. edn., 1². 325–6; 'Vita S. Romani presbyteri et confessoris', *AB* 5 (1886), 177–91, esp. (ch. 15) 191. See also 'Vita sancti Romani sacerdotis Blaviensis', ed. G. Vielhaber, *AB* 26 (1907), 52–6.

[137] Gregory of Tours, 'Liber in gloria confessorum' (ch. 44), 325.

[138] R. Collins, *The Basques* (Oxford, 1986), 179 n. 88.

areas, and in particular to the 'apostolic' shrines of neighbouring Aquitaine, Gascony's contribution to the sum of cults in Gaul was very modest— Celtic Brittany is a comparable region. The *Guide* is thus indirect but cogent evidence that Gascony, in itself, had little to offer the long- or middle-distance pilgrim.

At first sight, therefore, it would seem that Gascon arms-bearers of the late eleventh century could not have been fully exposed to the traditions and opportunities of pilgrimage enjoyed by their Limousin contemporaries, and that consequently one element of the preconditioning for crusading was absent. In fact this was not the case, for what Gascony lacked in local cult centres was more than compensated by the geographical accident which gave it a unique place in the development of pilgrimage to Santiago de Compostela.

It is very difficult to establish precisely when the shrine of St James in Galicia became widely accepted as the third great pilgrimage centre of Christendom alongside Jerusalem and Rome. The episcopate of Bishop (subsequently Archbishop) Diego Gelmírez of Compostela (1100–40) was an important turning-point because this long-lived and able prelate promoted the interests of his see and its cult with particular vigour and success.[139] But there is also solid evidence that Compostela was already attracting significant numbers of pilgrims in the eleventh century. In a passage celebrating Duke William V of Aquitaine's achievements and piety, for example, Adhemar of Chabannes noted that it was William's habit since his youth to go on annual pilgrimage to Rome or, if this were not practicable, to 'compensate' with a journey to Compostela.[140] Because it involved a similar distance, required the crossing of a high mountain range, and was directed towards the shrine of an Apostle,[141] the journey to Compostela, from a French perspective, had obvious points of comparison with that to Rome. To that extent Compostela's attraction was initially derivative: without the existence of other long-distance pilgrimage traditions, it is unlikely that remote Galicia would have appealed to anyone other than Spaniards. By the second half of the eleventh and the early twelfth centuries, however, as Christian conquests pushed the

---

[139] L. Vázquez de Parga, J. M. Lacarra, and J. Uría Ríu, *Las peregrinaciones a Santiago de Compostela* (Madrid, 1948–9), i. 53–9; Fletcher, *Saint James's Catapult, passim.*

[140] Adhemar, *Chronique* (ch. 41), 163. Fletcher, *Saint James's Catapult*, 93 points out that Adhemar's reference to pilgrimage in the eulogy is his one significant addition to his model, Einhard's encomium of Charlemagne.

[141] For the beliefs surrounding St James the Greater, see T. D. Kendrick, *St James in Spain* (London, 1960), 13–18. See also Matt. 20: 20–8; Mark 10: 35–45; Acts 12: 2. But see Fletcher, *Saint James's Catapult*, 61 n. 17.

Muslim frontier away from the pilgrimage road leading from the Pyrenees, Compostela began to establish its own self-perpetuating traditions. A good example of this process is the journey there of William V's great-grandson Duke William X in 1137. In an incident which made a great impression on contemporaries, William expired on Good Friday in the cathedral, possibly right in front of the high altar.[142]

Other pieces of evidence also suggest that interest in Compostela grew around the latter half of the eleventh century. For example, the train of events which led to the foundation of La Sauve-Majeure in 1079 was set in motion when St Gerard of Corbie joined up with a group of knights from the Laonnais travelling on pilgrimage to Spain; this Spanish connection was possibly one reason why, early in its history, La Sauve sought out properties south of the Pyrenees.[143] One of the first recorded English pilgrims to Compostela, Ansgot of La Haye, founded a priory dependency of La Sauve at Burwell in Lincolnshire (1094 × 1123) in gratitude for the reception which the monks had given him on his return journey.[144] Similarly, the abbey of Charroux acquired important dependencies in north-eastern France because of contacts made with Engelrand of Lillers and Count Baldwin of Guines, who went to Compostela together shortly before 1079 and may have gone again in 1082 × 1084.[145] The twelfth-century collection of miracles of St James which comprises book II of the *Codex Calixtinus* includes an account of a group pilgrimage led by Count Pons of Toulouse (1037–61) and an unnamed brother.[146] A further miracle story features the vassals of Girin of Donzy, a lord from the Lyonnais who flourished around the final quarter of the eleventh

---

[142] *La Chronique de Saint-Maixent 751–1140*, ed. and trans. J. Verdon (Les classiques de l'histoire de France au Moyen Âge, 33; Paris, 1979), 194; 'Sigeberti Gemblacensis Auctarium Laudunense', *MGH SS* 6. 446; *Chronique de Morigny* (III. 2), 65–6; OV vi. 480–2; Suger, *Vie de Louis VI le Gros* (ch. 34), ed. and trans. H. Waquet (Les classiques de l'histoire de France au Moyen Âge, 11; Paris, 1929), 280; Richard the Poitevin, 'Chronicon', extr. *RHGF* 12. 413–14.

[143] See LSM, pp. 385–92.

[144] Ibid. pp. 298, 462; D. W. Lomax, 'The First English Pilgrims to Santiago de Compostela', in H. Mayr-Harting and R. I. Moore (edd.), *Studies in Medieval History Presented to R. H. C. Davis* (London, 1985), 167–8.

[145] *Chartes et documents pour servir à l'histoire de l'abbaye de Charroux*, ed. P. de Monsabert (Archives historiques du Poitou, 39; Poitiers, 1910), 6, 16, 18, 19; Lambert of Ardres, 'Historia comitum Ghisnensium' (ch. 29), *MGH SS* 24. 575–6; G. Chapeau, 'Un pèlerinage noble à Charroux au XIe siècle', *Bulletin de la Société des Antiquaires de l'Ouest³*, 13 (1943), 250–71, esp. 254–5, 265–71.

[146] *Liber Sancti Jacobi* (II. 18), 282–3. For the prototype of the story, see *Memorials of St. Anselm*, 208–9.

century.[147] Other stories in the same collection, involving Lorrainers, Germans, a Poitevin, a Frenchman, a Catalan, and Italians, are expressly dated to 1080, 1090, 1102, 1104, 1105, and the early years of the twelfth century.[148] When Hugh of Die, archbishop of Lyons and papal legate, first heard of Urban II's plans to preach the crusade, he had just completed a pilgrimage to Compostela at the head of a large party of laymen and clergy.[149]

Gascony was exceptionally well placed to experience the full force of the growing pilgrim traffic to Spain because the land routes used by nearly all non-Iberian pilgrims converged to the north of the western Pyrenees. Attention has usually been fixed on the four great itineraries described by the *Pilgrim Guide*: from St-Gilles via Toulouse; from Le Puy via Conques and Moissac; from Vézelay via the Limousin and Périgord; and the road which led from Tours through western Aquitaine and Bordeaux.[150] It should be noted, however, that the *Guide*'s scheme was not intended to be comprehensive. One obvious illustration of this is the fact that passage via La Sauve required a detour from the *Guide*'s Bordeaux–Landes route. What is more, the *Guide* represents a fairly late stage in the development of the pilgrimage routes, which gradually became determined by the creation of hospices and *sauvetés* (areas of special jurisdictional safety) serving as staging-posts for the pilgrim traffic. This process reached maturity by the mid-twelfth century,[151] and even then roads other than the principal four remained in use.[152] Nevertheless, the limited number of convenient passes through the western Pyrenees—the four main routes used only two, Roncesvalles and Somport —effectively channelled pilgrims past or near most of the substantial religious communities and urban centres in southern Gascony. By the final third of the eleventh century Gascony must have been the scene of the greatest concentration of pilgrim traffic in the Latin West with the exceptions of the Spanish stage of the pilgrim road to Galicia and the northern approaches to Rome. Alongside other contacts with Spain such

---

[147] *Liber Sancti Jacobi* (II. 16), 276–8. See also *Memorials of St. Anselm*, 196–200. For Girin, see *Cartulaire de l'abbaye de Savigny suivi du petit cartulaire de l'abbaye d'Ainay*, ed. A. Bernard (Paris, 1853), i. 815, 817, 919.

[148] *Liber Sancti Jacobi* (II. 4–12, 14–15, 22), 265–74, 274–6, 286–7.

[149] Hugh of Flavigny, 'Chronicon', 474.

[150] *Guide* (chs. 1, 7, 8), 2–4, 16–24, 34–80; Vázquez de Parga *et al.*, *Peregrinaciones*, ii. 43–59 and map at 63.

[151] See C. Higounet, 'Les Chemins de Saint-Jacques et les sauvetés de Gascogne', *AM* 63 (1951), 295–300.

[152] Vázquez de Parga *et al.*, *Peregrinaciones*, ii. 59–67.

as trade and migration, pilgrimage broke down Gascony's isolation within Gaul: witness, for example, the flow of pilgrimage enthusiasm back along the pilgrim route from Gascony to Rocamadour. Consequently it is reasonable to suppose that Gascon laymen had ample opportunities to become steeped in pilgrimage, and by 1095–6 they could not have been handicapped by any lack of exposure to religious models and habits which contributed to the response to the crusade.

A Case-Study: St-Léonard, Noblat

An examination of pilgrimage in the eleventh century is apt to be diffuse because cult centres were scattered throughout many parts of the Christian world. It is therefore a valuable exercise to concentrate upon the experiences of one shrine in order to assess how a cult might influence the pious behaviour of the faithful, especially arms-bearers. The cult centre examined here, St-Léonard, Noblat, eleven miles east of Limoges, is particularly interesting because it became intimately associated with crusading ideas very soon after the First Crusade.

St Leonard was a sixth-century recluse who founded a hermitage near Noblat. His cult must have been originally very localized and modest, for virtually nothing is known of it before the eleventh century.[153] The first firm evidence for the growth of Leonard's reputation comes from Adhemar of Chabannes. In a passage of his chronicle placed between a description of the celebration of the Invention of the head of St John the Baptist at Angély (*c.* 1014) and the death of Bishop Gerald of Limoges (1021 or 1022), Adhemar reported that 'at that time' SS Leonard and Antoninus (in Quercy) began to coruscate with miracles, prompting the faithful to flock to their shrines from all quarters.[154] It was not uncommon for cults to appear as if from nowhere and then disappear with equal suddenness.[155] What undoubtedly ensured that St Leonard's popularity would endure was the creation of Jordan of Laron as bishop of Limoges in 1022–3. Jordan had been provost of St-Léonard.[156] This fact lends some weight to a late tradition that after the church of St-Léonard had been destroyed during the Viking invasions it had come into the hands of the lords

---

[153] Aubrun, *Ancien Diocèse*, 107, 325 and n. 37.
[154] Adhemar, *Chronique* (ch. 56), 181.
[155] Ward, *Miracles and the Medieval Mind*, 127–31.
[156] Adhemar, *Chronique* (ch. 57), 182.

of Laron, Jordan's kinsmen.[157] The relationship between Noblat and
the Larons was so close that when, in *c.*1045 × 1051, Bishop Jordan
negotiated with Duke William VII Aigret of Aquitaine the customs by
which bishops of Limoges should be elected, Noblat was one of two
castles whose castellans had the right to be consulted.[158]

Jordan set about promoting the cult of St Leonard early in his
episcopate. A letter of 1025 or 1026 addressed to Fulbert of Chartres by
his friend Hildegar, who was based in Poitiers, included a request made
on Bishop Jordan's behalf that Fulbert find a Life of the saint. Hildegar
implied that the cult of St Leonard was a novelty, for he expressed himself
with reserve, observing that Leonard was only believed to be buried
in the Limousin, and seeming uncertain whether a Life could in fact be
traced.[159] There is no evidence that Fulbert found a copy, if such existed.
Adhemar of Chabannes provides early evidence, however, that Bishop
Jordan persisted in his support of St Leonard. In a sermon recalling the
Council of Limoges in November 1028, which pronounced the Peace of
God, Adhemar wrote that there had been a large gathering of relics from
throughout Aquitaine. For four months after the council the Limousin
suffered very poor weather. Jordan was asked why St Leonard's relics had
not been brought to the assembly, and was told 'by certain persons' that
St Martial was refusing his intercession until St Leonard was paid the
honour he was owed by the clergy and people. In March 1029 St
Leonard's relics were brought to Limoges and processed around the city's
major shrines; the weather improved immediately; and the bishop ordered
a renewal of the Peace of God. Adhemar's account is undoubtedly fanciful
in places—it is unlikely that Bishop Jordan could have been as
disingenuous as he suggests—but much of what he said is credible. As a
keen proponent of the Peace, it was natural that Jordan would seek to
associate St Leonard's cult with the Peace programme. Furthermore,
Adhemar's statement that Jordan went to fetch Leonard's relics from
Noblat 'having brought together a large gathering of the greater-born'
looks like an oblique reference to the bishop's exploitation of his family
connections in order to publicize the cult.[160]

[157] G. Tenant de la Tour, *L'Homme et la terre de Charlemagne à Saint Louis: essai sur les
origines et les caractères d'une féodalité* (Paris, 1943), 327.

[158] *GC* 2, Inst., col. 172 (no. 12).

[159] Fulbert of Chartres, *The Letters and Poems*, ed. and trans. F. Behrends (Oxford, 1976),
no. 114, pp. 204–6.

[160] Adhemar of Chabannes, 'Aus ungedruckten Predigten' (no. 3), ed. E. Sackur, *Die
Cluniacenser in ihrer kirchlichen und allgemeingeschichtlichen Wirksamkeit* (Halle, 1892–4), ii.
482–4.

Adhemar's enthusiasm for St Leonard reveals that, at this early stage in its development, the cult posed no threat to the pre-eminence of St Martial in the Limousin. The community of canons at St-Léonard developed steadily and unspectacularly. In 1062 Bishop Itier of Limoges restored it to what was believed to be its pristine condition and confirmed it in possession of twelve prebends; and Bishop Peter (1100–5) imposed upon the community a regular life based on the Augustinian Rule.[161] By the early twelfth century the number of churches held by St-Léonard was modest but not insignificant.[162] The canons of Noblat do not appear to have attracted to themselves a reputation for particular holiness. It rather became a matter of policy—it is reasonable to suspect Bishop Jordan's initial guiding influence—to associate St Leonard with a certain type of miracle and, consequently, with a particular type of pilgrim, it being recognized that St Leonard freed captives, especially nobles and knights, from their fetters or dungeons. This emphasis is already apparent in the first known Life of St Leonard, which was written in the Limousin some time after *c*.1030 and, quite probably, before Bishop Jordan's death in 1051.[163]

According to this Life, St Leonard had been a Frankish younger contemporary of King Clovis. He renounced the world, journeyed to the Limousin by way of Orleans and Bourges, and founded a hermitage which developed into a small community of religious with the support of Leonard's former royal masters.[164] The Life apppears to have been largely a recasting and elaboration of local legends. With one important exception, to which we shall return, the author made no attempt to graft Leonard's history on to the better-known stories of other Merovingian saints. Nor did he make the sort of extravagent claims which were advanced by the monks of St-Martial for their own patron's role in the early history of Christianity. In fact the author's attitude to St Martial was reverential but restrained; he noted that it was Leonard's habit to visit St-Martial often, but referred to Martial as *pontifex*, not Apostle.[165]

[161] J. Becquet, 'Les Chanoines réguliers en Limousin aux XIe et XIIe siècles', *Analecta Praemonstratensia*, 36 (1960), pièce just. no. 1, pp. 229–31; id., 'Chanoines réguliers en Limousin au XIIe siècle: sanctuaires régularisés et dépendances étrangères', *BSAHL* 101 (1974), 78–9, 80–4.

[162] Aubrun, *Ancien Diocèse*, 191 n. 32.

[163] 'Vita S. Leonardi', in 'Vita et miracula S. Leonardi Nobiliacensia', *AASS* (Nov.) 3. 149–155. See Poncelet's comments, ibid. 139–41, disproving the views on the dating of the Life advanced by F. Arbellot, *Vie de saint Léonard solitaire en Limousin: ses miracles et son culte* (Paris, 1863), 239–42.

[164] 'Vita S. Leonardi' (chs. 1, 3–5, 8–9), 150, 151–2, 153.

[165] Ibid. (ch. 9), 153.

Because the Life was not greatly burdened by the need to correlate its contents with other Lives or historical events, it provides an unusually clear picture of how the stories associated with a saint could be exploited to establish the ground rules for a cult. The Life's various emphases thus reflect what the custodians of St Leonard's shrine around the middle of the eleventh century hoped to achieve for their patron. Some of the Life's themes were commonplaces of medieval hagiography aimed principally at a clerical audience: Leonard's stated purpose to serve God rather than his king, and the simplicity of his fledgling community's lifestyle, with its rejection of worldly riches, are good examples of this.[166] Other elements in the Life, however, clearly related to, by providing some historical basis for, the ways in which St Leonard might appeal to the lay faithful. For example, it was common for hagiographers to refer to their subject's high birth, but the author of Leonard's Life was particularly insistent on this point. According to him, King Clovis was Leonard's kinsman and godfather; Leonard's family enjoyed influence in the royal palace second only to the king, which suggests that they were at least counts; and Leonard could have expected to be given high rank and honour in Clovis's court had he been content to remain a secular cleric.[167] The turning-point in Leonard's life as a religious, moreover, came when a chance encounter with the *familia* of a king (who is unnamed, suggesting that the author was not writing from a copy of Gregory of Tours's *Histories* or other work of Merovingian history) led to the king granting him land on which to build an oratory. In addition, Leonard's counsel to the king that the valuable gifts which he intended to give to the hermit should be distributed instead to 'the poor, widows, orphans, and the suffering' recalls clearly the Peace programme's message to secular princes.[168]

The Life also establishes how St Leonard's specialisms in certain types of miracle originated. No saint, it is true, performed one sort of miracle to the exclusion of all others. The Life noted that, before leaving for Aquitaine and later during his passage through Berry, Leonard evangelized the laity and won renown for curing the sick; he continued his ministry in the Limousin; and at the time of writing, miracle cures for the blind, lepers, and paralytics—examples ultimately based upon thaumaturgic miracles recorded in the Gospels—were being effected at St Leonard's shrine.[169] There were two particular states of distress, however,

[166] Ibid. (chs. 3, 9, 11, 13) 151, 153, 154, 155.
[167] Ibid. (chs. 1, 3) 150, 151.
[168] Ibid. (chs. 5–8) 152–3.
[169] Ibid. (chs. 3, 4, 11, 14) 151–2, 154, 155.

for which St Leonard's intercession was particularly valued. First, according to the Life, he had helped a Frankish queen to survive a painful
childbirth.[170] Secondly, St Leonard interceded for captives. The author of
the Life could not point to a specific episode akin to the queen's labour on
which to base this assertion.[171] It is significant that he seems to have had
difficulty in justifying Leonard's patronage of prisoners, for the means he
used to do so smack of special pleading. He stated that Clovis was
converted by St Remigius—a very well-known story—and added that it
had been this saint's habit to ask the Frankish kings to release all their
captives whenever they passed through Reims; the author claimed that
this was still the practice of the kings of France.[172] In imitation of St
Remigius, whom the author introduced as Leonard's inspiration, Leonard
asked the king to release any prisoner whom he might visit. His request
granted, Leonard busied himself touring all the gaols he could find as a
witness to Christ's sayings 'I was in prison, and ye came unto me'
(Matthew 25: 36), and 'Inasmuch as ye have done it unto one of the least
of these my brethren, ye have done it unto me' (Matthew 25: 40).[173] In
this instance the author may not have been working simply from old
legends, for he seems to have extrapolated a tenuous connection from the
fact that Leonard's oratory was believed to have been dedicated by him
to Our Lady and St Remigius.[174] The author's convoluted argument
suggests that St Leonard's patronage of captives had emerged only
recently. Yet, equally, the Life's description of the saint's help to the
prisoners was placed in an important position, between the accounts of
his birth and family connections and his decision to quit the world,
which suggests that this recalled the facet of the cult which had quickly
achieved prominence.

The emphasis upon St Leonard the friend of captives is further evident
in the earliest surviving collection of his miracles, which was composed

---

[170] Ibid. (chs. 5–8) 152–3.

[171] See ibid. (ch. 11) 154, where Leonard's support of captives during his lifetime is
cited simply as a particular function of his sanctity.

[172] Ibid. (ch. 2), 150. According to Poncelet (ibid. n. 5), the corpus of legends
concerning St Remigius makes no mention of this arrangement with the Merovingian kings.
It is possible, however, that the author was aware of a famous letter written by Remigius to
Clovis which listed the duties and qualities of a good king. Towards the end of the letter
Remigius urged: 'Paternas quascumque opes possides, captiuos exinde liberabis, et a iugo
seruitis absolues': Remigius of Reims, 'Epistolae ad Chlodoueum Regem, et alios', ed. A.
Du Chesne, *Historiae Francorum scriptores coaetani* (Paris, 1636–49), i. 849.

[173] 'Vita S. Leonardi' (ch. 2), 150–1.

[174] Ibid. (ch. 14), 155.

before 1106 × 1111, probably by the author of the Life.[175] It contains accounts of eight miracles or groups of miracles, of which seven involve the freeing of prisoners. The exception (ch. 6) resumes the theme of Leonard's thaumaturgic powers found in the Life by describing how the saint's relics effected cures of a blind boy, a possessed woman, and a deaf man at the large gathering of churchmen and relics held to celebrate the Invention of the head of St John the Baptist at Angély in *c*.1014.[176] It was noted earlier that Adhemar of Chabannes believed that Leonard's cult blossomed around this time, so the presence of the relics at Angély is perfectly possible. The purpose of this story was to establish that Leonard was worthy to rank among the major saints of Aquitaine. His custodians could not, therefore, afford to ignore his attraction as a thaumaturge.

Predominantly, however, St Leonard appeared in the miracle stories as a friend of those who had been thrown into prison or clapped in irons. At this stage the cult involved men of knightly or noble rank in two ways. First, the beneficiary of the miracle who exhibited devotion to St Leonard might be of low status, in which case arms-bearers were typically cast as the villains of the piece. For example, an inhabitant of Noblat, rich enough to be held to ransom for 1,000*s*. but described only as 'a certain man', was captured by 'a most unjust tyrant', a castellan. The background to this episode, it was stated, was endemic local warfare 'as *milites* from this diocese [Limoges] squabbled amongst themselves'. Various stratagems devised by the castellan to secure his prisoner failed: St Leonard overturned the guard-house which had been built over the captive's pit, and the *milites* ordered to watch over him were thrown into confusion.[177] In another story a pilgrim returning from St-Léonard was captured and held to ransom by a lord based at a castle in the Auvergne. Warned by St Leonard in a vision to release his prisoner, the lord consulted with his *milites* and decided to do nothing. When a second vision was ignored, St Leonard intervened directly, leading the captive to safety and causing the castle to collapse suddenly. Many died horribly, but the lord was spared with both his legs broken so that his suffering should be prolonged.[178] In other accounts St Leonard appeared as a dove to a *servus* held in a pit and led him to safety, and helped a prisoner to escape from a castle situated near the Rhône.[179] On another occasion, a Poitevin peasant was waylaid by robber-knights who subjected him to

---

[175] 'Liber prior miraculorum', in 'Vita et miracula S. Leonardi Nobiliacensia', *AASS* (Nov.) 3. 155–9. See Poncelet's comments, ibid. 140.

[176] Ibid. 158.   [177] Ibid. (ch. 3), 156–7.   [178] Ibid. (ch. 4), 157.

[179] Ibid. (chs. 7–8), 158–9.

torture in order to extort money. The peasant invoked God, St Leonard, and St Martial (a further sign of the good relations between the two Limousin cult centres) and was miraculously freed; one of the foot-chains used on him he gave as an offering at St-Martial, the other at Noblat. In this instance the author of the miracle story was particularly hostile to the motives and methods of the captors, who were described as 'very faithless *milites*' and 'sons of the Devil' who lurked in woods to indulge in rapine and theft.[180]

These stories demonstrate that part of the appeal of Leonard's cult was generated by an attempt to discipline arms-bearers' conduct. This is not to say, however, that the behaviour of lords and knights was considered invariably wicked, as another case involving a humble beneficiary illustrates. A *servus* of St-Léonard was arrested by an unnamed viscount of Limoges and cruelly bound by the throat to a stake which the viscount had caused to be erected in his castle near the abbey of St-Martial. The device seems to have had some local notoriety: it was known as the *maura*, the 'Black Lady', and the author of the miracle implied that it was in frequent use. Nevertheless, the author did not fully object to the viscount's right to use such rough methods. He observed that, as a prince, the viscount exercised 'legal' authority over the people in the city, and that the stake had been put up 'to terrify and to punish the criminals'; as a means of cowing the local populace, he noted, it was very effective. The *servus* was freed by St Leonard's intervention principally because he was innocent of the crime imputed to him, and only indirectly in order to ensure that the device was not used again. Moreover, when news of the miracle spread a large crowd processed from Limoges to Noblat to give praise to St Leonard, but there is no record that divine vengeance was visited upon the viscount.[181]

Another story from the first collection of miracles further suggests that St Leonard's cult was not developed simply as an expression of protest against arms-bearers' violence, for here the beneficiary was himself of high social status. An *eques* from Nantes, finding himself chained in prison, invoked St Leonard, whereupon the saint secured his freedom.[182] The preface to the first miracle collection also makes it clear that, already in the eleventh century, the cult of St Leonard was appealing to arms-bearers

---

[180] Ibid. (ch. 9), 159. See also the reference in the Life to 'saecularibus rapinis': 'Vita S. Leonardi' (ch. 11), 154.

[181] 'Liber prior miraculorum' (ch. 2), 156 (amending 'delinquentiam' to 'delinquentium').

[182] Ibid. (ch. 5), 157–8.

not only as contrite villains but also as direct beneficiaries. Its author praised St Leonard for daily working miracles, as his shrine bore witness: 'For there hang in his basilica several iron chains, and the swords and lances of many whom he has freed from the fury of war.'[183] These were weapons closely associated with combat between mounted warriors. The first miracle collection does not mention St Leonard as a battle-helper, but this must have been an emerging feature of his cult as nobles and knights assimilated the dangers of combat and incarceration.

There is good evidence that by the first decade of the twelfth century at the very latest the process of directing St Leonard's cult at arms-bearers was reaching maturity. The cult received an important stimulus in this period with the arrival at Noblat in early 1106 of Bohemond, prince of Antioch, who had come to France to publicize his plans for a crusade. The sequence of events leading up to his visit had been remarkable. In the summer of 1100, on an ambitious long-distance campaign towards Melitene, Bohemond had been captured by the Turkish emir Danishmend and imprisoned at Niksar, deep in Muslim territory near the Black Sea.[184] After nearly three years' captivity he was freed after promising a large ransom and an alliance with Danishmend against the latter's rival, Kilij Arslan.[185] When Bohemond later went to France in 1106 he made a point of visiting Noblat, claiming that it was St Leonard's intercession which had made his release possible.

There survives a twelfth-century collection of seven of St Leonard's miracles.[186] The second miracle story in the collection, the longest, is an elaborate account of Bohemond's captivity and liberation.[187] It was long believed that this miracle was composed for the canons of St-Léonard by Bishop Walram of Naumburg, in Saxony, during a stay he made at

---

[183] Ibid. (ch. 1), 156.

[184] Fulcher of Chartres, *Historia Hierosolymitana* (1095–1127) (I. 35), ed. H. Hagenmeyer (Heidelberg, 1913), 344–8; Ralph of Caen, 'Gesta Tancredi in expeditione Hierosolymitana' (ch. 141), *RHC Occ.* 3. 704–5; Albert of Aachen, 'Historia Hierosolymitana' (VII. 27–8), *RHC Occ.* 4. 524–5; OV v. 354–6; Matthew of Edessa, 'Récit de la Première Croisade' (ch. 18), *RHC Documents arméniens*, 1. 51–3. For later criticism of Bohemond's rashness, William of Tyre, *Chronicon* (IX. 21), ed. R. B. C. Huygens (Corpus Christianorum, Continuatio Mediaeualis, 63/63A; Turnhout, 1986), i. 447–8.

[185] Guibert of Nogent, 'Gesta Dei per Francos' (VII. 37), *RHC Occ.* 4. 254; Ralph of Caen, 'Gesta Tancredi' (ch. 147), 709; Albert of Aachen, 'Historia Hierosolymitana' (IX. 35–6), 611–13; Matthew of Edessa, 'Récit' (ch. 28), 69–70.

[186] 'Liber alter miraculorum', in 'Vita et miracula S. Leonardi Nobiliacensia', *AASS* (Nov.) 3. 159–73.

[187] Ibid. (ch. 2), 160–8.

Noblat sometime between 1106 and his death in 1111. Walram was a man of letters who had written anti-Gregorian tracts at the height of the Investiture Contest but then transferred his loyalty to the pro-papal party within the German church, possibly after King Henry IV's death in 1106.[188] His visit to Noblat certainly stimulated interest in the cult of St Leonard, for he produced a revised version of the Life (now incomplete) with three appended miracles, which he presented with some relics to a Saxon noblewoman.[189] It has been effectively argued, however, that Walram only transcribed the Bohemond legend from material he found at Noblat.[190] The Bohemond story in the collection of seven miracles must therefore have been composed very soon after Bohemond's visit to Noblat, which means that it may be treated as close to the version of events which Bohemond himself recounted to the canons there. Furthermore, Walram's three miracles are based upon information contained in the first, second, and sixth stories in the collection of seven (in the case of the sixth story, it and Walram's account may have drawn on a common source), which means that the latter collection was at least predominantly based upon stories being told at Noblat before 1111.[191]

The first of the seven stories provides important evidence that Bohemond's devotion to St Leonard in 1106 owed something to the general popularity of the cult and not just his immediate wish for publicity. This account concerns a noble Norman, referred to only as Richard, who lapsed into arrogance and was punished by being captured by Muslims during or shortly after the First Crusade. He was then ransomed by the Byzantine emperor Alexius I and held captive at Constantinople until St Leonard (to whose protection Richard had committed himself, and whose shrine he had vowed to visit) warned the emperor in visions that he should let his prisoner go.[192] The identity of this Richard is vaguely expressed in the story, but it is almost certain that he was none other than Bohemond's kinsman Richard of the Principality, prince of

---

[188] On Walram's literary output see P. Ewald, *Walram von Naumburg: Zur Geschichte der publicistischen Literatur des XI. Jahrhunderts* (Bonn, 1874); I. S. Robinson, *Authority and Resistance in the Investiture Contest: The Polemical Literature of the Late Eleventh Century* (Manchester, 1978), 93–4, 103–4, 123. See also the comments of W. Schwenkenbecher in his introd. to 'Liber de unitate ecclesiae conservanda', *MGH Libelli de lite*, 2. 180–1.

[189] Walram of Naumburg, 'Vita et miracula S. Leonardi', *AASS* (Nov.) 3. 173–82.

[190] A. Poncelet, 'Boémond et S. Léonard', *AB* 31 (1912), 30–3, revising his comments in *AASS* (Nov.) 3. 141–2.

[191] Walram of Naumburg, 'Vita et miracula S. Leonardi', 176–82; Poncelet, 'Boémond et S. Léonard', 33.

[192] 'Liber alter miraculorum' (ch. 1), 159–60.

Salerno. Richard had been captured alongside Bohemond and probably imprisoned with him, at least at first.[193] Richard's fate thereafter is not wholly clear. Orderic Vitalis, in his account of Bohemond's incarceration, stated that Richard remained with him throughout his captivity until an alliance was agreed with Danishmend, and then led a mission to Antioch to collect a force which would escort Bohemond back into Latin territory.[194] But Matthew of Edessa, whose testimony is independent of the Noblat legends and Orderic, believed that Richard was ransomed by Danishmend to Alexius I in an agreement separate from, and apparently earlier than, the negotiations which led to Bohemond's release.[195] Matthew's version of events may be confirmed by Orderic's belief that Bohemond, shortly after his own liberation, sent Richard ahead to France to bear his silver shackles to Noblat as an *ex voto* offering.[196] The first miracle story in the collection of seven also suggests that the canons of Noblat learned of Richard's release before they received firsthand news of Bohemond's experiences. It made no mention of Bohemond. Significantly, too, its tone was anti-Byzantine but not as virulently so as Bohemond's miracle story, for it recognized that the Greeks did have a fair grievance against the Normans in the Levant.[197] The story also reveals how the crusading vocation could easily be assimilated with devotion to a western saint. Richard prayed to St Leonard in prison by stating that he was the saint's and Christ's *miles*, who had journeyed to serve the Lord and to defend Christians. St Leonard, moreover, told Alexius I in one vision that Richard had come to the East to protect *christianitas* and risk death for Christ's sake.[198]

The second miracle in the collection of seven, that concerning Bohemond, is much more intricately constructed, though it also resumes themes found in Richard's story. Just as Richard had been defeated because of his pride, so St Leonard told Bohemond that his captivity was punishment for his sins. On the other hand, the main reasons advanced for Bohemond's capture are the sheer weight of the Turkish numbers when he was overwhelmed and their rather devious tactics in battle (which are described with some care, suggesting a knowledgeable

---

[193] Albert of Aachen, 'Historia Hierosolymitana' (VII. 28), 525; Matthew of Edessa, 'Récit' (ch. 18), 51–2.
[194] OV v. 354, 372, 374–6.
[195] 'Récit' (ch. 28), 70.
[196] OV v. 376.
[197] 'Liber alter miraculorum' (ch. 1), 160; cf. ibid. (ch. 2), 164. It is noteworthy that in 1106 Bohemond was proposing a campaign against the Greeks.
[198] Ibid. (ch. 1), 160; cf. ibid. (ch. 2), 167.

informant).[199] According to this version of events, St Leonard encouraged Danishmend's wife in a vision to work for Bohemond's release, and Danishmend was finally convinced to let him go when St Leonard appeared to him also.[200] Bohemond emerges from the story in a very favourable light, for his warlike qualities and his zeal for the Christian faith are both vigorously celebrated.[201] The account also emphasizes the part played by the canons of Noblat, who prayed daily for Bohemond's release and processed barefoot to the high altar every Friday. The proof that their intercession was effective was supplied by a vision in which St Leonard appeared to Richard and ordered him to visit Bohemond with reassurances that help was at hand.[202] (This is interesting evidence that efforts were made to integrate the Richard story within the later and more famous miracle.) A combination of the fabulous and the plausible—the account, for example, of Bohemond's convoluted treaty negotiations with Danishmend has some realistic elements[203]—the miracle story occupies a mid-point between the factual descriptions of Bohemond's release found in some chronicles and the heavily embroidered legend recorded by Orderic Vitalis.

Orderic wrote his version of events some time later (*c.*1135), but he seems to have attended Bohemond's marriage to Constance, the daughter of King Philip of France, at Chartres in 1106 (no more than three months after Bohemond's visit to Noblat), where he heard stories of the captivity.[204] His long account has some important similarities with the Noblat legend. For example, it was set against the background of Danishmend's rivalry with Kilij Arslan; it celebrated the Franks' qualities of martial prowess and uprightness; and it stressed the important role played by a Muslim noblewoman, in this instance Danishmend's daughter Melaz, who was prepared to help Bohemond because she secretly wished to convert to Christianity.[205] Orderic's account, however, also contains elements closer to *topoi* of vernacular literature. Bohemond's men, he wrote, seized Danishmend's person and fortress by a cunning ruse after returning from a battle in which Bohemond had distinguished himself in single combat; and, when safely back in Christian territory, Bohemond married Melaz, now baptized, to Roger of Salerno (who was Richard of the Principality's son).[206] It is reasonable to suppose that Orderic's version of events preserves stories put about by Bohemond for knightly

[199] Ibid. (ch. 2), 161–2, 165.   [200] Ibid. 163, 166–7.
[201] Ibid. 164–5, 166–7, 168.   [202] Ibid. 163–4.
[203] Ibid. 167–8.   [204] OV, vol. v, pp. xi–xii, xviii–xix; vi. 70.
[205] Ibid. v. 358–72, 376–8.   [206] Ibid. 364–6, 376–8.

consumption as he tried to generate interest in his new crusade plans (the stories possibly being amplified in vernacular songs).[207] It is therefore interesting to note that St Leonard still features quite prominently in Orderic's account. Bohemond, he believed, regretfully refused to marry Melaz himself, giving as one excuse the vow which had been made in prison to visit Noblat; and in another part of his *Ecclesiastical History* he observed that Bohemond's principal purpose in coming to France in 1106 was to perform that vow.[208] Similarly, Ralph of Caen and William of Malmesbury, two writers well placed to be exposed to the sort of Norman stories which influenced Orderic (Ralph was in Antioch soon after 1106), recorded that St Leonard had been responsible for Bohemond's release, and hinted that Bohemond's vow dated from his time in prison.[209]

Patently, then, Bohemond thought it worth while to publicize his links with St Leonard both when he appealed to the pride of the canons of Noblat and when he tried to stimulate interest in crusading among arms-bearers. It is noteworthy that the guardians of Leonard's shrine reciprocated Bohemond's enthusiasm by celebrating his crusade vocation and assimilating it with their cult. One passage in Bohemond's miracle story lauded the crusade's victories at Nicaea, Antioch, Jerusalem, and Ascalon as God's miracle operated through his few servants against the many.[210] Before his capture, it was recorded, Bohemond had encouraged his troops by reminding them that they were fighting 'the Lord's wars' for faith and justice; St Michael would help them in battle and lead the fallen to Paradise (the emphasis on this saint suggesting further that the account came from a Norman); and it was better to die for the Lord than to survive, since death earned eternal blessedness. In addition, St Leonard told Danishmend in a vision that 'Bohemond is Christ's and my *miles*'.[211] Ideas such as these, asserting the importance of divine favour shown to the crusaders and magnifying the spiritual rewards which they could receive, developed during the course of the First Crusade itself.[212] The

[207] See ibid. vi. 68–72.
[208] Ibid. v. 378; vi. 68.
[209] Ralph of Caen, 'Gesta Tancedi' (ch. 152), 713; William of Malmesbury, *De Gesti. Regum Anglorum Libri Quinque* (ch. 387), ed. W. Stubbs (RS 90; London, 1887–9), ii 454. See also 'Historia peregrinorum euntium Jerusolymam' (ch. 140), *RHC Occ.* 3. 228. But cf. Guibert of Nogent, 'Gesta Dei per Francos' (VII. 37), 254; Albert of Aachen, 'Historia Hierosolymitana' (IX. 35–6, 47), 611–13, 620.
[210] 'Liber alter miraculorum' (ch. 2), 164.
[211] Ibid. 162, 167.
[212] Riley-Smith, *First Crusade*, 107, 111, 114–19. But cf. H. E. J. Cowdrey, 'Martyrdom and the First Crusade', in P. W. Edbury (ed.), *Crusade and Settlement: Paper. read at the First Conference of the Society for the Study of the Crusades and the Latin East and*

fact that they were received so readily by the canons of Noblat into the corpus of stories surrounding their patron saint demonstrates how crusading enthusiasm and local cults could feed off each other.

There is further evidence that St Leonard's cult benefited from being closely identified with the concerns of arms-bearers. The first miracle which Walram of Naumburg appended to his Life was either based on the sixth miracle in the collection of seven or drew on a common source.[213] This sixth miracle deals with a brutal private war (*guerra*) in the diocese of Viviers (Aps). A well-born *miles* whose life was in danger became terrified at the prospect of God's judgement because of his many sins. His priest advised him to submit himself to the lordship of St Leonard, whose power no constraints or prisons could withstand; the saint, he was assured, would protect him during ambushes and, if necessary, in prison. Later the *miles* dreamt that he had been captured and imprisoned, prompting him to reaffirm his devotion to St Leonard. In due course he was indeed taken prisoner, but St Leonard miraculously opened the pit in which he was held, and procured his enemy's horse with which to make good his escape. The return of the horse to its owner ended the feud.[214]

Most of the remaining miracles in the twelfth-century collection similarly reasserted St Leonard's patronage of arms-bearers. Two *milites* from the Rouergue escaped miraculously from castles; and a vendetta between two powerful nobles from the Auvergne led to the capture and torture of one by the other, whereupon St Leonard's intervention resulted in the conclusion of peace, again confirmed, interestingly, by the exchange of horses.[215] Only the final surviving miracle, which concerns a rich burgess of Noblat who was cured of leprosy by St Leonard's intercession after surrendering all his possessions and becoming a hermit, points to a broader social range of devotees to the saint as a thaumaturge (and even then very locally).[216] At the end of our period the compiler of the *Pilgrim Guide* devoted a long passage to St-Léonard, which he

presented to R. C. Smail (Cardiff, 1985), 49–53; J. Flori, 'Mort et martyre des guerriers vers 1100: l'exemple de la Première Croisade', *CCM* 34 (1991), 121–39.

213 Poncelet, 'Boémond et S. Léonard', 33.
214 'Liber alter miraculorum' (ch. 3), 169.
215 Ibid. (chs. 4–6), 169–72.
216 Ibid. (ch. 7), 172–3. In the absence of a concluding passage after the 7th miracle, it is possible that the collection was originally longer. The final miracle's author noted that the events he described took place 'in nostris temporibus'. The story's *terminus ad quem* is 1183, by which time Geoffrey of Vigeois had probably seen what is now the surviving MS of the collection: 'Chronica' (ch. 33), 297 (which also records that Bohemond visited St-Martial in 1106). See Poncelet, 'Boémond et S. Léonard', 31–2.

had almost certainly visited. St Leonard was referred to as 'Leonardus Lemovicensis', which was possibly meant as a snub to St-Martial, which the compiler studiously ignored, but is also an indication of Leonard's close identification with his region. The compiler claimed that St Leonard's fame as a friend of captives extended world-wide; and he described what must have been a remarkable scene in the church at Noblat, where 'thousands' of chains, torture instruments, and items of weaponry hung on all sides from walls and wooden poles especially erected for the purpose. He also mentioned Bohemond by name as one of the most noted beneficiaries of the saint's help—proof of the enduring importance to the canons of Noblat of Bohemond's story and the shrine's association with crusading.[217]

It is important to note that St Leonard's patronage of captives was not unique. Other saints were also believed to tender help to prisoners.[218] The means chosen to promote Leonard's appeal simply accentuated one of the many benefits—thaumaturgic, protective, retributive, or dispensatory—which the faithful expected from saints' intercession. Leonard's popularity is, moreover, a particularly clear example of a general trend evident around the time of the First Crusade to give encouragement to arms-bearers to involve themselves in devotion to saints. Using the Limousin and Gascony as models, it has been seen that accidents of geography and early Christian history created regional variations in the concentration of important shrines, the appeal of certain saints, and the rhythms of pilgrimaging (the Limousin experience probably being more representative of Gaul as a whole). The two regions vary, however, only to the extent that they reveal differing facets of a general enthusiasm for pilgrimage. Pilgrimage's appeal extended from devotion to minor local cults to the attraction of journeying great distances to famous shrines. Consequently it was a normative feature of religious life by the eleventh century, expressed through language, rituals, and customs with which most of the faithful would have been perfectly familiar. Far from being an esoteric devotional practice sustained by its own self-contained traditions and motivations, pilgrimage in all its forms was accommodated and

---

[217] *Guide* (ch. 8), 52–6.

[218] e.g. *Liber Sancti Jacobi* (II. 1, 11, 14, 20, 22), 261–2, 273, 274–5 (chains hanging in church), 285–6, 286–7; *Guide* (ch. 8), 76 (St-Eutrope, Saintes); *Miracles de saint Benoît* (IX. 1), 359–61; OV iii. 346–60. See also Sigal, *Homme*, 268–70.

stimulated by laymen's contacts with religious communities; and it was informed by the same preoccupations with sin, punishment, and Judgement. Pilgrimage therefore helps us to locate the appeal of crusading in the habitual and commonplace behavioural routines of pious arms-bearers.

6

The Response of Laymen to the First Crusade

THIS study has comprised two main sections. The former has attempted to establish that efforts to trace the origins of the crusade through ideas associated with the Peace of God and the Spanish *Reconquista* bear insufficient relation to the realities of nobles' and knights' lives in later eleventh-century south-western France. The latter section, in contrast, has suggested that in the regions under study many arms-bearers were fully and intimately involved with religious communities, and that their enthusiasm stimulated ideas of recruitment, spiritual association, benefaction, and pilgrimage which had resonances in crusading thought. It remains to examine how the crusade message was put to the laity and how eagerly it was received in order to establish whether the impressive body of circumstantial evidence linking lay–religious contacts and the crusade can be supplemented by specific examples.

Problems of Evidence and Chronology

The response of the laity in the Bas-Limousin and Gascony to the appeal of the First Crusade must be understood in the light of how the crusade was preached. After formally launching the crusade at the Council of Clermont on 27 November 1095, Pope Urban II embarked on a journey through French-speaking lands which was to occupy him for the following nine months. It would be misleading to regard Urban's travels solely as a preaching tour, for throughout the pope and his party pursued church reform, issued privileges, and consecrated various ecclesiastical sites.[1] Yet there can also be no doubt that the preaching of the cross and the organization of the crusade occupied a good deal of the papal energy.

[1] R. Crozet, 'Le Voyage d'Urbain II en France (1095–1096) et son importance au point de vue archéologique', *AM* 49 (1937), 42–69; id., 'Le Voyage d'Urbain II et ses négociations avec le clergé de France (1095–1096)', *Revue historique*, 179 (1937), 271–310; A. Becker, *Papst Urban II. (1088–1099)* (MGH Schriften, 19; Stuttgart, 1964–88), ii. 435–57 and map at end of volume.

Before the pope's impact in France in 1095–6 is considered in detail, however, two preliminary points should be made: one concerns the problems of the evidence; the other the relationship between the 1096 and 1101 expeditions of the First Crusade. First, it must be emphasized that, although observers were struck by the large numbers and wide provenance of those who responded to the crusade appeal,[2] our knowledge of identifiable individuals who went to the East is limited by the nature of both the narrative and documentary sources. This is particularly so with respect to south-western French crusaders. No event in the central medieval period spawned as great a corpus of narrative material as did the First Crusade, but most of the accounts of the expedition were written by northern Frenchmen—the Norman author of the *Gesta Francorum* who came from southern Italy should be included in this category—and reflected both conscious regional bias and variations in the availability of information about crusaders from different areas. For example, our knowledge of the performance during the latter stages of the 1096–9 expedition of southerners such as Raymond Pilet, William of Montpellier, and the Limousins Raymond of Turenne and Gulpher of Lastours is a result of the fortuitous decision made in late 1098 by the author of the *Gesta Francorum* to abandon the following of Bohemond of Taranto, to whom he had previously been very loyal, and to join Raymond of St-Gilles, the crusade leader who was most sympathetic to popular agitation to press on towards Jerusalem without delay.[3] Among the other early narrative authors it might be expected that Peter Tudebode, who probably came from Civray near Charroux, would have been well placed to note the contribution of crusaders from Aquitaine. In fact the amount of additional information he provides is disappointing. Recently the established view that Tudebode's work was mostly a slavish reworking of the *Gesta Francorum* has been challenged, and the problem has yet to be resolved satisfactorily.[4] For our purposes, however, the question of the relationship between the two texts, and thus of Tudebode's originality, is

---

[2] e.g. Ekkehard of Aura, 'Hierosolymita' (ch. 6), *RHC Occ.* 5. 15–16; Hugh of Fleury, 'Liber qui modernorum regum Francorum continet actus', *MGH SS* 9. 393; Bernold of St Blasien, 'Chronicon', *MGH SS* 5. 464, 466; Albert of Aachen, 'Historia Hierosolymitana' (I. 2, 5), *RHC Occ.* 4. 272, 274; 'Anonymi Florinensis brevis narratio belli sacri', *RHC Occ.* 5. 371; 'Annales S. Dionysii Remenses', *MGH SS* 13. 83.

[3] *Gesta Francorum et aliorum Hierosolimitanorum*, ed. and trans. R. Hill (London, 1962), pp. xi–xii, xiii, xxxvi, 78–9, 83–4, 87–9; but cf. ibid. 26.

[4] Peter Tudebode, *Historia de Hierosolymitano Itinere*, ed. J. H. and L. L. Hill (Documents relatifs à l'histoire des croisades, 12; Paris, 1977), 8–12, 16–24. For Tudebode's origins, ibid. 12–13.

less important than the fact that his specifically Aquitanian or southern French contribution to the narrative tradition was modest.[5] The history of the crusade by the Auvergnat eye-witness Raymond of Aguilers was not independent of the *Gesta*/Tudebode tradition but added some important details about Raymond of St-Gilles's crusade. Unfortunately, Raymond of Aguilers, an impressionable priest on the fringes of Raymond of St-Gilles's entourage, was temperamentally suited to describing visions and miracles rather than cataloguing names.[6] Given the problems which northern writers could experience in finding information about southerners—Guibert of Nogent, for example, was very confused about as important a prince as Gaston of Béarn and confessed to finding it difficult to come by even rudimentary information about Adhemar of Le Puy, the papal legate on the expedition[7]—and given too the evidence that there was north–south hostility within the crusade army exacerbated by linguistic and cultural differences,[8] it is not surprising that southern, and particularly south-western, French crusaders were poorly served in the narrative sources.

This problem is compounded by the limitations of the surviving documentary evidence. Many charters which recorded the names of crusaders must now be lost. Others, documenting pledges which were redeemed, soon lost their evidential value and were not preserved. Still more charters may have been abridged by later cartulary scribes who excised untopical references to the crusade. Some of the difficulties involved in identifying crusaders are illustrated by the case of Viscount Adhemar III of Limoges. Many pieces of circumstantial evidence suggest that he was the sort of man who could have gone on crusade. There were powerful traditions of support of reformed religious communities within his kindred.[9] His great-grandfather Guy I and his grandfather Adhemar I had been on pilgrimage to Jerusalem; Adhemar III himself exhibited

---

[5] See ibid. 32, 78, 97, 110, 116, 123, 129.

[6] Raymond of Aguilers, *Le 'Liber'*, ed. J. H. and L. L. Hill (Documents relatifs à l'histoire des croisades, 9; Paris, 1969), 10, 12–20.

[7] 'Gesta Dei per Francos' (preface; II. 5, 13; VII. 8), *RHC Occ.* 4. 121, 140, 148, 228.

[8] Ibid. (II. 18), 150; Ralph of Caen, 'Gesta Tancredi in expeditione Hierosolymitana' (chs. 15, 61, 99, 109), *RHC Occ.* 3. 617, 651, 676, 682–3. For criticism of southerners on the 1101 expedition, Albert of Aachen, 'Historia Hierosolymitana' (VIII. 11), 566. For northerners' limited range of terms for southerners, see Robert the Monk, 'Historia Iherosolimitana' (VII. 9), *RHC Occ.* 3. 828.

[9] Mart. I. 29 (1015 × 1036); C. de Lasteyrie, *L'Abbaye de Saint-Martial de Limoges: étude historique, économique et archéologique, précédée de recherches nouvelles sur la vie du saint* (Paris, 1901), pièce just. no. 6, pp. 426–7 (1062); Uz. 43 (1068), 460 (1048); V. 101 (1092 × 1110).

interest in pilgrimage after the First Crusade; and his grandson Guy IV was to die on the Second Crusade having earlier exploited his family's close ties with St-Martial, Limoges in order to raise funding for his journey.[10] In 1095–1101 Adhemar III was not old (he died at Cluny in 1139).[11] Furthermore, he most probably met Pope Urban during the latter's long stay at Limoges a month after the crusade was launched, and it is reasonable to suppose that he was courted with the same combination of ceremonial honour and deference to family prestige with which the pope flattered Count Fulk of Anjou a few weeks later.[12] (Fulk did not go on the crusade, however.) Adhemar does not appear among the names of local crusaders supplied by the Limousin chronicler Geoffrey of Vigeois, who wrote many years later but used documentary material as sources, was meticulous in recording the careers of his region's leading families, and took a keen interest in crusading history and the Latin East.[13] Another late source (*c.*1180), but one based on the account of a participant on the First Crusade, records the presence of 'li quens de Limoges' in the crusade army at Nicaea in 1097.[14] No known charter reference, however, makes any mention of Adhemar's participation on the crusade. Without such a document, or a more secure reference in an early narrative source, the cryptic mention of the count of Limoges cannot establish that Adhemar went to the East.

Another case demonstrates that even apparently solid documentary evidence can be tantalizingly obscure. In about 1096 Ramnulf of Courcelles, in the presence of his three eldest sons, granted half of a mill to the abbey of St-Jean d'Angély, in the Saintonge, for the large sum of 12*l.* (which represented a released debt) and a measure of corn. This was more than a straightforward cash transaction, however, for Ramnulf also offered a son as an oblate. Pope Urban visited St-Jean in April 1096,

---

[10] Adhemar, *Chronique* (ch. 40), 162; 'Vita Beati Gaufredi' (ch. 5), ed. A. Bosvieux, *Mémoires de la Société des Sciences Naturelles et Archéologiques de la Creuse*, 3 (1862), 92; Geoffrey of Vigeois, 'Chronica' (chs. 41, 53), ed. P. Labbe, *Novae Bibliothecae Manuscriptorum Librorum* (Paris, 1657), ii. 300, 307; *Catalogue général des manuscrits des bibliothèques publiques de France: Départements*, vol. 35 (= Carpentras, vol. 2), comp. Duhamel and Liabastres (Paris, 1899), 635.

[11] *Les Documents nécrologiques de l'abbaye de Saint-Pierre de Solignac*, ed. J.-L. Lemaître and J. Dufour (Recueil des historiens de la France, Obituaires, 1; Paris, 1984), 569.

[12] Fulk IV of Anjou, 'Fragmentum Historiae Andegavensis', ed. P. Marchegay and A. Salmon, *Chroniques des comtes d'Anjou* (Paris, 1871), 380–1.

[13] See Geoffrey of Vigeois, 'Chronica' (chs. 6, 23–5, 30, 32–3, 41), 281–2, 290–1, 296, 297, 300.

[14] *La Chanson d'Antioche*, v. 1186, ed. S. Duparc-Quioc (Documents relatifs à l'histoire des croisades, 11; Paris, 1976–8), i. 71.

and the abbey assisted many departing first crusaders.[15] The date of Ramnulf's grant fits a possible crusade connection; so too the combination of cash deal, spiritual contacts, and family solidarity which was characteristic of crusaders' charters. In this instance Ramnulf may have had a vested interest in not leaving home, for he agreed that he and the monks should divide the proceeds of the mill and co-operate in the construction of new mills, sharing authority over the miller. Nevertheless, Ramnulf's charter may stand for many others which make no mention of the crusade but which may well have been connected with it.[16]

It is also important to clarify the relationship between the chronology of the First Crusade and the involvement of crusaders from south-western France. The crusade was not preached definitively by the church's hierarchy in 1095–6 and then left to its own devices. The crusade appeal's momentum was rather maintained by papal letters and legatine missions stretching over a number of years. Of particular importance in Aquitaine was the Council of Poitiers in November 1100, presided over by the papal legates John of Sta Anastasia and Benedict of Sta Eudoxia.[17] The cross was preached there, an event which made a strong impression on one eye-witness who had also seen Pope Urban preach the crusade at Limoges nearly five years earlier.[18] Shortly before the council Duke William IX of Aquitaine had taken the cross at Limoges with many leading men from his duchy. By this action he effectively proclaimed his intention to lead an Aquitanian rather than a narrowly Poitevin contingent on the forthcoming expedition. At least two important Limousin lords left in 1101, as did others from Aquitaine such as Herbert of Thouars.[19] It is important, however, to dismiss any notion that the 1096 and 1101 expeditions comprised forces from distinct territorial blocs determined by the provenance of their leaders. Gascons went on the first wave of the crusade and gravitated towards the forces of Raymond of

---

[15] Becker, *Papst Urban II.*, ii. 448.

[16] *Cartulaire de Saint-Jean d'Angély*, ed. G. Musset (Archives historiques de la Saintonge et de l'Aunis, 30 and 33; Paris, 1901–3), i. 46.

[17] *La Chronique de Saint-Maixent 751–1140*, ed. and trans. J. Verdon (Les classiques de l'histoire de France au Moyen Âge, 33; Paris, 1979), 172.

[18] 'Vita Beati Gaufredi' (ch. 5), 91–2.

[19] 'Cartae et chronica prioratus de Casa Vicecomitis', ed. P. Marchegay and É. Mabille, *Chroniques des églises d'Anjou* (Paris, 1869), 340–3.

St-Gilles,[20] though some may have left in 1097.[21] Limousin lords distinguished themselves in Syria and Palestine in 1098–9.[22] One, Viscount Raymond of Turenne from the southern Limousin marches with Quercy, belonged as much to the outer edge of the political orbit of the counts of Toulouse as to that of the dukes of Aquitaine, and linguistically was much closer to the former.[23]

It is also necessary to avoid a simplistic contrast between the expedition of 1096–9 as a grim slog by pious *milites Christi* and that of 1101 as a chivalric jaunt by *milites*. Such a distinction rests on two unsure foundations: the fact that the 1101 campaigns failed miserably, forcing commentators to explain the defeat by casting around for signs of improper behaviour and contrasts with the supposed conduct of the victors of 1096–9;[24] and the attractive personality of Duke William IX of Aquitaine, who came to personify the reasons why the 1101 expedition floundered,[25] but whose supposed proto-chivalric tastes and reputation (which research is now revising) should not be treated as emblematic of a different mood among the 1101 crusaders.[26] Moreover, many of those who left home in 1101 had probably taken the cross some years earlier; it is known that the non-performance of crusading vows became a matter

---

[20] Raymond of Aguilers, *'Liber'*, 52; Fulcher of Chartres, *Historia Hierosolymitana (1095–1127)* (i. 6, 23), ed. H. Hagenmeyer (Heidelberg, 1913), 157–8, 255; Ralph of Caen, 'Gesta Tancredi' (ch. 99), 676. For the suggestion that Gaston of Béarn attracted a following of crusaders from throughout Aquitaine, Peter Tudebode, *Historia*, 110.

[21] Auch, 57 (probably dated no earlier than 1097 by the presence of Archbishop Raymond of Auch).

[22] *Gesta Francorum*, 79, 83–4, 87–8.

[23] See the map of the dialects of Gaul in the Central Middle Ages in P. Rickard, *A History of the French Language*, 2nd edn. (London, 1989), 40.

[24] J. S. C. Riley-Smith, *The First Crusade and the Idea of Crusading* (London, 1986), 132–4.

[25] Guibert of Nogent, 'Gesta Dei per Francos' (VII. 23), 243; William of Malmesbury, *De Gestis Regum Anglorum Libri Quinque* (ch. 383), ed. W. Stubbs (RS 90; London, 1887–9), ii. 447–8; Geoffrey of Vigeois, 'Chronica' (ch. 32), 297. For a more sympathetic treatment of William's failure, see Richard the Poitevin, 'Chronicon', extr. *RHGF* 12. 412. See also OV v. 324, 328, 338, 340.

[26] See J. L. Cate, 'A Gay Crusader', *Byzantion*, 16 (1942–3), 503–26, esp. 509–19. See also G. T. Beech, 'Contemporary Views of William the Troubadour, IXth Duke of Aquitaine (1086–1126)', in N. Bulst and J.-P. Genet (edd.), *Medieval Lives and the Historian: Studies in Medieval Prosopography* (Kalamazoo, Mich., 1986), 73–89, esp. 79–81, 82–3; J. Martindale, '"Cavalaria et Orgueill": Duke William IX of Aquitaine and the Historian', in C. Harper-Bill and R. Harvey (edd.), *The Ideals and Practice of Medieval Knighthood II: Papers from the Third Strawberry Hill Conference 1986* (Woodbridge, 1988), 87–116. For the persistence of crusading enthusiasm in William's family, see J. P. Phillips, 'A Note on the Origins of Raymond of Poitiers', *EHR* 106 (1991), 66.

of concern both for those on the 1096 expedition and the church's hierarchy.[27] For these reasons it is necessary to treat the period of crusade enthusiasm between 1095 and 1101 as a continuum. Obviously the ideas of the 1101 crusaders were influenced by the knowledge that Jerusalem was in Christian hands, so that the pilgrimage element in crusading ideology became even more prominent than before.[28] The difference, however, was a subtle one of degree. The motivations of crusaders who left in 1096, 1101, or any intervening year were fundamentally the same. Consequently the manner in which the crusade message was first put to the laity assumes great significance, and to it we must now turn.

The Preaching of the Cross

Limoges was the first important urban centre which the pope visited after the Council of Clermont, and he remained there for a little over a fortnight (23 December 1095–*c*.6 January 1096). The accounts of his stay demonstrate that he made a powerful impression on the area. Arriving at Limoges, Urban and his large party established themselves in the episcopal *cité* and embarked on a remarkable sequence of public liturgical ceremonial. On the Feast of the Nativity Urban celebrated Morrow Mass in the convent of La Règle, then climbed to the *enceinte* of St-Martial, where he celebrated High Mass in St-Martial's basilica of St-Sauveur. There he preached to the people before returning to the episcopal quarter wearing the papal crown. He dedicated the cathedral of St-Étienne on 29 December and may have used this occasion, too, to preach the cross. Probably during the first week of his stay Urban also consecrated the abbey church of St-Martin, the most important ecclesiastical centre beyond the *enceintes* of St-Étienne and St-Martial. On 30 December the pope moved to St-Martial, which became his base for the latter half of his stay. On 31 December he dedicated the recently rebuilt basilica of St-Sauveur, himself anointing the high altar, translating its relics, and celebrating Mass 'in the presence of an innumerable throng of the people'. It is likely that he then preached the cross once more.[29]

[27] *Die Kreuzzugsbriefe aus den Jahren 1088–1100*, ed. H. Hagenmeyer (Innsbruck, 1901), nos. 9, 15, 16, 19, pp. 148, 160, 165, 175.

[28] Riley-Smith, *First Crusade*, 126–9.

[29] 'Notitiae duae Lemovicenses de praedicatione crucis in Aquitania', *RHC Occ.* 5, nos. 1, 2, pp. 350, 351–3; 'Vita Beati Gaufredi' (chs. 5, 6), 91, 93; Geoffrey of Vigeois, 'Chronica' (ch. 27), 293–4; Becker, *Papst Urban II.*, ii. 442–3.

The significance of the pope's ability to draw large crowds to liturgical ceremonies is revealed by his actions at the abbey of Charroux, his next important staging-post after leaving Limoges. An account by a local monk who was almost certainly an eye-witness states that Urban arrived with his 'most holy entourage', attracting a large crowd of nobles and ordinary folk drawn by reverence for the papal office and the novelty of the pope's presence in their locality. On 10 January 1096 Urban consecrated Charroux's high altar and celebrated Mass. The monk's account makes no mention of the crusade or its preaching (the author seems to have been principally impressed by the liturgical ceremonial and the honour which this did to Charroux). Yet he noted with approval that the crowds who witnessed the consecration believed that their presence there would earn them 'a great indulgence of their sins'.[30] Precisely what this may have involved is revealed by Count Fulk IV Le Réchin of Anjou's account of Urban's actions at Angers in the following February. Here the pope preached the cross, dedicated the church of St-Nicholas, and ordained that the anniversary of the dedication be thenceforth kept as a feast, attendance at which would earn the faithful a remission of one-seventh of their penances.[31] Similar arrangements were made when the pope consecrated an altar at the abbey of La Trinité, Vendôme on 26 February 1096.[32]

Although there is no direct evidence that Urban instituted such remissions of penances at Limoges—it is only recorded that he 'blessed' the crowds after the consecration of St-Sauveur[33]—it is very likely that he did so. This would have meant that the cross was preached there in an ambiance of penitence accentuated by the prospect of spiritual rewards and focused by devotion to local religious centres with which the laity was familiar.

Urban, moreover, enhanced his impact in Limoges by contriving to spread his attention between the episcopal *cité* and St-Martial's quarter (relations between which, as we shall see, were very cool) and by exploiting the rhythms of the local liturgical cycle. His presence in Limoges over the Feast of the Nativity must have created opportunities for par-

---

[30] *Chartes et documents pour servir à l'histoire de l'abbaye de Charroux*, ed. P. de Monsabert (Archives historiques du Poitou, 39; Poitiers, 1910), pp. 25–7.

[31] 'Fragmentum Historiae Andegavensis', 380–1; Urban II, 'Epistolae et privilegia', *PL* 151, no. 175, col. 448.

[32] *Cartulaire de l'abbaye cardinale de la Trinité de Vendôme*, ed. C. Métais (Paris, 1893–1904), ii, p. 93 n. 1; 'Chronicon Vindocinense seu de Aquaria', ed. P. Marchegay and É. Mabille, *Chroniques des églises d'Anjou* (Paris, 1869), 171.

[33] Geoffrey of Vigeois, 'Chronica' (ch. 27), 294.

ticularly powerful preaching of the cross centred on the person of Christ.
The only comparable opportunity to build crusade preaching so squarely
around the liturgical calendar was Lent and Easter. In mid-Lent Urban
held his first great council after Clermont at Tours, where the cross was
preached.[34] Easter was spent at Saintes.[35] It cannot be merely the result
of chance survivals that the cartulary of St-Jean d'Angély, which was near
Saintes and which Urban briefly visited, contains references to at least a
dozen early crusaders.[36]

The pope's passage through Gascony was much swifter than that
through the Limousin and is not as well documented. Reaching Bordeaux
from the north (by boat, there being no bridge over the Garonne at
Bordeaux at that time), Urban consecrated the cathedral of St-André on 1
May 1096.[37] No more than three days later he was at Bazas, where he
also consecrated the cathedral.[38] By May 1096 there had been ample time
for the crusade message to reach Gascony independently of the papal
party, but it is still reasonable to suppose that Urban's (albeit fleeting)
presence in the Bordelais and Bazadais was an important stimulus to
crusade recruitment there.[39] Significantly, there is evidence that Archbishop Amatus of Bordeaux was closely associated with the pope in the
preaching of the cross.[40]

Local Religious Institutions and the Response
to the Crusade Appeal

The quasi-regal grandeur, penitential atmosphere, and sheer novelty value
of the pope's passage through south-western France helped to generate
enthusiasm for the crusade. The dissemination of the crusade message was

[34] Becker, *Papst Urban II.*, ii. 446–7.

[35] Ibid. 448–9.

[36] *Cartulaire de Saint-Jean d'Angély*, i. 86, 120, 212, 319, 322, 326; ii. 416, 421–2,
448, 450. Most of the documents are datable only to the abbacy of Ansculf (1091 × 1095–
1103) but almost certainly involve crusaders. No. 322 dates from 1076 × 1101 but refers to
a 'via'. Two documents (421–2) may refer to 1101 crusaders who departed with Duke
William IX of Aquitaine, for whom see ibid. ii. 420.

[37] Becker, *Papst Urban II.*, ii. 449.

[38] J.-G. Dupuy, 'Chronique de Bazas', ed. E. Piganeau, *AHG* 15 (1874), 24. For the
speed of Urban's movements in early May, see Becker, *Papst Urban II.*, ii. 449–50.

[39] See LR 131[129]. For a possible crusader from the area to the east of the Bazadais, see
P. D. Du Buisson, *Historiae Monasterii S. Severi Libri X*, ed. J.-F. Pédegert and A. Lugat
(Aire, 1876), ii. 192; R. Mussot-Goulard, *Les Princes de Gascogne* (Marsolan, 1982), 235.

[40] Peter Tudebode, *Historia*, 31–2.

not dependent on Urban alone, however. The pope is best seen as the mobile nodal point of waves of preaching effort. One example of how this process could have operated, linking the papal party to grass-roots contacts, can be reconstructed from the documents of the priory of Aureil, which was near Limoges but not on the pope's itinerary. Of particular interest is an undated document which records that Bishops Geoffrey of Maguelonne and Humbald of Limoges consecrated the priory's cemetery.[41] This event is most probably to be dated to the period between the Council of Clermont and the pope's arrival at Limoges.[42] There is no evidence that Bishop Geoffrey was in the papal party at Limoges, but it is recorded that he attended the Council of Clermont and the third great council of Urban's French tour, at Nîmes in July 1096.[43] A supporter of the regular life for canons and in contact with the canons regular of St-Ruf, Avignon (with whom Pope Urban and Prior Gaucher of Aureil also had ties),[44] Geoffrey would have served as a natural link between the papal party and Aureil. Prior Gaucher of Aureil, for his part, was well suited to be receptive to Urban's crusade initiative. Still young and vigorous (he died in 1140), he pursued a form of reformed religious life favoured by the pope. The title of a lost chapter of Gaucher's Life (ch. 18) records that he received from Urban the authority to minister to sinners, which means that he would have been able to advise intending crusaders on their penitential needs.[45]

The final link in the chain of crusading communication appears in the remarkable number of crusaders who appear in Aureil's cartulary. In every instance the relevant documents can only be dated to the earlier part of Gaucher's rule (*c.* 1081 × 1085−1140), but they are so closely inter-connected that it is highly unlikely that they refer to individual pilgrimages distinct from the crusade; whether they refer to the 1096 or 1101 expeditions or to intermediate departures is uncertain. One charter-notice records that a man named Peter Gauzbert, wishing to go to

---

[41] A. 53.

[42] See 'La Vie de saint Gaucher, fondateur des chanoines réguliers d'Aureil en Limousin', ed. J. Becquet, *RM* 54 (1964), 35 and n. 46 where the ed. suggests that the consecration took place some years before 1095. But it is difficult to imagine circumstances in which Bishop Geoffrey, from Languedoc, would have been in the Limousin before then. There is no evidence that he ever held legatine powers in south-western France.

[43] Urban II, 'Epistolae et privilegia', no. 158, cols. 431−2; Mansi, 20. 940.

[44] Urban II, 'Epistolae et privilegia', no. 136, cols. 408−10; 'Vie de saint Gaucher', 35−6 and (ch. 13) 52.

[45] 'Vie de saint Gaucher', 45. The date of Urban's licence to Gaucher is not recorded, but there is no reason to suppose that it post-dated 1095−6.

Jerusalem, gave to the canons one half-*bordaria* if he died 'in via', and added that if his brother, who had already departed for the East, should also die, the whole property should pass to Aureil.[46] This transaction was witnessed by a man named Arnald Bernard who is almost certainly to be identified with an Arnald Bernard of Jaunac attested in 1101.[47] Arnald Bernard was also present when Aymeric of Ponroy sold to Aureil one half-*mansus* and rights in the other half before leaving for Jerusalem, and when Fulcher of St-Genest, a cleric, granted his rights in the church of Eyjeaux when he set out for the Holy City.[48] Arnald Bernard also witnessed the charter which recorded Peter of Vizium's grant of properties to Aureil if he died 'in via de Jerusalem' on condition that if he returned he should retain the usufruct for a modest annual payment.[49] Peter of Vizium was probably the brother of Itier of Vizium, who went to Jerusalem after granting to Aureil his rights in a *villa*.[50] Around the same time ('in illis diebus') two brothers named Boso and Joscelin of La Chèze granted their rights in the same *villa* when Boso and a third brother, Rainald, went to Jerusalem, probably with Itier.[51] In the same year as these three men left for the East, Adhemar of Mouliéras and his brother Gerald added to Aureil's interest in the *villa* by granting an adjacent area of woodland.[52] Adhemar's and Gerald's brother William of Mouliéras pledged another property to a canon of Aureil for 60s. 'long before he went to Jerusalem'. When he did leave for the East, and fell ill on the journey, he gave that property and part of its tithe to Aureil. (Subsequently his daughter took the veil there.)[53] The canon who advanced money to William had the same name, Geoffrey of La Brugère, as a canon who witnessed the gift of a *mansus* by Gerald of St-Junien on the day he set out for Jerusalem.[54] Brunet of Le Treuil gave to Aureil allodial land on condition that, if he

---

[46] A. 259.   [47] Ibid. 76[1].

[48] Ibid. 128, 56[2]. The latter transaction took place sometime after ('subsequenti tempore') events datable to c.1081 × 1086: ibid. 56[1].

[49] Ibid. 35, 260.   [50] Ibid. 236[1].   [51] Ibid. 236[2].   [52] Ibid. 236[4].

[53] Ibid. 163; cf. ibid. 170[2]. William's instructions were witnessed by his servant (*cliens*), a priest named Gerald of 'Beiriu' and Humbert of Dognon. Dognon may be settlements of that name in comm. St-Maurice, cant. La Souterraine in the northern Limousin, or comm. Châtenet-en-Dognon, cant. St-Léonard. The latter is nearer Aureil and Mouliéras, but a Humbert of Dognon is attested in c.1086 × 1091 associated with Count Boso III of La Marche, which makes identification with the more northerly Dognon preferable. A Gerald of 'Beirut' appears as a canon of Aureil around the first quarter of the 12th C.: ibid. 289, 304.

[54] Ibid. 165. A Gerald of St-Junien is attested in Prior Gaucher's lifetime, but a canon named Geoffrey of La Brugère is attested around the mid-12th C., so it cannot be certain that Gerald was a first crusader: ibid. 183, 129, and p. 87 n. 1.

wished to go to the East, the prior and canons would furnish his
equipment (*apparatus*). When Brunet decided to invoke this promise, the
canons arranged that a boy be received as a canon in his place (Brunet had
obviously begun moves to enter Aureil in the interim) and that the boy's
father, Peter Rigald of St-Genest, divert some of the entry gift to
Brunet's use on his journey.[55]

Brunet's original grant was witnessed by a man named Bernard of Le
Breuil, who also witnessed a grant to Aureil by Peter II of Pierrebuffière,
who went on the 1101 crusade.[56] Aymeric of Ponroy held lands of the
lords of Lastours, who were well represented on the First Crusade.[57] The
Aureil documents therefore seem to record one or more subgroups of
crusaders associated with the expeditions of the most important crusading
figures of the area.[58] By 1096 Aureil had existed as an established
community for less than two decades. Its resources and regional prestige
must still have been modest compared to, for example, St-Jean d'Angély.
But its cartulary contains references to about the same number of
crucesignati as that of St-Jean. Aureil assisted a network of local crusaders
with such vigour that it is impossible to believe that Gaucher and his
followers did not actively promulgate the crusade message, pursuing a
policy of combining existing family ties to local kindreds with the
development of their own propertied interest.[59]

The value of canons regular in stimulating crusade enthusiasm is
further demonstrated by the Life of Geoffrey of Le Chalard. In a portion
of the Life which recorded certain of Geoffrey's own reminiscences, it
was stated that Geoffrey heard Urban II preach at Limoges and decided
to take the cross. According to the Life's author, Geoffrey was then
approached by Gulpher of Lastours, who had probably also heard Urban
at Limoges and suggested that the two men journey to the East together.
Gulpher was insistent, promising to serve and provide for Geoffrey and
expressing the conviction that Geoffrey would act as a sort of talisman,
guaranteeing their safety. In the event, Geoffrey was dissuaded from
leaving the Limousin by the anxiety of his followers and his own misgivings about the security of his community in his absence. Gulpher did

---

[55] Ibid. 47.

[56] Ibid. 40.

[57] Ibid. 128[1]. Aymeric of Ponroy is recorded in 1093: *Documents nécrologiques*, 603–4.

[58] The Lastours and Pierrebuffières were themselves closely connected. See Uz. 611
(1097 × 1108); Sol. fo. 30[v] (late 11th C.); V. 59 (1092 × 1108).

[59] The statement in A. 80 (of uncertain date) that the canons were wary of acting as
mortgagees may reflect unfavourable experiences with debtor-crusaders and their families.

go on crusade, however, and won great renown. Nevertheless, it is clear that Gulpher, even allowing for the Life's imputation to him of well intentioned naïvety, had treated Geoffrey as a valued spiritual mentor whose domestic familiarity (Gulpher's *caput* lay some eight miles north of Le Chalard) could be translated into crusading support.[60] In this context it is surely significant that Geoffrey was a close associate of Gaucher of Aureil.[61]

Impressive as the efforts of progressive canons regular seem to have been, however, they alone could not have generated the level of crusading enthusiasm which is revealed in the Limousin sources. In fact crusade preaching, official or otherwise, was undertaken by all levels of the local church. This was so even though leadership from the very top of the Limousin ecclesiastical hierarchy was hampered by scandal and discontinuity during much of the period of early crusade fervour. When Urban II arrived at Limoges in December 1095 he was presented with a bitter running dispute between the two *enceintes* centred on St-Martial and St-Étienne which had also drawn in the vicecomital family, the Benedictine monasteries of the Bas-Limousin, and, one must suppose, the communities of canons such as Aureil, Le Chalard, and St-Léonard which derived support from Limoges's bishop and cathedral clergy. Guy of Laron (d. April 1086) had been succeeded as bishop of Limoges by Humbald of St-Sévère, a Berrichon. In 1087 the abbots of St-Martial, Solignac, Tulle, Uzerche, and Vigeois had complained to Archbishop Richard II of Bourges that Humbald's election had been uncanonical. It is hard to detect precisely what impediment the abbots had in mind. Humbald was an outsider unlikely to have been well known in Limoges; at the time of his election he may not have been a priest; and, at least according to the abbots' hyperbole, he had deployed a force of archers to protect the episcopal quarter when his election had led to public disorder.[62] Humbald, however, could not have been entirely unsuitable, for the abbots' letter reveals that Archbishop Richard, hostile at first, was coming round to his side. The 'uncanonical' election seems to have been mainly a vigorous act of independence on the part of the canons of St-

---

[60] 'Vita Beati Gaufredi' (chs. 5, 6), 90–1, 93–6. For the attribution of part of this section to Geoffrey, see ibid. (chs. 4, 6) 90, 93. The portion of the Life dealing with Urban II's preaching of the cross at Limoges, as well as, *inter alia*, the launching of the 1101 expedition, is also edited as Geoffrey of Le Chalard, 'Dictamen de primordiis ecclesiae Castaliensis', *RHC Occ.* 5. 348–9.

[61] 'Vita Beati Gaufredi' (ch. 13), 116.

[62] Lasteyrie, *Abbaye de Saint-Martial*, pièce just. no. 8, pp. 429–31.

Étienne contrary to the claims of the abbot of St-Martial to be consulted in the choice of bishop.[63] According to Geoffrey of Vigeois's rather garbled, pro-monastic version of the dispute, Humbald, by resorting to a piece of sharp practice, sabotaged an appeal which Abbot Adhemar of St-Martial had made to Rome, and then fabricated a papal confirmation of his election. This fooled the monastic party until Urban came to Limoges, exposed the forgery, and deposed Humbald immediately.[64] The process of deposition must in fact have been more protracted than this, however, for Humbald is recorded as bishop issuing a charter on 24 February 1096, by which time Urban had long since left the Limousin and was in distant Anjou.[65] But on 12 April, at Saintes, the pope issued a privilege to Abbot Adhemar, Humbald's principal antagonist, in which the abbot was praised fulsomely for his work in restoring St-Sauveur and introducing Cluniac reform to St-Martial. Adhemar was granted the right to assume pastoral responsibilities during episcopal vacancies or absences by co-opting another bishop of his choosing, to liaise with the senior clergy of St-Étienne in the election of a bishop 'according to the manner of ancient custom', and to be consulted by the bishop of Limoges on important church business.[66] Urban seems deliberately to have avoided defining the abbot of St-Martial's precise role in episcopal elections beyond stating that the abbot's advice should carry great weight; this suggests that it was still a live issue.[67] But, equally, the pope would not have written to Adhemar in such favourable terms and lavished praise on his abbey if Humbald's days as bishop had not been over or, at least, numbered.[68] Furthermore, in the first weeks of 1097 Pope Urban wrote from Rome to his legate Archbishop Hugh of Lyons addressing the latter's complaint that Archbishop Amatus of Bordeaux had exceeded his own legatine powers by consecrating what would appear to have been a new bishop in Limoges. Urban sympathized with Hugh in principle, but added that,

---

[63] See J. Becquet, 'Les Évêques de Limoges aux Xᵉ, XIᵉ et XIIᵉ siècles', *BSAHL* 106. 108–11; A. Sohn, *Der Abbatiat Ademars von Saint-Martial de Limoges (1063–1114): Ein Beitrag zur Geschichte des cluniacensischen Klösterverbandes* (Beiträge zur Geschichte des alten Mönchtums und des Benediktinertums, 37; Münster, 1989), 246–7.

[64] 'Chronica' (chs. 26, 28), 291–2, 295; cf. Sohn, *Abbatiat Ademars*, 248–54, 256–9, 265. See also 'Historia monasterii Usercensis', extr. *RHGF* 14. 338. For contemporary monastic criticism of Humbald, see Uz. 474 (1091).

[65] SEL 75[68]; see also T. 476 (1096).

[66] Urban II, 'Epistolae et privilegia', no. 189, cols. 462–4.

[67] For such powers claimed by some abbots in central and western France, see Becquet, 'Évêques', 106. 109–10; Sohn, *Abbatiat Ademars*, 271 n. 160.

[68] For a slightly different view, Sohn, *Abbatiat Ademars*, 264, 265.

after consulting with Abbot Hugh of Cluny, he had decided to let the consecration stand.[69] Amatus of Bordeaux seems to have been consistently with the papal party until July 1096.[70] Allowing for time for Archbishop Hugh's complaint to have reached the pope and for the pope to have consulted Hugh of Cluny, who was probably in France for most of the latter half of 1096,[71] it would seem that the consecration of Humbald's successor had taken place in the late summer or early autumn of that year.[72]

Thus in the months immediately following Urban's visit to Limoges the Limousin ecclesiastical hierarchy lacked a single focus of firm leadership. Further disruptions followed. Humbald's successor, the former prior of St-Martial William of Huriel (his antecedents providing further evidence that Abbot Adhemar had effectively secured Humbald's deposition or resignation), died in early 1100,[73] and his office was not filled until November of that year, when Peter of Bordeaux was elected at the Council of Poitiers.[74] The problem of continuity at the centre may have been eased to some extent by Bishop Rainald of Périgueux (1081 or 1082 to 1101 or 1102), whose diocese had a long common border with that of Limoges. Rainald was from Thiviers in north-eastern Périgord near the Limousin. He was an active supporter of Geoffrey of Le Chalard and his community, and he consecrated Le Chalard's priory church on 18 October 1100.[75] He also consecrated the church of St-Junien, probably in the same year.[76] Rainald had close ties with the southern Limousin, for he

---

[69] 'Epistolae et privilegia', no. 216, cols. 488–9. Text amended by J. Becquet, 'Le Bullaire du Limousin', *BSAHL* 100 (1973), no. 25, p. 122. The date of early 1097 suggested in *JL* 5678 is accepted by Becker, *Papst Urban II.*, i. 104–5.

[70] See the various references in Becker, *Papst Urban II.*, ii. 443–53.

[71] H. Diener, 'Das Itinerar des Abtes Hugo von Cluny', in G. Tellenbach (ed.), *Neue Forschungen über Cluny und die Cluniacenser* (Freiburg, 1959), 370–1.

[72] See the comment of Geoffrey of Vigeois, 'Chronica' (ch. 28), 295 that William did not succeed Humbald immediately. See also Uz. 335, which suggests that William was elected shortly before 28 Aug.

[73] *Documents nécrologiques*, 519; Becquet, 'Évêques', 106. 113–14.

[74] *Chronique de Saint-Maixent*, 172.

[75] 'Vita Beati Gaufredi' (chs. 1, 4, 5), 77, 81, 89, 92.

[76] Stephen Maleu, *Chronique*, ed. F. Arbellot (St-Junien, 1847), 41–3. The dating of the ceremony to 20 Oct. 1102 (p. 41) cannot be correct. Becquet, 'Évêques', 107. 112 argues for 1100, but he incorrectly dates the consecration to 21 Oct. The approx. 30 miles from Le Chalard to St-Junien could have been covered in 2 days—just about. Maleu (*Chronique*, 41), however, places the consecration during the pontificate of Bishop Peter of Limoges. If correct—the chronology and dating in this late source are often very weak—this can only be Oct. 1101. The presence of Rainald in the Limousin at that time, however, is not without its problems. For Stephen Maleu, see J.-L. Lemaître, 'Note sur le texte de la *Chronique* d'Étienne Maleu, chanoine de Saint-Junien', *RM* 60 (1982), 175–9.

encouraged Uzerche to acquire dependencies in his diocese, and joined Bishop William of Limoges at the consecration of various altars at Uzerche in January 1098.[77] Rainald attended the Councils of Clermont and Poitiers, and himself left for the East, probably in 1101, meeting his death there.[78] Nevertheless, although it is reasonable to suppose that Bishop Rainald lent some continuity to the episcopal direction of crusade recruitment in the Limousin, it would have been impossible for him to have given the area his full attention. Similarly, we have seen that Archbishop Amatus of Bordeaux was not consistently in his own diocese or province during Pope Urban's stay in France. Bishops were a potentially important link in the chain of authority by which the crusade was preached—as the pope intended—but they were not indispensable, for other churchmen were prepared to lend their full support to the crusade initiative.

An important element of the church which assisted in the promulgation of the crusade message was Benedictine monasticism. This was particularly so in the southern Limousin and Gascony, where black-monk influence was strong relative to that of canons regular and ascetics. In strict theory monks worked at a disadvantage, for they might technically be barred from full involvement in the cure of souls, potentially a serious impediment to the counselling of intending crusaders. The Council of Poitiers in 1078, for example, had enacted (canon 5) that no abbot or monk could grant penances without episcopal licence.[79] At the Council of Poitiers in 1100, at which the needs of the crusade were vigorously pursued, monks were prohibited from assuming such duties of the secular

---

[77] Uz. 36 (possibly 1094), 23 and 960 [*recte* 1099], 626 (1081 × 1097); see also ibid. 38 [*recte* 1108], 962 (1151); 'Historia monasterii Usercensis', 338–9; Geoffrey of Vigeois, 'Chronica' (ch. 28), 295 (where Rainald is incorrectly named Arnald).

[78] 'Notitia de Petragoricensibus episcopis, qui donis suis primordia canonicorum S. Asterii adjuvere', *RHGF* 14. 222; 'Fragmentum de Petragoricensibus Episcopis', ed. P. Labbe, *Novae Bibliothecae Manuscriptorum Librorum* (Paris, 1657), ii. 738; Geoffrey of Vigeois, 'Chronica' (ch. 32), 297; 'Charte de Pierre, évêque de Limoges, administrateur de l'évêché de Périgueux, de l'an 1101', ed. Marquis de Bourdeille, *Bulletin de la Société Historique et Archéologique du Périgord*, 55 (1928), 158; J. Becquet, 'Un acte de l'évêque Pierre de Limoges (1101)', *BSAHL* 112 (1985), 18–19. A document from Charroux places Rainald in Aquitaine as late as 27 Dec. 1101 (*Chartes de Charroux*, 24), but this is very possibly a conflation of a charter which Rainald granted to Charroux in Jan. 1096 and a confirmation made by his successor William on the latter date: see M. Laharie, 'Évêques et société en Périgord du Xᵉ au milieu du XIIᵉ siècle', *AM* 94 (1982), 345 n. 10. See, *contra*, Becquet, 'Acte', 14–17, revising his comments in 'La Mort d'un évêque de Périgueux à la Première Croisade: Raynaud de Thiviers', *Bulletin de la Société Historique et Archéologique du Périgord*, 87 (1960), 66–9.

[79] Mansi, 20. 498.

clergy as preaching and granting penance (canon 11); this was in sharp contrast to the powers conferred upon canons regular (canon 10).[80] The very fact, however, that the council fathers deemed such enactments necessary, and the close involvement of monasteries with acts of pilgrimage, demonstrate that monks often assumed the task of acting as spiritual advisers and confessors to intending crusaders.[81] Certainly Pope Urban must have anticipated such a role for the abbot of St-Martial, given the important position granted to him in the governance of the Limousin church.[82] A party of monks from Tulle under Abbot William and Prior Bernard attended the pope at Limoges and must have heard about the crusade.[83] Urban himself visited Uzerche on his journey to Limoges.[84] The number of crusaders recorded in Uzerche's cartulary makes it likely, indeed, that the pope discussed his crusade plans there and intended the monastery to be a recruitment centre in the southern Limousin.

Uzerche's response to the demands of the crusade appeal is obscured by problems surrounding the chronology of its abbots. Immediately after reporting Pope Urban's presence at Limoges and the deposition of Bishop Humbald, the mid-twelfth-century chronicle of Uzerche observed that 'at length' (*tandem*) Abbot Gerald died at the abbey of St-Martial. His date of death is given as 1096 and the length of his abbacy as twenty-eight years (Gerald was probably elected abbot in June 1068).[85] Geoffrey of Vigeois, who most probably used the Uzerche narrative, reported the same sequence of events but linked Urban's 'deposition' of Humbald at Limoges to Gerald's death with the phrase 'While these events were taking place' ('Dum ista peraguntur'), adding that Gerald went to Limoges on the Feast of the Nativity in order to conduct some business on behalf of his abbey, and died there on 15 January.[86] Both sources state

---

[80] Ibid. 1123–4. But for episcopal resistance in the Limousin to canons' assumption of pastoral duties, see Ivo of Chartres, *Correspondance*, ed. and trans. J. Leclercq (Les classiques de l'histoire de France au Moyen Âge, 22; Paris, 1949), no. 69, pp. 304–8.

[81] In general terms, the assumption of broad pastoral duties by monks was implicit in their frequent possession of tithes: see G. Constable, *Monastic Tithes from their Origins to the Twelfth Century* (Cambridge, 1964), esp. 61–3, 144–53, 165–85.

[82] Cf. Sohn, *Abbatiat Ademars*, 109–10.

[83] T. 502, 600.

[84] Becker, *Papst Urban II.*, ii. 442.

[85] 'Historia monasterii Usercensis', 338.

[86] 'Chronica' (ch. 28), 295; cf. *Documents nécrologiques*, 141, 464; 'Additions à l'obituaire de S. Martial', ed. A. Leroux, E. Molinier, and A. Thomas, *Documents historiques bas-latins, provençaux et français concernant principalement la Marche et le Limousin* (Limoges, 1883–5), i. 67.

that Gerald was succeeded by Gauzbert Malafaida (d. 1108), who ruled Uzerche for twelve years, and add that Gauzbert was present as abbot when Uzerche's altars were consecrated by the bishops of Limoges and Périgueux. This last event is dated 22 January 1097 'in the second year after the death of Lord Gerald'.[87] The late narrative tradition therefore suggests that Abbot Gerald followed Urban to Limoges, fell ill, and died in January 1096. It has been argued, however, that the dates supplied by the narratives might be rendered as 1097 and 1098 (NS) respectively.[88] In fact around this time the dating conventions used in the Limousin—as adopted, one must suppose, in documents used by the narratives—were not entirely consistent.[89] More conclusive is the fact that Gauzbert Malafaida appears simply as a sacrist in a Vigeois document which dates from after the election of Bishop William of Limoges, which we have seen should be assigned to the latter half of 1096.[90] Abbot Gerald, moreover, was alive when Gerald Malafaida of Noailles, departing for Jerusalem, surrendered customs which he had exacted from the lands of Uzerche's dependency at Exandon.[91] If this man and this journey are identified with a reference in the *Historia peregrinorum* to a 'Girardus Malafaida' in the company of Limousins and other southern French at Antioch in March 1098, then it seems most probable that it was Abbot Gerald who presided over the first year of Uzerche's response to the crusade.[92]

The contribution to the crusade of Gerald's successor Abbot Gauzbert therefore becomes very interesting, for it demonstrates crusade enthusiasm retaining its momentum after the immediate impact of Pope Urban's

---

[87] Geoffrey of Vigeois, 'Chronica' (ch. 28), 295; 'Historia monasterii Usercensis', 338–9. It is worth noting that 22 Jan. fell on a Sunday, the most common day for a dedication ceremony, in neither 1097 nor 1098.

[88] Becquet, 'Évêques', 106. 114.

[89] T. 502 ('1096' = Jan. 1096 NS); *Chartes, chroniques et mémoriaux pour servir à l'histoire de la Marche et du Limousin*, ed. A. Leroux and A. Bosvieux (Tulle, 1886), charte 20, pp. 25–6 (17 Feb. '1108' = 1109 NS).

[90] V. 169. But see Uz. 531 (1097?).

[91] Uz. 52[7].

[92] 'Historia peregrinorum euntium Jerusolymam' (ch. 55), *RHC Occ*. 3. 193. Girardus Malafaida's appearance in this late source (*c.*1140) is odd, however, for he is the only addition to the list of names supplied by the *Historia*'s principal source: Peter Tudebode, *Historia*, 78. But a Giralt de Malafalda appears associated with Gulpher of Lastours in *La Gran Conquista de Ultramar* (I. 219), ed. L. Cooper (Publicaciones del Instituto Caro y Cuervo, 51–4; Bogotá, 1979), i. 408. This work drew upon the now mostly lost Limousin *Chanson d'Antioche provençale*. No other near-definite crusader appears in a Uzerche document from Abbot Gerald's lifetime. A précis of a lost charter records that Arnald Rufus of Nontron, Gerald's nephew or great-nephew, went to Jerusalem. This notice is appended to a reference to events involving Arnald's mother in 1113 X 1133. Consequently, the Jerusalem journey may be later than the First Crusade: Uz. 943; cf. ibid. 293.

French tour. During Gauzbert's abbacy three members of the Durnais family went to Jerusalem, probably as crusaders.[93] So too, possibly, did Gauzbert's own brother, Peter of Noailles.[94] Gauzbert, moreover, assisted the crusader Guy of Bré before the 1101 expedition. His support for crusaders may lie behind a cryptic passage in the chronicle of Uzerche which (though otherwise favourable to the abbot) states that, 'Finding our house with an abundance of resources, he began to lavish things on *milites*, giving them the horses from the stables and other items to the great detriment of the church.' It was also noted that Gauzbert's abbacy was a boom period for the monastery's acquisition of property.[95] Taken together these two comments on his rule may point to a brisk trade through purchase, gift, or pledge between Uzerche and intending crusaders.

Religious communities cannot have monopolized the land market, there being exchanges between laymen which are largely hidden from the surviving, church-dominated record. For example, the sales and secured loans within the immediate family of two early crusading brothers came to the attention of the canons of Auch, and were recorded by them, only after most of the family's collateral lines had died out and the single surviving representative, Bertrand of St-Jean, found himself in financial difficulties and sold his rights in his pledged property to Ste-Marie.[96] No doubt other crusaders financed themselves as Bertrand's forebears had done. Nevertheless, as the experience of intending pilgrims had demonstrated in the past, religious communities could be approached for material and financial assistance; and enough surviving documents record would-be crusaders exploiting this source of help to suggest that it was a widespread practice.[97] The case of William of Mouliéras noted above demonstrates how crusade finance could be agreed on the basis of earlier indebtedness to religious. Similarly, William of Le Breuil (probably not a

---

[93] Uz. 675.

[94] Ibid. 998, 1000. Peter left for the East after his brother Gerald Malafaida had returned home and become a monk: see ibid. 52[8]. A Vigeois charter records that Peter left for Jerusalem in 1111 × 1124 during Gerald's lifetime: V. 211. Uz. 998 and 1000 may therefore refer to this journey.

[95] 'Historia monasterii Usercensis', 338. The horses seem to have come from stables established by Abbot Gerald in a drive to improve the abbey's material resources. See C. du F. Du Cange, *Glossarium Mediae et Infimae Latinitatis*, rev. G. A. L. Henschel (Paris, 1840–50), i. 765, s.v. 'bravaria', citing this source.

[96] Auch, 64.

[97] G. Constable, 'The Financing of the Crusades in the Twelfth Century', in B. Z. Kedar, H. E. Mayer, and R. C. Smail (edd.), *Outremer: Studies in the History of the Crusading Kingdom of Jerusalem Presented to Joshua Prawer* (Jerusalem, 1982), 70–80.

kinsman of the Le Breuil crusaders featuring in the Aureil cartulary)[98] decided to go 'in hoste de Ierusalem' sometime after he had pledged to Vigeois and his local priest for 50*s.* his share in a property which his brothers had already sold to the same cleric. The priest gave William a little less than 18*s.* in return for a guarantee that the pledge become a sale to himself and the abbey.[99] After William Alboin had granted to Vigeois his rights in the church of Donzenac, his nephew Gauzbert Alboin bought a horse for 50*s.* from Abbot Peter, his uncle, with money raised by pledging his own rights in the church. Gauzbert Alboin later contested the terms of William Alboin's grant, insisting that he be allowed to enjoy the usufruct of the property. Abbot Peter reluctantly agreed on condition that the rights which Gauzbert Alboin had pledged cease to be redeemable. Then Gauzbert defaulted by not returning the money he had borrowed or, one imagines, the horse. Finally, when he decided to go to Jerusalem, he recognized that he had been at fault and granted Abbot Peter his own and William Alboin's shares of the church as well as a lock and certain renders in return for 20*s.* and a swineherd worth 5*s.*[100] In other instances disputes were not pressing but formed a convenient pretext for cash transactions. For example, the *miles* Raymond of Curemonte pledged to Tulle one quarter of the church of Branceilles, his tower, and other properties when he left for the East with Raymond of Turenne in 1096. In return he received 200*s.* and a mule worth 100*s.* His charter recorded that the church had once been Tulle's allodial holding, but the monks' desire to regain their claimed rights could not have been urgent, for Raymond was granted the right to redeem his pledge if he returned home. The church features in Pope Paschal II's confirmation of Tulle's properties in 1105, which suggests that by the time of Raymond's transaction the abbey had already substantially realized its claim to most of Branceilles and could afford to bide its time for Raymond's share.[101]

The capacity of religious communities to fund those departing on crusade is clearly illustrated by a long charter-notice from the priory of La Réole in the Bazadais which records three definite and two probable first crusaders. A 'miles acerrimus' named Amanieu of Loubens who left for

---

[98] See A. Lecler, *Dictionnaire topographique, archéologique et historique de la Creuse* (Limoges, 1902), 88–9.

[99] V. 107; cf. ibid. 108. The editor's dating of *c.*1097 is only approximate. A witness to William's charter, Peter Blanquet, is attested in 1111 × 1124: ibid. 218.

[100] Ibid. 104 (1092 × 1110). Were the swineherd and his pigs meant to go on the crusade?

[101] T. 517, 644, 3 at p. 10; see also ibid. 14 at p. 29.

the East pledged his rights in a weekly fair held next to the priory on condition that, if he were to die in Jerusalem or settle there, no kinsman might claim rights of redemption. Amanieu and his brother Bernard, whose participation in the crusade is to be inferred, also pledged half of a church and their rights in three artisans whom they held in fief from the monastery. Amanieu's companion on the crusade, Bertrand of Taillecavat, pledged to La Réole a quarter of two churches and adjacent properties, limiting rights of redemption to his wife, brother, or sons. The crusader Gerald of Landerron gave the priory allodial lands for the souls of his parents and himself and also pledged a *villa*. Raymond of Gensac, whose presence on the crusade is not explicitly mentioned but who stood surety for Gerald of Landerron and whose own transaction took place at the same time as that of Amanieu and Bernard of Loubens, pledged all his patrimonial rights in the settlement next to the priory. In total 3,300s. changed hands.[102] Another document records that Raymond of Gensac also pledged for 450s. his rights in Pierrefitte, a property near La Réole which he had already granted to the priory. This transaction was witnessed by some of the same men who were present at Raymond's other grant and that of the Loubens, and most probably took place at the same time.[103] In addition, around the time that Gerald of Landerron left on crusade Viscount Bernard I of Bezaume-Benauges mortgaged his rights in four properties to La Réole in return for 160s.; it is not recorded whether Bernard went on crusade, but there is good evidence that one of his kinsmen, William Amanieu, did.[104]

On one level the priory of La Réole was able to disburse such large sums because it acquired valuable rights which enormously strengthened its jurisdictional and economic power over its adjacent settlement and local properties. Given the possibility that the priory's debtors might die on crusade, the difficulty which returning crusaders or their kindred

---

[102] LR 100[93–7].

[103] Ibid. 85[60], 86[61–2].

[104] Ibid. 149[147]; Baldric of Bourgueil, 'Historia Jerosolimitana' (I. 8), *RHC Occ.* 4. 17; William of Tyre, *Chronicon* (I. 17; II. 17; VI. 17), ed. R. B. C. Huygens (Corpus Christianorum, Continuatio Mediaeualis, 63/63A; Turnhout, 1986), i. 139, 182, 331; J. de Jaurgain, *La Vasconie* (Pau, 1898–1902), ii. 116–17, whose suggestion that the crusader William Amanieu was the son of William Amanieu I, attested in the 1080s, and the father of Bernard I, is unlikely on chronological grounds and partly rests upon a conflation of William Amanieu I's brother named Bernard (who is probably our Bernard I) and Bernard Ez I of Albret (whose son Amanieu III went on the First Crusade: ibid. 118 and Peter Tudebode, *Historia*, 129). See C. Higounet, 'En Bordelais: "Principes castella tenentes"', in P. Contamine (ed.), *La Noblesse au Moyen Âge XIᵉ–XVᵉ siècles: essais à la mémoire de Robert Boutruche* (Paris, 1976), 98 and the dating clause of LR 85[60].

might face in raising the redemption value, and the clauses restricting who was entitled to redeem, the priory had every chance of acquiring the rights permanently. To this extent La Réole's support of crusaders was calculating and self-interested.[105] The value to the priory of exploiting the crusade to extend its power over its immediate locality was demonstrated in 1103, when the monks successfully contested Bernard of Bezaume-Benauges's imposition of tolls in their *bourg* by pleading before Duke William IX and an impressive court comprising many of the senior nobles of southern Gascony.[106]

On another level, however, the monks' extraction of prized rights from intending crusaders was perfectly compatible with offers of practical support and favourable expressions of the crusade's purpose, which together reveal that the monks were motivated by more than a desire for material gain. In a remarkable example of how crusaders' temporal privileges might find practical expression on a local level, Gerald of Landerron committed his castles and lands to Prior Auger of La Réole, who was his brother. Auger was also entrusted with the custody of Gerald's sons until they came of age, and was guaranteed for life the enjoyment of the *villa* which Gerald had pledged to the priory. During Gerald's absence Auger erected a church in the castle of Landerron, apparently on his own initiative, and on his return from the East Gerald granted it to La Réole, thus cementing still further his close ties to the monastic community.[107]

The details of the crusaders' money transactions were couched by La Réole's scribes in language which demonstrates that the monks recognized and encouraged the crusade's devotional aspect. It was written of Amanieu of Loubens and Gerald of Landerron that they went on crusade moved by the Holy Spirit and that they abandoned their inheritances, a reference to Matthew 16: 24 which was both an early and prominent element of crusade jargon and used of converts to professed religion. Similarly, it was recorded that Amanieu of Loubens and Bertrand of Taillecavat had 'converted' to pilgrimage, and that Bertrand went to Jerusalem in order to redeem his soul. Amanieu was conceived as being on a divine mission to kill the enemies of the Christian faith and cleanse the scene of Christ's Passion.[108]

---

[105] See P. A. Lewis, 'Mortgages in the Bordelais and Bazadais', *Viator*, 10 (1979), 27–8, 33.

[106] LR 88[64].

[107] Ibid. 100[95], 131[129].

[108] Ibid. 100[93–5].

The favour shown by religious to crusaders' intentions was expressed in liturgical support. Before Viscount Peter of Castillon left on the 1096 expedition he granted the abbey of La Sauve-Majeure a boat which the monks might ply on the River Dordogne as far downstream as Civrac without tolls or any other dues. A valuable grant merited a valuable consideration, and Peter received 150*s.* loaded on to a mule. To this extent Peter's motives were practical and immediate. The transaction, however, took place in a powerful spiritual ambiance. It was noted that Peter's grant redounded 'to the praise and glory of God's name'; he was presented as acting for the salvation of his soul and in order to seek divine assistance; and he received the *beneficium* of La Sauve, meaning that he was guaranteed to feature prominently in the monks' intercession.[109] Similarly, sometime between 1092 and his departure on the First Crusade, and very possibly in connection with the latter, Guy III of Lastours pledged properties and fodder rights to Vigeois in return for 60*s.*, on condition that the pledge should pass permanently to the abbey on his death for the soul of himself and his parents, and that his name be enrolled 'amongst the brethren'.[110] The crusaders Peter Gauzbert, Brunet of Le Treuil, Itier of Vizium, and Adhemar Catard granted property to Aureil for the souls of themselves and their parents; Boso of La Chèze did so for the soul of his father-in-law, from whom he had received the rights which he now granted to the canons.[111]

The concern of many crusaders to be remembered in the intercession of local religious communities was typically rooted in past contacts. There is a danger here of using a circular argument, for crusaders with traditional ties to religious bodies are most likely to feature in those communities' surviving documents. Furthermore, it is impossible to draw up a list, to serve as a control, of those men who definitely did not take the cross. Even so, the sample of crusading names is sufficiently large to point to a pattern of support for religious institutions over generations. Such support need not have been perfectly consistent—William IX of Aquitaine, for example, fell out bitterly with the papal legates at the Council of Poitiers,[112] and few kindreds, however pious, did not occasionally challenge certain alienations of property to religious com-

---

[109] LSM, pp. 255–6. Cf. *Cartulaire de Saint-Jean d'Angély*, ii. 416 (1095 × 1101), 422 (1100 or 1101).

[110] V. 153.

[111] A. 47, 52, 78, 236[1–2], 259 (all 1096 × 1101).

[112] 'Gesta in concilio Pictavensi, circa excommunicationem Philippi I Francorum Regis', *RHGF* 14. 108–9; Hugh of Flavigny, 'Chronicon', *MGH SS* 8. 493.

munities[113]—but some measure of regular contact with the church is a recurrent feature of the spiritual careers of many crusaders' families.

The first crusader Gaston of Béarn (his half-brother Centulle II of Bigorre probably went with him to the East) belonged to a kindred with a strong tradition of support for reformed religious communities in southern Gascony.[114] Raymond of Turenne, like his father Boso before him, was an enthusiastic benefactor of Tulle and Uzerche; shortly after his return from the crusade he also supported the introduction of Cluniac reform at St-Sour, Terrasson, alongside the future crusader Bishop Rainald of Périgueux; and he made a grant of lands to Vigeois in a charter whose dating clause included a powerful evocation of the recent restoration of the Holy Sepulchre from infidel control, an interesting indication of how Raymond's crusade enthusiasm could prompt an extension of his links to local monasteries and be valued by the monks.[115] As Geoffrey of Le Chalard's flirtation with crusading demonstrates, the lords of Lastours were known for their support of professed religion. Members of their kindred were benefactors of religious communities,[116] and the family's reputation was to be confirmed in 1114 when the former crusader Gulpher and his brother Gerald provided Gerald of Sales with land with which to found Dalon, part of a reformed monastic congregation which was to expand impressively before being absorbed by the Cistercians.[117] Peter II of Pierrebuffière, who went on the 1101 expedition, belonged to a family with close ties to Uzerche, Vigeois, Aureil, Solignac, and St-Étienne, Limoges.[118] Raymond of Gensac was a member of a group of nobles from Entre-Deux-Mers who had sworn to defend the interests of La Sauve-Majeure in return for regular and intense liturgical commemoration and favourable burial rights.[119] Among

---

[113] For a dispute involving a future first crusader, see T. 325 (1084 × 1091).

[114] See above, 135, 158.

[115] Uz. 265–6, 271, 277–8, 282; 'Varia chronicorum fragmenta ab ann. DCCC.XLVIII. ad ann. 1658', ed. H. Duplès-Agier, *Chroniques de Saint-Martial de Limoges* (Paris, 1874), 187; Geoffrey of Vigeois, 'Chronica' (ch. 31), 297; V. 106 (see also ibid. 253).

[116] B. 14 (1062 × 1072), 15 (1073 × *c*.1076); Uz. 504 (1061), 611 (1097 × 1108), 768 (*c*.1060), 1070 (1037 × *c*.1050).

[117] 'Liber fundationis et donationum abbatiae B. Mariae Dalonis', extr. *RHGF* 14. 161–2.

[118] Uz. 152 (1037), 322 (*c*.1000 × 1025), 325 (1097 × 1108), 460 (1048), 611 (1097 × 1108), 670 (1068 × 1086), 838–9 (1086 × 1096), 1076 (1044); V. 58 (1092 × 1110); A. 40 (*c*.1100), 95⁶ (*c*.1100); SEL 121[102] (1052). For Solignac, see below.

[119] LSM, pp. 10–11, 14, 165. See also ibid. 22.

humbler and therefore less-documented crusading families, there is evidence that the Durnais had traditions of support of Uzerche and Vigeois.[120] The Charieyras, four of whom probably went on the Saragossa crusade in 1118, had given to Tulle and Uzerche, and at least one member of their kindred was a monk at Vigeois.[121] Raymond of Cure-monte was most probably a kinsman of past donors to Uzerche.[122] These men were thus in a very similar position to those who have already been noted combining an enthusiasm for pilgrimage with support of religious communities. Such a combination, extended from the experience of pilgrimage to the crusade, can be demonstrated clearly in a detailed study of an active crusading family, the Brés.

A Case-Study: The Brés

The first recorded members of the Bré family, from the western central Limousin, are Fruinus, attested in 1011 and 1025, and his son Bernard I, attested in 1025, 1036, and 1044. Both seem to have been *fideles* of the viscounts of Limoges.[123] It appears that by the early eleventh century the kindred already enjoyed well-established and close relations with Vigeois, which was to remain its favoured religious community into the twelfth century, for Geoffrey of Vigeois records that a Firminus (i.e. Fruinus) granted the chapel of Bré itself to the abbey.[124] Direct documentary evidence for the identity of Bernard I's progeny is lacking, but it is most probable that he was the father of three brothers, Gerald Bernard, Bernard II, and Peter I.[125] It is in these brothers' generation that evidence for their family's support of local monasteries becomes impressive.

The formative event in the Brés' religious careers was the reform of Vigeois in 1082. According to Geoffrey of Vigeois, conventual life at the abbey had been disrupted during the abbacy of Peter I Mirabel (c.1060–c.1080) by a serious fire which exposed the monastery and its property to

---

[120] Uz. 153 (1068 × 1097), 690 (1073 × 1086); V. 120 (1092 × 1110).

[121] T. 361 (1091); Uz. 750 (1068 × 1097); V. 140 (1092 × 1110), 156 (1100 × 1104), 172 (1108 × 1111), 217 (c.1100), 280 (1124 × 1164), 303 (1111 × 1124). For the crusaders, V. 220.

[122] Uz. 563 (1068 × 1097). See also B. 111 (1102 × 1111).

[123] Uz. 47, 172, 315, 716; see also ibid. 323, 664.

[124] 'Chronica' (ch. 9), 283.

[125] The relationship between the brothers is established by Uz. 571 (1053 × 1067 and 1068 × 1097) and 1153 (1062); see also ibid. 1307. Geoffrey of Vigeois, 'Chronica' (ch. 18), 288 records that Gerald Bernard's father was named Bernard.

spoliation during an outbreak of disorder in the area.[126] Geoffrey's account is vague in places and may be embroidered by legend. He recounts that a *miles* from Pierrebuffière, who was related to Vigeois's sacrist, was captured, sparking off reprisals and provoking local princes to devastate the abbey's lands. Whatever the roots of the monks' involvement in what appears to have been a bitter feud,[127] it is clear that the abbey suffered badly, for in the charter of 1082 in which Gerald Bernard of Bré granted the *abbatia* of Vigeois to St-Martial he was made to promise that neither he nor anyone acting on his behalf would engage in arson, theft, warfare, or judicial duels on the abbey's lands, nor would he continue to receive renders and fines. Gerald Bernard also surrendered the 'evil customs' which he had raised on Vigeois's land, and licensed his vassals to give their rights to the abbey.[128] Geoffrey of Vigeois, who doubtless knew this document but also had additional information, confirms that Vigeois's resources had become an important element of Gerald Bernard's lordship: he records that Gerald Bernard's surrender to the Cluniacs was not total, for he retained rights in Vigeois's *vicaria* held as a fief from the abbot (which suggests that the Brés had earlier come to dominate the abbey's resources as lay abbots). It is also clear, however, that the reform of Vigeois was largely Gerald Bernard's initiative.[129] This is confirmed by the identity of Vigeois's first Cluniac abbot, Gerald of Lestrade, who before entering St-Martial had been a canon of St-Yrieix, a community which lay very close to the Brés' *caput*.[130] Relations between the Brés and Vigeois were not invariably warm thereafter. Two of Gerald Bernard's sons, Boso and Guy I, contested a modest gift of part of a tithe which had been made by their brother Bernard III Juvenior.[131] Generally, however, the Brés were active supporters, endowing Vigeois with further lands after 1082.[132]

---

[126] 'Chronica' (ch. 17), 288.

[127] It is possible that these events were linked to a bitter feud in *c*.1062 between the lords of Pierrebuffière and Viscount Adhemar II of Limoges, whose family had an interest in the area near Vigeois by reason of its castles at Ségur and Salon: Sol. fo. 19^{r-v}. If so, regular life at Vigeois may have been disrupted for as long as twenty years, which would explain the paucity of charters from between *c*.1060 and *c*.1080 surviving in its cartulary. V. 12 is dated by the ed. to this period but more securely belongs to 1092 × 1110. The suggested dates of V. 7 (1073 × 1086) and 16 (1062 × 1072) are only approximate. See also the first transaction in ibid. 23 (*c*.1068 × *c*.1074), in which no abbot of Vigeois is recorded.

[128] *Chartes, chroniques*, charte 11.

[129] Geoffrey of Vigeois, 'Chronica' (ch. 18), 288.

[130] Ibid. (ch. 17) 288.

[131] V. 170 (1092 × 1101).

[132] See ibid. 139.

In the years after the Council of Clermont the Brés became one of the most enthusiastic crusading families in the Limousin. Because of the recurrence of the family name Bernard among their number there cannot be certainty about the identity of some of the Brés who went on crusade. A charter-notice from Vigeois records that a Bernard, 'unus ex Breenensium principibus', granted two *mansi* which Gerald Bernard of Bré had previously given to the abbey *post obitum* and which Bernard had himself promised to surrender on his own death. The later transaction took place when Bernard was wishing to set out for Jerusalem, the notice adding that he died on the journey.[133] The document may refer to Gerald Bernard's brother Bernard II, or to his son Bernard III Juvenior, either of whom may have had an interest in Gerald Bernard's two *mansi*. An undated notice from Aureil records that Gerald Bernard gave to the canons there a pool, possibly (the document is fragmentary) when his son Bernard III left for Jerusalem.[134] Gerald Bernard's date of death is unknown—the Vigeois document noted before only implies that he was dead by the time of the First Crusade—so it is uncertain whether this document refers to a departure on crusade. Bernard II is recorded in 1062, 1068 × 1097, and probably 1082 × 1091, so it is not impossible that he was still alive in 1096–1101.[135] Other factors, however, favour the identification of Bernard III with the 'unus ex Breenensium principibus'. The crusading Bernard's wife was named Titburgis. A Titbugis is recorded alongside her sons Gerald and Bernard in two précis of lost charters from Uzerche dated 1097 × 1108. Her son Bernard appears elsewhere in the same period, and two brothers, Guy II and Bernard IV, are recorded in 1111 × 1124 conceding the grant by their father Bernard of the two *mansi* mentioned before.[136] On the assumption that there were not two women named Titburgis married to men named Bernard of Bré, we may treat Guy II, Bernard IV, and Gerald as brothers. Guy II (who went to Jerusalem sometime after 1107) is recorded in 1124 × 1164 (probably *c.*1130 × *c.*1150) granting properties to Vigeois for the soul of his brother Bernard, which may suggest that the latter had died fairly recently.[137] The likelihood that at least two of the brothers were alive in

---

[133] V. 86. The charter is datable to 1092 × 1110, but the reference to the *iter* almost certainly recalls the First Crusade.

[134] A. 244.

[135] Uz. 571, 1034, 1153; V. 239.

[136] Uz. 709–10, 933; V. 191.

[137] V. 319. The ed.'s dating of the charter to 1147 is conjectural. See also Uz. 1004 (1135 × 1149). The Vigeois document was witnessed by Fruinus of Bré, a monk recorded in 1130 × 1143, 1146, and 1135 × 1149: V. 137–8, 316.

the second quarter of the twelfth century suggests that they were more likely to have been the sons of Bernard III Juvenior than of Bernard II, and that consequently it was the former who went on the First Crusade. There are further problems with the evidence, however. A modern précis of a lost charter from Uzerche records that a Bernard of Bré, son of Bernard, went to Jerusalem, where he died.[138] This could refer to Bernard II, an otherwise unrecorded son of Bernard II, or Bernard IV. Overall, the Brés' confused genealogy means that it is impossible to be definite about who went on crusade. It is tolerably clear, however, that early and powerful crusading traditions developed within their kindred.[139]

Fortunately more precise information is available concerning another member of the Bré family, Guy I, whose presence on crusade is securely documented.[140] Guy's experience is significant for it highlights a number of crusaders' preoccupations and how they were addressed through contacts with religious communities. In the first place, Guy needed money to fund his journey. It is quite possible that he raised cash from Roger the Merchant, a mysterious figure who normally features in the documents closely associated with the Brés, and from whom Guy had borrowed in the past.[141] Sometime before he left on crusade Guy also pledged property for 30*s*. to the 'men of Montneyger', and sold and mortgaged property to Peter Pipiola of Allassac, who was probably a layman.[142] Even in the context of cash deals between laymen, however, Guy was concerned to associate a religious community, for in the last transaction

[138] Uz. 717, 1144. The former notice is appended to a reference to 'Bernard of Bré and his brother Guy', but it is unclear what the copyist intended by this.

[139] G. Tenant de la Tour, *L'Homme et la terre de Charlemagne à Saint Louis: essai sur les origines et les caractères d'une féodalité* (Paris, 1943), 362 n. 1 supposes that two Bernards of Bré went on the 1096 crusade, i.e. Bernard II on the basis of V. 86, and the Bernard son of Bernard recorded in Uz. 717. From the foregoing it is apparent that the presence of two Bernards on the First Crusade cannot be established with confidence.

[140] No surviving document precisely dates Guy's departure for the East. The most important source (V. 113) is undated but must refer to the First Crusade. A Guy of Bré seems to be attested in the Limousin in January or February (on one of the Feasts of St Peter *ad cathedra*) of a year during the episcopate of Bishop Peter of Limoges, who was elected in Nov. 1100: Uz. 365; cf. ibid. 361, 364. Bishop Peter died no later than 1106 and probably in 1105: Becquet, 'Évêques', 107. 112. Given that Guy II of Bré is not securely recorded before 1124 × 1164, it is most probable that Uz. 365 refers to Guy I, who would have gone on crusade shortly thereafter, possibly with William IX of Aquitaine who left France in the spring of 1101.

[141] V. 6 (1082 × 1101), 52 (1082 × 1097), 68 (1092 × 1101), 93 (1092 × 1110), 113 (1101 or 1102), 170 (1092 × 1110); A. 244 (late 11th C.). Roger may be the Roger of Uzerche recorded in 1082 × 1091: V. 51.

[142] V. 113; Uz. 361 (1097 × 1101), 364–5.

he arranged with Peter Pipiola that the pledged property pass to Uzerche on the latter's death. There is evidence, moreover, that Guy raised money directly from monks. When he lay dying at Latakia he ordered his companions to convey his instructions that whoever inherited his lands should redeem all the property which was pledged 'to the saints', a phrase suggesting that Guy had agreed secured loans with more than one religious community.[143]

Guy's second preoccupation was the welfare of his wife and his sole child, a daughter named Stephana. A notice from Uzerche records that Guy, as he left for Jerusalem, settled his daughter and lands on Oliver of Lastours. Whether Oliver was specifically the individual chosen is not wholly certain: this document was written up some years later, after Oliver had married Stephana, and the instructions which Guy transmitted on his death-bed, in the form in which they were recorded at Vigeois, refer only to an unnamed prospective husband of the daughter. The date of the marriage cannot be established precisely, other than that it took place between the news of Guy's death reaching the Limousin and 1110. It would seem most likely, however, that Guy had betrothed Stephana to Oliver before his departure. The match would have dispelled doubt about the girl's future, and Oliver was a good choice in the context of early crusading enthusiasm among the lords of the central Limousin, being the son of Guy III of Lastours, who died on the 1096 expedition, and the nephew of the 1096 crusader Gulpher of Lastours, who survived the campaign and was presumably back in the Limousin by 1100–1.[144] Given Guy's anxieties about the future of his immediate family, it is striking how, on his death-bed, he attempted to balance its material security with a desire to add to the (as shall be seen) already generous grants he had made to Vigeois. He gave instructions that, if his daughter were to die without issue, the land which he had settled on his wife at Estivaux was to pass to the abbey on the two women's deaths; so too a number of vines, *bordariae*, and renders which included property which Guy had received from his own mother. In effect Guy was inviting the monks of Vigeois, through their reversionary interest in widely dispersed properties, to associate themselves closely with his wife's and daughter's future.[145]

---

[143] V. 113.

[144] Uz. 361; V. 113–14; Geoffrey of Vigeois, 'Chronica' (ch. 6), 281, 282.

[145] V. 113. For the location of the properties, ibid. 69 nn. 5–9. The unidentified *mansus* 'de Las Cumbas' may be that 'de la Cumba' which appears in a charter from 1082 × 1086, apparently (the document is truncated) granted by Gerald Bernard, Guy's father: ibid. 6.

Guy's final concern was to gain the favour of religious communities in his region so that he might benefit from their intercession. This aim was addressed in two stages. First, he made over properties to Vigeois and Uzerche before his departure. As the evidence stands, the grants do not appear to have been conspicuously generous. Guy gave to Vigeois, in the event of his death, a tithe which he had already pledged to the abbey (in other words, he renegotiated the terms of an outstanding loan) as well as rights in a *mansus* which was already held as security by Roger the Merchant. As we have seen, Guy also arranged that Uzerche should have the reversion on a property pledged to a third party.[146] As Guy lay dying and reflected on the fate of his soul, however, his thoughts turned to his favoured religious communities back home. He remembered that he had not always behaved correctly towards Vigeois (though in fact the minor dispute over Bernard III's gift of a tithe noted earlier is the only recorded instance of ill will between Guy and the monks), and, on the advice of two Limousin companions and in the presence of more than five others, he granted the abbey various rights in no fewer than ten locations in the central Limousin (excluding those properties bound up with the settlement on his wife and daughter).[147] The history of some of these properties is not documented; four (Nespouls, Sioussac, Montcoulomb, and La Cipière) are not mentioned elsewhere in the Vigeois cartulary. Yet it is reasonably clear that Guy attempted to combine donations of new rights with a rounding-off and confirmation of properties already in Vigeois's hands. One *mansus*, at Anglars, and probably a *bordaria* at Vall were held by Guy from Viscount Adhemar III of Limoges.[148] Guy's rights at Murat may have been held from Viscount Bernard of Comborn.[149] Lands which Guy granted at Charliac were close to properties given to Vigeois by his aunt Agnes.[150] Similarly, two properties (Charliac and La Mazière) were very close to lands earlier granted to Vigeois by a Bernard of Bré, either Guy's uncle or brother.[151]

Vigeois was not the only community on Guy's mind. He confirmed Aureil in possession of the pool given by his father.[152] A brief notice in the cartulary of Solignac records that on his death-bed Guy also granted

---

[146] Ibid. 113; Uz. 361, 364–5.

[147] V. 113.

[148] Ibid. 101. This charter can only be dated to 1092 × 1110, but it probably records Adhemar's confirmation of Guy's donation.

[149] Ibid. 203 (1092 × 1110 and 1111 × *c*.1120).

[150] Ibid. 139 (1092 × 1108). Agnes's grant most probably preceded Guy's.

[151] Ibid. 239 (1082 × 1091).

[152] A. 244.

one property to the monks there.[153] The Brés' *caput* was approximately equidistant from Vigeois and Solignac, so it is not impossible that Guy had some family connections with the latter abbey. The Brés, however, do not feature as benefactors elsewhere in the cartulary of Solignac. A family which most certainly does is the Pierrebuffières, whose castle was adjacent to one of Solignac's principal dependencies, Ste-Croix.[154] Peter II of Pierrebuffière was the senior layman recorded at Guy's side when he died, advising him on his dispositions to Vigeois.[155] It would therefore seem that Guy's grant to Solignac was an expression of crusading bonds between himself and Peter. Yet one has only to compare the extent of Guy's respective grants to Vigeois and Solignac to realize the importance of traditional family ties to religious communities. In a stressful state and thousands of miles from home in a strange land, Guy turned his thoughts to his family and to monks. Guy's kindred's record of support for religious communities, particularly Vigeois and Uzerche, his own arrangements for his departure on crusade, and the preoccupations of his dying moments, all form a continuum in which the interests of family and monastery were intimately linked.

It would be misleading to argue that only men with traditions of support of religious communities could have experienced the impulse to go on crusade. The evidence is lacking that would explain why non-crusaders did not go, or why some men who took the cross did not perform their vows. Furthermore, and allowing for the paucity of the sources, it is highly unlikely that crusaders formed anything more than a small, if significant, minority among the identifiable adult benefactors of religious communities between *c.*1080 and *c.*1110. But the object of this study has not been to isolate reasons why one particular individual went on crusade and another did not, for such a task would leave too much to conjecture. Obliged by the evidence to generalize from a few specific cases, what we can say is that the fund of religious ideas upon which laymen drew when they entered or endowed religious communities was essentially the same as the beliefs which generated crusade enthusiasm. Patently, no self-

---

[153] Sol. fo. 23ᵛ. The property was at 'Siussac' which is almost certainly to be identified with Sioussac near Vigeois, where, as we have seen, Guy granted a *bordaria* to Vigeois. Two of the three witnesses in the Solignac notice occur in V. 113. The third, Adhemar of Felez, may be the Adhemar, priest of St-Germain, recorded in the Vigeois document.

[154] Sol. fos. 17ʳ⁻ᵛ, 19ʳ⁻ᵛ, 26ᵛ⁻27ʳ, 27ʳ, 30ᵛ, 32ᵛ⁻33ʳ, 35ʳ⁻36ʳ. The lords of Pierrebuffière also appear in many other documents as consenters and witnesses.

[155] V. 113.

contained body of crusade traditions and values could have existed in 1095–6 or have been anything more than embryonic in 1100–1. Common sense dictates that an idea as novel as the crusade vocation (laymen, remember, were not itching to redirect *Reconquista* fever elsewhere or bursting to break the constraints of the Peace movement) owed its appeal to elements within it which were natural extensions of familiar values. Pope Urban II's exploitation of local ecclesiastical centres when preaching the cross, the willingness of religious communities to fund crusaders by recourse to established patterns of loan, purchase, and dispute settlement, and the easy manner in which crusaders drew upon religious communities' spiritual as well as material resources, all point to the root of the value-system which made the crusade vocation possible: the interaction of laymen and professed religious, particularly monks, in the years around 1095.

Conclusion

IT is impossible to frame a concise statement of crusaders' motivations which would apply accurately to all periods, regions, and social classes. Crusading proved to be a durable institution because it was able to adapt to new religious and ethical impulses, to changes within the church, and to evolving social conditions (as demonstrated by, for example, the friars' preaching of the cross in thirteenth-century towns). In such areas as the theology of the indulgence, preaching, organization, and strategic planning, the crusade gradually became a more sophisticated instrument; and the ideas of the participants developed accordingly. To take only a few examples of the dynamics within the crusading movement: current research is establishing that from very early in the movement's history family traditions—transmissible through both the male and female lines—contributed significantly to many individuals' interest in crusading;[1] in the thirteenth and fourteenth centuries the papacy developed an impressive system of taxation, redemptions, and other sources of revenue, which meant that not all those who supported crusades were expected to go on campaign, and that some who fought did so for pay as well as for spiritual rewards;[2] and in the later medieval period crusading enthusiasm came to be influenced (but not supplanted) by chivalric ethics, as the experience of Chaucer's Knight reveals.[3] The capacity of crusade institutions and ideas to respond to changing conditions suggests that a thorough examination of crusaders' motivations across the whole of the movement's history would need to address a wide range of social, political, economic, spiritual, and other factors.

In the present study such an ambitious approach has not been attempted with respect to the First Crusade for two reasons: there is a danger that the causes of the expedition, itself a novelty, become conceived as nothing less than a commentary upon virtually every feature of Latin Christian

[1] J. S. C. Riley-Smith, 'Family Traditions and Participation in the Second Crusade', in M. Gervers (ed.), *The Second Crusade and the Cistercians* (New York, 1992), 101–8.

[2] N. Housley, *The Italian Crusades: The Papal–Angevin Alliance and the Crusades against Christian Lay Powers 1254–1343* (Oxford, 1982), 173–90, 207–51; id., *The Avignon Papacy and the Crusades 1305–1378* (Oxford, 1986), 129–30, 131–2, 134–43, 154–5, 162–98.

[3] J. S. C. Riley-Smith, *The Crusades: A Short History* (London, 1987), 208–9.

society; and certain specific problems—in particular the significance of the Peace of God and the *Reconquista*—are so embedded in crusade scholarship that they merited some detailed treatment. It is a valid exercise to bring some order to the mass of reasons why men went on crusade by concentrating upon ideas and impulses which operated on or near the level of consciousness. The motives which an averagely intelligent crusader might have been able to put into words are only part of the story, of course, but they have a particular interest because crusading was voluntary. It was an undertaking which had to be made attractive to the faithful through spiritual and temporal privileges.

This straightforward observation creates certain problems of definition and interpretation, for few actions can be described as wholly voluntary. In the Central Middle Ages men and women of knightly or noble rank seldom enjoyed personal autonomy. Usually their actions were guided by many considerations: the obvious restraints of material resources and time, but also other factors such as the expectations of the kin-group (which largely absorbed the individual's identity in many settings), the demands of honour, and the social conventions which accompanied status. (Proof of this is the reverence shown to men and women who resolutely pursued spiritual careers in the face of domestic responsibilities and sometimes even hostility from relatives and peers.)[4] The constraints which acted upon arms-bearers mean that it is impossible to imagine crusaders as footloose voluptuaries indulging their whims, as free-spirited adventurers in search of excitement.

This does not mean, however, that crusaders were nothing more than the reluctant or unwitting servants of convention. There must have been a point when they made a conscious decision either to go on crusade or at least to be receptive to pressures which might push them in that direction. Whenever, for example, crusading was enjoined as a penance, it cannot, on one level, be treated as voluntary; yet it is important to remember that the crusader had chosen to submit himself to penitential discipline in the first place. The element of volition in the crusading vocation, however it manifested itself in individual cases, was particularly important in the response to the First Crusade, when the crusade message was still a novelty and family traditions and social norms linked to crusading had not had time to develop fully.

From the perspective of an arms-bearer living in south-western

---

[4] See D. Weinstein and R. M. Bell, *Saints and Society: The Two Worlds of Western Christendom* 1000–1700 (Chicago, 1982), 59–63, 74–5.

France at the end of the eleventh century, the crusade appeal tapped his ideological potential on some levels but not on others. The ethical content of the Peace of God, such as it was, was of little direct relevance. The Peace in Aquitaine had been pitched predominantly at the very highest levels of the lay aristocracy, the princes and more important lords who controlled castles and were closely related to the ecclesiastical leaders who framed the Peace decrees. Furthermore, the Aquitanian Peace had been in abeyance for more than sixty years (near the very limit of living memory) in 1095–6; and in an area such as southern Gascony it had never made an impact. When Pope Urban II planned the recruitment and organization of the crusade he drew on his own experiences of Peace ideas and, where appropriate, on local traditions. This he did to optimize the expedition's chances of success. There was no deeper, direct link between the crusade vocation and any ethical programme which could be construed from Peace enactments. Quite probably Urban and other senior clergy were alive to the possibility that Latin Europe might be pacified somewhat by the diversion of some arms-bearers' energies abroad; but this could have been nothing more than an incidental strategic aim. The crusade was no 'war on war', nor did it represent a deliberate conceptual ploy by ecclesiastics attempting to pursue the Peace of God programme by other means.

In south-western France only a small fraction of the audience of the crusade appeal would have been able to draw parallels between the new expedition and direct experience of campaigning against infidels in Spain. Even then the parallels could have been nothing more than approximate. The circumstances behind the recruitment and organization of the Barbastro campaign in 1064, although they might seem to anticipate the crusade, are in reality very obscure and prone to misinterpretation. The one consistent theme linking Barbastro to later instances of French participation in peninsular warfare, notably in 1087, is a desire for material gain and the assertion of family prestige. Between the Barbastro campaign and the eve of the First Crusade the French contribution to the *Reconquista* was dominated by a kindred network which was so tightly drawn that it points to a system of control among Spanish rulers in keeping with their own policies: policies which were much more subtle than all-out holy war. Geography and, to some extent, language meant that an area such as southern Gascony enjoyed close ties with parts of northern Spain, but there is no reason to suppose that Gascons were particularly receptive to the crusade message for that reason. Furthermore, crusade ideals did not cross over into the Spanish theatre to a significant degree before the 1110s, and when they did so the influence of external agents such as the papacy and former first crusaders from France was pronounced, and probably decisive. The French ex-crusaders were influential because their experience combined the traditional nexus of trans-Pyrenean military contacts—the kinship bond—with the newer crusade enthusiasms. For these reasons the eleventh-century *Reconquista* cannot be seen as a trial run for the crusade movement, preparing men's minds for the summons of 1095–6.

The Peace of God and the *Reconquista* have often been considered significant by historians of the early crusades because they suggest that laymen could, in certain circumstances, act in harmony with progressive ecclesiastics' emerging ideas about the morality and correct purposes of violence. Many laymen were indeed attuned to the concerns of clerics, but not principally on this sort of abstract level. Rather, arms-bearers exchanged ideas about immediate and personal issues—the fate of a kinsman's soul, for example, or the demands of penances—with the monks and canons who staffed local religious communities. The ways in which religious institutions in south-western France recruited members, and associated practices such as confraternity and burial, demonstrate that the links between churches and their local aristocratic community could be very close. Oblation, adult conversion, and reception *ad succurrendum* were all, on one level, social acts informed by conventions. But they were also the means to express genuinely pious impulses. The fact that support of religious communities was typically bound up with family traditions points to the depth, not the superficiality, of those impulses.

It would be mistaken to underestimate the vitality of laymen's religious sentiments around the time of the First Crusade. One of the most important features of the piety of eleventh-century arms-bearers (at least in its formal and thus documented aspects) was that it was associative, passive to the extent that it was inspired and sustained by the spiritual resources of a monastic or clerical élite. Associative behaviour is not necessarily insincere or shallow, however. Although the crusade message encouraged laymen to assume more direct responsibility for their salvation, crusading was not considered the alternative to traditional forms of pious expression. Why should the first crusader Gaston of Béarn have supported Ste-Christine and other religious communities after 1100 and founded Sauvelade in 1127, and why should Raymond of Turenne have continued to endow Tulle's dependency at Auriol and other churches, unless old practices continued to meet present spiritual needs? As far as the inevitably distorted evidence can reveal, first crusaders did not return from the East convinced that, spiritually speaking, they had 'done their bit' and could now dispense with the ministrations of the church. Some in fact entered religious communities;[5] more brought back relics to present to the traditional foci of their piety back home.[6] The fact that crusading complemented rather than supplanted the established devotional norms suggests that the appeal of the crusade message was rooted in the very value-system through which those norms were sustained.

The charters of religious communities, although the picture which they present cannot be complete, contain important amounts of information about why arms-bearers supported monks and canons. The simple fact that the doctrinal statements made in the charters were usually tradition-bound and selected from a limited range of themes is itself of great significance, for it points to a common currency of ideas passing between the laity and professed religious. Chief among the themes expounded were: a powerful sense of sinfulness which could be more intense than the aggregated feelings of guilt inspired by a number of separate sins; a concern for the spiritual welfare of kinsfolk, and an appreciation that families had a transcendental existence in which the living members bore some responsibility for the dead; an instinct to associate one's spiritual prospects as closely as possible with those of monks and other religious; both respect for and some inchoate unease with the church's penitential disciplines; a lively fear of damnation made more potent by a tendency to literalize the torments of the damned; and a belief in an intermediate other-world, not fully distinguished from Hell, which offered the hope of salvation to those compromised by immersion in this world, and without which it would have been idle to provide for intercession for dead relatives. Narrative accounts of visions and miracles from south-western France and elsewhere complement the picture of lay religiosity which is revealed by the documentary evidence.

Many elements of the lay religious culture which emerge from the charters and narratives—the fear of Hell, for example, and concern about penances—feature in the sources for the response to the First Crusade. A further bridge between crusade ideas and established forms of piety is provided by pilgrimage, enthusiasm for which was often channelled through contacts and family traditions linking arms-bearers to local religious communities. Pilgrimage was also important because it profoundly influenced the way in which the crusade was conceived and

---

[5] J. S. C. Riley-Smith, *The First Crusade and the Idea of Crusading* (London, 1986), 121.
[6] Ibid. 122–3.

conducted. The Peace of God, the *Reconquista*, and pilgrimage have tended to form a trinity in discussions of the crusade's origins. From the perspective of a first crusader from south-western France, the third element was likely to be by far the most significant, not only because of pilgrimage's immediate relevance to the rites, symbolism, and goal of the crusade, but also because it was an expression of the lay pious culture which drew its inspiration from religious bodies.

This study may conclude by entering and addressing two caveats. First, it must be emphasized that the field of enquiry has concentrated upon religious ideas and crusade enthusiasm as manifested in three parts of south-western France. It cannot be stated with total confidence that conclusions drawn from the sources for these areas can be applied to the rest of Latin Europe or even every French-speaking territory. The Peace of God is one obvious example of how local experiences may have varied from region to region; the impact of reformed monasticism another. Nevertheless, the striking similarities between the three areas under review—the Limousin most in the 'mainstream' of religious and cultural life in eleventh-century France, southern Gascony the most remote and backward, the Bordelais/Bazadais in an intermediate position—suggest that the religious experiences of regional aristocracies throughout France varied subtly in degree, not in kind. This can be no more than a suggestion, but it has provided the basis for the inclusion in the present study of sources from beyond south-western France whenever they seemed to complement Aquitanian material: for example, Orderic Vitalis's account of Herlequin's Hunt, the narrative accounts of the First Crusade written by northern Frenchmen, and certain Anglo-Norman texts bearing on the monastic vocation.

The second caveat is that it is inevitable that attention should be concentrated on the careers of the better-documented arms-bearing families. The historian's knowledge of the actions, even the existence, of a family can depend on the chance survival of a handful of charters; and no doubt the evidence of many pious kindreds is lost for ever. A still more serious problem is posed by those individuals or families who did not generate written records in the first place because they chose not to endow or enter religious communities. Any picture of lay piety which ignores the existence of the lazy and the indifferent is vulnerable to distortion. Nevertheless, it can be argued that the number of surviving charters is sufficiently large, and the sample of recorded lay benefactors thus sufficiently randomized, to permit generalizations about popular religious ideas and devotional practices around the time of the First Crusade.

Some caution is necessary. Even if it did not involve a subjective value-judgement, it would be unwise to give a figure for the proportion of noble and knightly families in south-western France which had traditions of support for monasteries or other ecclesiastical institutions by 1095. Instead, it is more realistic to posit a sliding scale of commitment (with variations between the members of single families) stretching from the energetic patron of reform, through the intermittently generous benefactor, to the occasional supporter of the church and the reluctant consenter to a third party's pious dispositions. No popular religious culture is embraced by all people at all times with equal enthusiasm. But, on the other hand, the existence of a sliding scale of piety in eleventh-century Aquitaine and Gascony is itself significant, for it was defined and perpetuated by the ideas and enthusiasms of those at its upper limits: men (and women) who were the most vigorous (and so conspicuous) supporters of a system of religious values which extended beyond them to touch the lives of most of their peers. It is difficult to imagine that men such as Gaston of Béarn, Raymond of Turenne, Gulpher of Lastours, Guy of Bré, Gerald of Landerron, and William Amanieu of Benauges were so exceptional in their devotion that they departed from the norms which governed the society in which they lived. On the basis of what is known about their support for local ecclesiastical centres, they may be regarded as representative of the associative but potentially deep arms-bearing piety of their age. The foregoing has attempted to demonstrate that it is not surprising that all these men, and many others like them, also went on the First Crusade.

Bibliography

MANUSCRIPTS

Bordeaux, Bibliothèque Municipale, MS 745 ('Cartulaire de Bigorre').
Bordeaux, Bibliothèque Municipale, MS 769 ('Grand Cartulaire de la Sauve-Majeure').
London, BL Add. MS 8873.
Paris, Bibliothèque Nationale, MS lat. 12752.
Paris, Bibliothèque Nationale, MS lat. 18363 ('Cartulaire de Solignac').

PRINTED SOURCES
(INCLUDING LATER WORKS
CONTAINING SUBSTANTIAL AMOUNTS
OF MEDIEVAL MATERIAL)

Acta Pontificum Romanorum Inedita, ed. J. Pflugk-Harttung, 3 vols. (Tübingen, 1881; Stuttgart, 1884–6).
'Additions à l'obituaire de S. Martial', ed. A. Leroux, E. Molinier, and A. Thomas, *Documents historiques bas-latins, provençaux et français concernant principalement la Marche et le Limousin*, 2 vols. (Limoges, 1883–5), i, pp. 63–80.
ADHEMAR OF CHABANNES, 'Aus ungedruckten Predigten', ed. E. Sackur, *Die Cluniacenser in ihrer kirchlichen und allgemeingeschichtlichen Wirksamkeit*, 2 vols. (Halle, 1892–4), ii. 479–87.
—— *Chronique*, ed. J. Chavanon (Collection de textes pour servir à l'étude et à l'enseignement de l'histoire, 20; Paris, 1897).
—— 'Epistola de apostolatu Sancti Martialis', *PL* 141. 87–112.
—— 'Sermones tres', *PL* 141. 115–24.
—— 'Translatio beati Martialis de Monte Gaudio', ed. E. Sackur, *Die Cluniacenser in ihrer kirchlichen und allgemeingeschichtlichen Wirksamkeit*, 2 vols. (Halle, 1892–4), i. 392–6.
—— *et al.*, 'Commemoratio abbatum Lemovicensium basilice S. Marcialis, apostoli', ed. H. Duplès-Agier, *Chroniques de Saint-Martial de Limoges* (Paris, 1874), 1–27.
AIMO OF FLEURY, 'Vita Sancti Abbonis', *PL* 139. 375–414.
ALBERT OF AACHEN, 'Historia Hierosolymitana', *RHC Occ.* 4. 265–713.
ALEXANDER II, 'Epistolae et diplomata', *PL* 146. 1279–1430.
'Altera S. Adelelmi Vita', *ES* 27. 434–59.

AMATUS OF MONTECASSINO, *Storia de' Normanni*, ed. V. de Bartholomaeis (Rome, 1935).

'Ancienne vie anonyme de saint Martial (VIe siècle)', ed. J. C. E. Bourret, *Documents sur les origines chrétiennes du Rouergue: saint Martial* (Rodez, 1887–1902), 2–8.

ANDREW OF FLEURY, *Vie de Gauzlin, abbé de Fleury*, ed. and trans. R.-H. Bautier and G. Labory (Sources d'histoire médiévale, 2; Paris, 1969).

'Annales Compostellani', *ES* 23. 318–25.

'Annales Engolismenses', *MGH SS* 16. 485–7.

'Annales Lemovicenses', *MGH SS* 2. 251–2.

'Annales S. Dionysii Remenses', *MGH SS* 13. 82–4.

'Annales Toledanos, I', *ES* 23. 382–401.

'Anonymi Florinensis brevis narratio belli sacri', *RHC Occ.* 5. 371–3.

ANSELM OF CANTERBURY, *Opera Omnia*, ed. F. S. Schmitt, 6 vols. (Rome, 1940; Edinburgh, 1946–61).

BALDRIC OF BOURGUEIL, 'Historia Jerosolimitana', *RHC Occ.* 4. 9–111.

BALUZE, É., *Historiae Tutelensis Libri Tres* (Paris, 1717).

BERNARD ITIER, 'Chronicon', ed. H. Duplès-Agier, *Chroniques de Saint-Martial de Limoges* (Paris, 1874), 28–129.

BERNOLD OF ST BLASIEN, 'Chronicon', *MGH SS* 5. 400–67.

'Cartae et chronica prioratus de Casa Vicecomitis', ed. P. Marchegay and É. Mabille, *Chroniques des églises d'Anjou* (Paris, 1869), pp. 327–47.

Cartulaire de l'abbaye cardinale de la Trinité de Vendôme, ed. C. Métais, 5 vols. (Paris, 1893–1904).

Cartulaire de l'abbaye de Beaulieu (en Limousin), ed. M. Deloche (Paris, 1859).

Cartulaire de l'abbaye de Conques en Rouergue, ed. G. Desjardins (Paris, 1879).

Cartulaire de l'abbaye de la Sainte-Trinité de Tiron, ed. L. Merlet, 2 vols. (Chartres, 1883).

Cartulaire de l'abbaye de Saint-Amant-de-Boixe, ed. A. Debord (Poitiers, 1982).

Cartulaire de l'abbaye de Saint-Cyprien de Poitiers, ed. L. Rédet (Archives historiques du Poitou, 3; Poitiers, 1874).

Cartulaire de l'abbaye de Saint-Jean de Sorde, ed. P. Raymond (Paris, 1873).

Cartulaire de l'abbaye de Saint-Sernin de Toulouse (844–1200), ed. C. Douais (Paris, 1887).

Cartulaire de l'abbaye de Saint-Victor de Marseille, ed. M. Guérard, 2 vols. (Collection des cartulaires de France, 8–9; Paris, 1857).

Cartulaire de l'abbaye de Savigny suivi du petit cartulaire de l'abbaye d'Ainay, ed. A. Bernard, 1 vol. in 2 (Paris, 1853).

'Cartulaire de l'abbaye de Talmond', ed. L. de la Boutetière, *Mémoires de la Société des Antiquaires de l'Ouest*, 36 (1873), pp. 41–498.

Cartulaire de l'abbaye de Vigeois en Limousin (954–1167), ed. M. de Montégut (Limoges, 1907).

Cartulaire de l'abbaye des bénédictins de Saint-Savin en Lavedan (945–1175), ed. C. Durier (Cartulaire des Hautes-Pyrénées, 1; Paris, 1880).

Cartulaire de l'abbaye d'Uzerche, ed. J.-B. Champeval (Paris, 1901).

Cartulaire de l'abbaye royale de Notre-Dame de Saintes, ed. T. Grasilier (Cartulaires inédits de la Saintonge, 2; Niort, 1871).

'Cartulaire de l'abbaye Sainte-Croix de Bordeaux', ed. A. Ducaunnès-Duval, *AHG* 27 (1892), pp. 1–157.

Cartulaire de l'église collégiale Saint-Seurin de Bordeaux, ed. J.-A. Brutails (Bordeaux, 1897).

Cartulaire de Saint-Jean d'Angély, ed. G. Musset, 2 vols. (Archives historiques de la Saintonge et de l'Aunis, 30 and 33; Paris, 1901–3).

Cartulaire de Saint-Vincent-de-Lucq, ed. L. Barrau Dihigo and R. Poupardin (Pau, 1905).

Cartulaire de Sainte-Foi de Morlàas, ed. L. Cadier (Collection de pièces rares ou inédites concernant le Béarn, 1; Pau, 1884).

Cartulaire des abbayes de Tulle et de Roc-Amadour, ed. J.-B. Champeval (Brive, 1903).

'Cartulaire du prieuré d'Aureil', ed. G. de Senneville, *BSAHL* 48 (1900), pp. 1–289.

Cartulaire du prieuré de Paray-le-Monial, ed. C. U. J. Chevalier (Collection de cartulaires dauphinois, 8^2; Montbéliard, 1891).

Cartulaire du prieuré de Saint-Mont, ed. J. de Jaurgain (Archives historiques de la Gascogne[2], 7; Paris, 1904).

'Cartulaire du prieuré de Saint-Pierre de la Réole', ed. C. Grellet-Balguerie, *AHG* 5 (1863), pp. 99–186.

Cartulaires du chapitre de l'église métropolitaine Sainte-Marie d'Auch, ed. C. Lacave La Plagne Barris (Archives historiques de la Gascogne[2], 3; Paris, 1899).

El Cartulario de Roda, ed. J. F. Yela Utrilla (Estudios históricos, 1; Lérida, 1932).

Cartulario de San Juan de la Peña, ed. Antonio Ubieto Arteta, 2 vols. (Textos Medievales, 6 and 9; Valencia, 1962–3).

Cartulario de San Millán de la Cogolla, ed. L. Serrano (Madrid, 1930).

Cartulario de Santa Cruz de la Serós, ed. Antonio Ubieto Arteta (Textos Medievales, 19; Valencia, 1966).

Catalogue général des manuscrits des bibliothèques publiques de France: Départements, vol. 35 (= Carpentras, vol. 2), comp. Duhamel and Liabastres (Paris, 1899).

Catalogus codicum hagiographicorum latinorum antiquiorum saeculo XVI qui asservantur in Bibliotheca Nationali Parisiensi, ed. Société des Bollandistes, 3 vols. (Paris, 1889–93).

La Chanson d'Antioche, ed. S. Duparc-Quioc, 2 vols. (Documents relatifs à l'histoire des croisades, 11; Paris, 1976–8).

'Charte d'Almodis, comtesse de La Marche, en faveur de l'abbaye de l'Esterps (12 novembre 1098)', ed. G. Babinet de Rencogne, *Bulletin de la Société Archéologique et Historique de la Charente*[3], 4 (1864), 409–14.

'Charte de Pierre, évêque de Limoges, administrateur de l'évêché de Périgueux, de l'an 1101', ed. Marquis de Bourdeille, *Bulletin de la Société Historique et Archéologique du Périgord*, 55 (1928), 156–60.

'Chartes', ed. A. Leroux, E. Molinier, and A. Thomas, *Documents historiques bas-latins, provençaux et français concernant principalement la Marche et le Limousin*, 2 vols. (Limoges, 1883–5), i, pp. 121–258.

Chartes, chroniques et mémoriaux pour servir à l'histoire de la Marche et du Limousin, ed. A. Leroux and A. Bosvieux (Tulle, 1886).

Chartes de l'abbaye de Nouaillé de 678 à 1200, ed. P. de Monsabert (Archives historiques du Poitou, 49; Poitiers, 1936).

Chartes et documents pour servir à l'histoire de l'abbaye de Charroux, ed. P. de Monsabert (Archives historiques du Poitou, 39; Poitiers, 1910).

Chartes et documents pour servir à l'histoire de l'abbaye de Saint-Maixent, ed. A. Richard, 2 vols. (Archives historiques du Poitou, 16 and 18; Poitiers, 1886).

Chronica Adefonsi Imperatoris, ed. L. Sánchez Belda (Escuela de Estudios Medievales, Textos, 14; Madrid, 1950).

Chronica monasterii Casinensis, ed. H. Hoffmann (MGH SS 34; Hanover, 1980).

'Chronicon Aquitanicum', *MGH SS* 2. 252–3.

'Chronicon Dolensis Coenobii', extr. *RHGF* 11. 387–8.

'Chronicon Lusitanum', *ES* 14. 415–32.

'Chronicon Trenorciense', extr. *RHGF* 11. 112–13.

'Chronicon Vindocinense seu de Aquaria', ed. P. Marchegay and É. Mabille, *Chroniques des églises d'Anjou* (Paris, 1869), 153–77.

La Chronique de Morigny (1095–1152), ed. L. Mirot (Collection de textes pour servir à l'étude et à l'enseignement de l'histoire, 41; Paris, 1909).

La Chronique de Saint-Maixent 751–1140, ed. and trans. J. Verdon (Les classiques de l'histoire de France au Moyen Âge, 33; Paris, 1979).

Chronique de Saint-Pierre-le-Vif de Sens, dite de Clarius, ed. and trans. R.-H. Bautier and M. Gilles (Sources d'histoire médiévale, 3; Paris, 1979).

Colección diplomática de Irache, ed. J. M. Lacarra and A. J. Martín Duque, 2 vols. (Saragossa, 1965; Pamplona, 1986).

Colección diplomática de la catedral de Huesca, ed. A. Duran Gudiol, 2 vols. (Fuentes para la Historia del Pirineo, 5–6; Saragossa, 1965–9).

Colección diplomática de Pedro I de Aragón y Navarra, ed. Antonio Ubieto Arteta (Escuela de Estudios Medievales, Textos, 19; Saragossa, 1951).

Colección diplomática medieval de la Rioja (923–1225), ed. I. Rodríguez de Lama, 3 vols. (Logroño, 1976–9).

Conciliorum Oecumenicorum Decreta, ed. J. Alberigo, J. A. Dossetti Perikle, P. Joannou, C. Leonardi, and P. Prodi, 3rd edn. (Bologna, 1973).

The Councils of Urban II, i. *Decreta Claromontensia*, ed. R. Somerville (Annuarium Historiae Conciliorum, Suppl., 1; Amsterdam, 1972).

Crónica de San Juan de la Peña, ed. T. Ximenez de Embun (Biblioteca de Escritores Aragoneses, Sección Histórico-Doctrinal, 1; Saragossa, 1876).

Crónicas Anónimas de Sahagún, ed. Antonio Ubieto Arteta (Textos Medievales, 75; Saragossa, 1987).

Documentación medieval de Leire (siglos IX a XII), ed. A. J. Martín Duque (Pamplona, 1983).

La documentación pontificia hasta Inocencio III (965–1216), ed. D. Mansilla (Monumenta Hispaniae Vaticana, Sección Registros, 1; Rome, 1955).

Documentos correspondientes al reinado de Sancho Ramírez, ed. J. Salarrullana de Dios and E. Ibarra y Rodríguez, 2 vols. (Colección de documentos para el estudio de la historia de Aragón, 3 and 9; Saragossa, 1907–13).

Documentos medievales Artajoneses, ed. J. M. Jimeno Jurío (Pamplona, 1968).

Documentos para el estudio de la reconquista y repoblación del valle del Ebro, ed. J. M. Lacarra, 2 vols. (Textos Medievales, 62–3; Saragossa, 1982–5).

Les Documents nécrologiques de l'abbaye de Saint-Pierre de Solignac, ed. J.-L. Lemaître and J. Dufour (Recueil des historiens de la France, Obituaires, 1; Paris, 1984).

DU BUISSON, P. D., Historiae Monasterii S. Severi Libri X, ed. J.-F. Pédegert and A. Lugat, 2 vols. (Aire, 1876).

DUPUY, J.-G., 'Chronique de Bazas', ed. E. Piganeau, AHG 15 (1874), 1–67.

EKKEHARD OF AURA, 'Hierosolymita', RHC Occ. 5. 1–40.

Epistolae Pontificum Romanorum Ineditae, ed. S. Loewenfeld (Leipzig, 1885).

'Formulae Salicae Merkelianae', MGH Legum Sectio, 5. 239–64.

'Formulae Turonenses vulgo Sirmondicae dictae', MGH Legum Sectio, 5. 128–65.

Les Fors Anciens de Béarn, ed. and trans. P. Ourliac and M. Gilles (Paris, 1990).

'Fragment d'une Chanson d'Antioche en provençal', ed. and trans. P. Meyer, Archives de l'Orient Latin, 2 (1884), 467–509.

'Fragmenta chronicorum comitum Pictaviae, ducum Aquitaniae', ed. E. Martène and U. Durand, Veterum Scriptorum et Monumentorum, Historicorum, Dogmaticorum, Moralium, Amplissima Collectio, 9 vols. (Paris, 1724–33), v. 1147–60.

'Fragmentum de Petragoricensibus Episcopis', ed. P. Labbe, Novae Bibliothecae Manuscriptorum Librorum, 2 vols. (Paris, 1657), ii. 737–40.

FULBERT OF CHARTRES, The Letters and Poems, ed. and trans. F. Behrends (Oxford, 1976).

FULCHER OF CHARTRES, Historia Hierosolymitana (1095–1127), ed. H. Hagenmeyer (Heidelberg, 1913).

FULK IV OF ANJOU, 'Fragmentum Historiae Andegavensis', ed. P. Marchegay and A. Salmon, Chroniques des comtes d'Anjou (Paris, 1871), 373–83.

GEFFREI GAIMAR, L'Estoire des Engleis, ed. A. Bell (Anglo-Norman Texts, 14–16; Oxford, 1960).

GELASIUS II, 'Epistolae et privilegia', PL 163. 487–514.

GEOFFREY OF LE CHALARD, 'Dictamen de primordiis ecclesiae Castaliensis', RHC Occ. 5. 348–9.

GEOFFREY OF VIGEOIS, 'Chronica', ed. P. Labbe, Novae Bibliothecae Manuscriptorum Librorum, 2 vols. (Paris, 1657), ii. 279–329.

GERARD OF CORBIE, 'Notitia de fundatione monasterii Silvae-majoris', RHGF 14. 45–6.

Gesta Comitum Barcinonensium, ed. L. Barrau Dihigo and J. Massó Torrents (Cròniques Catalanes, 2; Barcelona, 1925).

Gesta Francorum et aliorum Hierosolimitanorum, ed. and trans. R. Hill (London, 1962).

'Gesta in concilio Pictavensi, circa excommunicationem Philippi I Francorum Regis', *RHGF* 14. 108–9.

GILBERT CRISPIN, 'Vita Domni Herluini Abbatis Beccensis', ed. J. A. Robinson, *Gilbert Crispin Abbot of Westminster* (Notes and Documents relating to Westminster Abbey, 3; Cambridge, 1911), 87–110.

La Gran Conquista de Ultramar, ed. L. Cooper, 4 vols. (Publicaciones del Instituto Caro y Cuervo, 51–4; Bogotá, 1979).

GREGORY VII, *Das Register*, ed. E. Caspar, 2nd edn., 1 vol. in 2 (MGH Epistolae Selectae, 4–5; Berlin, 1955).

GREGORY OF TOURS, 'Liber in gloria martyrum', *MGH Scriptores rerum Merovingicarum*, rev. edn., 1^2. 34–111.

—— 'Liber in gloria confessorum', *MGH Scriptores rerum Merovingicarum*, rev. edn., 1^2. 294–370.

—— *Libri Historiarum X*, *MGH Scriptores rerum Merovingicarum*, rev. edn., 1^1.

GUIBERT OF NOGENT, *Autobiographie*, ed. and trans. E.-R. Labande (Les classiques de l'histoire de France au Moyen Âge, 34; Paris, 1981).

—— 'Gesta Dei per Francos', *RHC Occ.* 4. 117–263.

Le Guide du pèlerin de Saint-Jacques de Compostelle, ed. and trans. J. Vielliard, 3rd edn. (Mâcon, 1963).

HAMELIN OF ST ALBANS, 'Liber de monachatu', extr. *Thesaurus Novus Anecdotorum*, ed. E. Martène and U. Durand, 5 vols. (Paris, 1717), v. 1453–6.

HELGAUD OF FLEURY, *Vie de Robert le Pieux*, ed. and trans. R.-H. Bautier and G. Labory (Sources d'histoire médiévale, 1; Paris, 1965).

HERMAN OF LAON, 'De miraculis B. Mariae Laudunensis libris tribus', extr. *RHGF* 12. 266–72.

Historia Compostellana, ed. E. Falque Rey (Corpus Christianorum, Continuatio Mediaeualis, 70; Turnhout, 1988).

'Historia monasterii Usercensis', extr. *RHGF* 14. 334–42.

'Historia peregrinorum euntium Jerusolymam', *RHC Occ.* 3. 165–229.

Historia Pontificum et Comitum Engolismensium, ed. J. Boussard (Paris, 1957).

Historia Silense, ed. J. Pérez de Urbel and A. Gonzalez Ruiz-Zorrilla (Escuela de Estudios Medievales, Textos, 30; Madrid, 1959).

'Historiae Francicae Fragmentum', extr. *RHGF* 10. 210–12; 11. 160–2; 12. 1–8.

HUGH OF FLAVIGNY, 'Chronicon', *MGH SS* 8. 280–503.

HUGH OF FLEURY, *Chronicon, quingentis ab hinc annis et quod excurrit, conscriptum*, ed. B. Rottendorff (Münster, 1638).

—— 'Liber qui modernorum regum Francorum continet actus', *MGH SS* 9. 376–95.

—— 'Opera historica', *MGH SS* 9. 337–406.

—— 'Vita S. Sacerdotis Episcopi Lemovicensis', *AASS* (May) 2. 14–22.

IVO OF CHARTRES, 'Epistolae', *PL* 162. 9–288.

—— *Correspondance*, ed. and trans. J. Leclercq (Les classiques de l'histoire de France au Moyen Âge, 22; Paris, 1949).

Jaca: documentos municipales 971–1269, ed. Antonio Ubieto Arteta (Textos Medievales, 43; Valencia, 1975).

JOHN OF SALERNO, 'Vita sancti Odonis', *PL* 133. 43–86.

Die Kreuzzugsbriefe aus den Jahren 1088–1100, ed. H. Hagenmeyer (Innsbruck, 1901).

LAMBERT OF ARDRES, 'Historia comitum Ghisnensium', *MGH SS* 24. 550–642.

LANFRANC, 'De celanda confessione', *PL* 150. 625–32.

LETHALD OF MICY, 'Delatio corporis S. Juniani in synodum Karrofensem', *PL* 137. 823–6.

'Liber alter miraculorum', in 'Vita et miracula S. Leonardi Nobiliacensia', *AASS* (Nov.) 3. 159–73.

'Liber Anselmi archiepiscopi de humanis moribus', ed. R. W. Southern and F. S. Schmitt, *Memorials of St. Anselm* (Auctores Britannici Medii Aevi, 1; London, 1969), 37–104.

'Liber de unitate ecclesiae conservanda', *MGH Libelli de lite*, 2. 173–284.

'Liber fundationis et donationum abbatiae B. Mariae Dalonis', extr. *RHGF* 14. 161–2.

Liber Miraculorum Sancte Fidis, ed. A. Bouillet (Paris, 1897).

'Liber prior miraculorum', in 'Vita et miracula S. Leonardi Nobiliacensia', *AASS* (Nov.) 3. 155–9.

Liber Sancti Jacobi: Codex Calixtinus, i. *Texto*, ed. W. M. Whitehill (Santiago de Compostela, 1944).

Liber Tramitis Aevi Odilonis Abbatis, ed. P. Dinter (Corpus Consuetudinum Monasticarum, 10; Siegburg, 1980).

MARBOD OF RENNES, 'Vita sancti Gualterii abbatis et canonici Stirpensis in dioecesi Galliarum Lemovicensi', *PL* 171. 1563–76.

MARCA, P. DE, *Histoire de Béarn* (Paris, 1640).

MARCULF, 'Formulae', *MGH Legum Sectio*, 5. 32–112.

MARTIN OF MONTIERNEUF, 'De constructione Monasterii novi Pictavis', ed. F. Villard, *Recueil des documents relatifs à l'abbaye de Montierneuf de Poitiers (1076–1319)* (Archives historiques du Poitou, 59; Poitiers, 1973), pp. 421–41.

MATTHEW OF EDESSA, 'Récit de la Première Croisade', *RHC Documents arméniens*, 1. 24–150.

Memorials of St. Anselm, ed. R. W. Southern and F. S. Schmitt (Auctores Britannici Medii Aevi, 1; London, 1969).

Les Miracles de Notre-Dame de Roc-Amadour au XIIe siècle, ed. and trans. E. Albe (Paris, 1907).

Les Miracles de saint Benoît, ed. E. de Certain (Paris, 1858).

Les Miracles de saint Privat suivis des opuscules d'Aldebert III, évêque de Mende, ed. C. Brunel (Collection de textes pour servir à l'étude et à l'enseignement de l'histoire, 46; Paris, 1912).

El monasterio de San Pelayo de Oviedo: historia y fuentes, ed. F. J. Fernández Conde, I. Torrente Fernández, and G. de la Noval Menéndez, 2 vols. (Oviedo, 1978–81).

MORET, J. DE, *Anales del reino de Navarra*, 12 vols. (Tolosa, 1890–2).

'Narratio Floriacensis de captis Antiochia et Hierosolyma et obsesso Dyrrachio', *RHC Occ.* 5. 356–62.

'Necrologio del monasterio de San Victorian', *ES* 48. 276–84.

'Notitia de Petragoricensibus episcopis, qui donis suis primordia canonicorum S. Asterii adjuvere', *RHGF* 14. 221–2.

'Notitiae duae Lemovicenses de praedicatione crucis in Aquitania', *RHC Occ.* 5. 350–3.

ODO OF CLUNY, 'Collationum Libri III', ed. M. Marrier and A. Du Chesne, *Bibliotheca Cluniacensis* (Paris, 1614), 159–262.

—— 'Vita sancti Geraldi Auriliacensis comitis', *PL* 133. 639–704.

ORDERIC VITALIS, *Historiae Ecclesiasticae*, ed. A. Le Prévost and L. Delisle, 5 vols. (Paris, 1838–55).

—— *The Ecclesiastical History*, ed. and trans. M. Chibnall, 6 vols. (Oxford, 1969–80).

'Papsturkunden in Florenz', ed. W. Wiederhold, *Nachrichten der K. Gesellschaft der Wissenschaften zu Göttingen, Phil.-hist. Kl.* (1901), 306–25.

Papsturkunden in Spanien, i. *Katalanien*, ed. P. Kehr (Abhandlungen der Gesellschaft der Wissenschaften zu Göttingen, Phil.-hist. Kl., NS 18^2; Berlin, 1926).

Papsturkunden in Spanien, ii. *Navarra und Aragon*, ed. P. Kehr (Abhandlungen der Gesellschaft der Wissenschaften zu Göttingen, Phil.-hist. Kl., NS 22^1; Berlin, 1928).

PELAYO OF OVIEDO, *Crónica*, ed. B. Sánchez Alonso (Textos latinos de la edad media española, 3; Madrid, 1924).

PETER DAMIAN, 'Opusculum decimum sextum', *PL* 145. 365–80.

PETER OF MAILLEZAIS, 'De antiquitate et commutatione in melius Malleacensis insulae', ed. P. Labbe, *Novae Bibliothecae Manuscriptorum Librorum*, 2 vols. (Paris, 1657), ii. 222–38.

PETER THE VENERABLE, *The Letters*, ed. G. Constable, 2 vols. (Harvard Historical Studies, 78; Cambridge, Mass., 1967).

PETER TUDEBODE, *Historia de Hierosolymitano Itinere*, ed. J. H. and L. L. Hill (Documents relatifs à l'histoire des croisades, 12; Paris, 1977).

'Le Plus Ancien Cartulaire de Saint-Mont (Gers) (XIe–XIIIe siècles)', ed. C. Samaran, *Bibliothèque de l'École des Chartes*, 110 (1952), pp. 5–56.

Le Pontifical romano-germanique du dixième siècle, ed. C. Vogel and R. Elze, 3 vols. (Studi e Testi, 226–7 and 269; Vatican City, 1963–72).

'Premier cartulaire de l'aumônerie de S. Martial', ed. A. Leroux, E. Molinier, and A. Thomas, *Documents historiques bas-latins, provençaux et français concernant principalement la Marche et le Limousin*, 2 vols. (Limoges, 1883–5), ii, pp. 1–17.

Primera Crónica General de España, ed. R. Menéndez Pidal, 3rd edn., 1 vol. in 2 (Fuentes Cronísticas de la Historia de España, 1; Madrid, 1977).

RALPH GLABER, 'Historiarum Libri Quinque', ed. and trans. J. France, in *Rodulfus Glaber Opera*, ed. J. France, N. Bulst, and P. Reynolds (Oxford, 1989), 1–253.

RALPH OF CAEN, 'Gesta Tancredi in expeditione Hierosolymitana', *RHC Occ.* 3. 603–716.

RAYMOND OF AGUILERS, *Le 'Liber'*, ed. J. H. and L. L. Hill (Documents relatifs à l'histoire des croisades, 9; Paris, 1969).

Recueil des chartes de l'abbaye de Cluny, ed. A. Bernard and A. Bruel, 6 vols. (Paris, 1876–1903).

REGINO OF PRÜM, 'De ecclesiasticis disciplinis et religione christiana', *PL* 132. 185–400.

La Règle de saint Benoît, ed. and trans. A. de Vogüé and J. Neufville, 3 vols. (Sources chrétiennes, 181–3; Paris, 1972).

REMIGIUS OF REIMS, 'Epistolae ad Chlodoveum Regem, et alios', ed. A. Du Chesne, *Historiae Francorum scriptores coaetani*, 5 vols. (Paris, 1636–49), i. 847–75.

RICHARD THE POITEVIN, 'Chronicon', extr. *RHGF* 7. 258–9; 10. 263–4; 11. 285–6; 12. 411–17.

ROBERT OF TORIGNY, *Chronica*, ed. R. Howlett (RS 82[4]; London, 1889).

ROBERT THE MONK, 'Historia Iherosolimitana', *RHC Occ.* 3. 721–882.

RODRIGO JIMÉNEZ DE RADA, *Historia de Rebus Hispaniae sive Historia Gothica*, ed. J. Fernández Valverde (Corpus Christianorum, Continuatio Mediaeualis, 72; Turnhout, 1987).

Saint-Denis de Nogent-le-Rotrou 1031–1789: histoire et cartulaire, ed. Vicomte de Souancé and C. Métais (Archives du diocèse de Chartres, 1; Vannes, 1899).

'Sancti Stephani Lemovicensis Cartularium', ed. J. de Font-Réaulx, *BSAHL* 69 (1922), pp. 5–258.

Le Siège de Barbastre: chanson de geste du XIIe siècle, ed. J. L. Perrier (Les classiques français du Moyen Âge, 54; Paris, 1926).

SIGEBERT OF GEMBLOUX, 'Chronica', *MGH SS* 6. 300–74.

'Sigeberti Gemblacensis Auctarium Laudunense', *MGH SS* 6. 445–7.

The Song of Roland, ed. and trans. G. J. Brault, 2 vols. (Phil., 1978).

STEPHEN MALEU, *Chronique*, ed. F. Arbellot (St-Junien, 1847).

SUGER, *Vie de Louis VI le Gros*, ed. and trans. H. Waquet (Les classiques de l'histoire de France au Moyen Âge, 11; Paris, 1929).

'Textes anciens sur la discipline monastique', ed. H. Rochais, *RM* 43 (1953), 41–7.

ULRICH OF CLUNY, 'Antiquiores consuetudines Cluniacensis monasterii', *PL* 149. 635–778.

URBAN II, 'Epistolae et privilegia', *PL* 151. 283–558.

'Varia chronicorum fragmenta ab ann. DCCC.XLVIII. ad ann. 1658', ed. H. Duplès-Agier, *Chroniques de Saint-Martial de Limoges* (Paris, 1874), 186–216.

'De vera et falsa Poenitentia', *PL* 40. 1113–30.

'La Vie ancienne de S. Front', ed. M. Coens, *AB* 48 (1930), 324–60.

'La Vie de saint Gaucher, fondateur des chanoines réguliers d'Aureil en Limousin', ed. J. Becquet, *RM* 54 (1964), 25–55.

'Vita Aridii abbatis Lemovicini', *MGH Scriptores rerum Merovingicarum*, 3. 576–609.

'Vita Beati Gaufredi', ed. A. Bosvieux, *Mémoires de la Société des Sciences Naturelles et Archéologiques de la Creuse*, 3 (1862), 75–160.

'Vita et miracula S. Leonardi Nobiliacensia', *AASS* (Nov.) 3. 148–73.

'Vita S. Fronti ep. Petragoricensis', ed. M. Coens, in 'La *Scriptura de sancto Fronto nova*, attribuée au chorévêque Gauzbert', *AB* 75 (1957) 351–65.

'Vita S. Frontonis, Episcopi Petragoricensis', *AASS* (Oct.) 11. 407–14.

'Vita S. Geraldi abbatis', *AASS* (Apr.) 1. 414–23.

'Vita S. Leonardi', in 'Vita et miracula S. Leonardi Nobiliacensia', *AASS* (Nov.) 3. 149–55.

'Vita S. Martialis Episcopi Lemovicensis et Galliarum Apostoli', ed. L. Surius, *De Probatis Sanctorum Vitis*, 12 vols. (Cologne, 1618), vi. 365–74.

'Vita S. Romani presbyteri et confessoris', *AB* 5 (1886), 177–91.

'Vita sancti Romani sacerdotis Blaviensis', ed. G. Vielhaber, *AB* 26 (1907), 52–8.

Vita Sanctissimi Martialis Apostoli, ed. W. de G. Birch (London, 1872).

WALRAM OF NAUMBURG, 'Vita et miracula S. Leonardi', *AASS* (Nov.) 3. 173–82.

WILLIAM OF JUMIÈGES, *Gesta Normannorum Ducum*, ed. J. Marx (Rouen, 1914).

WILLIAM OF MALMESBURY, *De Gestis Regum Anglorum Libri Quinque*, ed. W. Stubbs, 1 vol. in 2 (RS 90; London, 1887–9).

WILLIAM OF POITIERS, *Histoire de Guillaume le Conquérant*, ed. and trans. R. Foreville (Les classiques de l'histoire de France au Moyen Âge, 23; Paris, 1952).

WILLIAM OF TYRE, *Chronicon*, ed. R. B. C. Huygens, 1 vol. in 2 (Corpus Christianorum, Continuatio Mediaeualis, 63/63A; Turnhout, 1986).

SECONDARY WORKS

ALBE, E., 'La Vie et les miracles de S. Amator', *AB* 28 (1909), 57–90.

ALPHANDÉRY, P., and DUPRONT, A., *La Chrétienté et l'idée de croisade*, 2 vols. (L'évolution de l'humanité, 38; Paris, 1954–9).

ALTHOFF, G., 'Nunc fiant Christi milites, qui dudum extiterunt raptores: Zur Entstehung von Rittertum und Ritterethos', *Saeculum*, 32 (1981), 317–33.

ANCIAUX, P., *La Théologie du Sacrement de Pénitence au XII* siècle (Universitas Catholica Lovaniensis, Dissertationes ad Gradum Magistri in Facultate Theologica vel in Facultate Iuris Canonici consequendum conscriptae[2], 41; Louvain, 1949).

ANGENENDT, A., 'Theologie und Liturgie der mittelalterlichen Toten-Memoria', in K. Schmid and J. Wollasch (edd.), *Memoria: Der geschichtliche Zeugniswert des liturgischen Gedenkens im Mittelalter* (Münstersche Mittelalter-Schriften, 48; Munich, 1984), 79–199.

ANGOLD, M., *The Byzantine Empire 1025–1204: A Political History* (London, 1984).

ARBELLOT, F., *Vie de saint Léonard solitaire en Limousin: ses miracles et son culte* (Paris, 1863).

ARCHER, T. A., 'Giffard of Barbastre', *EHR* 18 (1903), 303–5.

ARIÈS, P., *The Hour of Our Death*, trans. H. Weaver (London, 1983).

ARNOLD, B., *German Knighthood 1050–1300* (Oxford, 1985).

AUBRUN, M., 'Le Prieur Geoffroy du Vigeois et sa chronique', *RM* 58 (1974), 313–26.

—— *L'Ancien Diocèse de Limoges des origines au milieu du XIᵉ siècle* (Publication de l'Institut d'Études du Massif Central, 21; Clermont-Ferrand, 1981).

AVRIL, J., 'La Pastorale des malades et des mourants aux XIIᵉ et XIIIᵉ siècles', in H. Braet and W. Verbeke (edd.), *Death in the Middle Ages* (Mediaevalia Lovaniensia¹, 9; Louvain, 1983), 88–106.

—— 'Observance monastique et spiritualité dans les préambules des actes (xᵉ– XIIIᵉ siècle)', *Revue d'histoire ecclésiastique*, 85 (1990), 5–29.

BACHRACH, B. S., 'Toward a Reappraisal of William the Great, Duke of Aquitaine (995–1030)', *Journal of Medieval History*, 5 (1979), 11–21.

BAKER, D., '*Vir Dei*: Secular Sanctity in the Early Tenth Century', in G. J. Cuming and D. Baker (edd.), *Popular Belief and Practice* (Studies in Church History, 8; Cambridge, 1972), 41–53.

BALAGUER, F., 'Los límites del obispado de Aragón y el Concilio de Jaca en 1063', *EEMCA* 4 (1951), 69–138.

—— 'La vizcondesa del Bearn doña Talesa y la rebelión contra Ramiro II en 1136', *EEMCA* 5 (1952), 83–114.

BATES, D., *Normandy before 1066* (London, 1982).

BECKER, A., *Papst Urban II. (1088–1099)*, 2 vols. (MGH Schriften, 19; Stuttgart, 1964–88).

BECQUET, J., 'La Mort d'un évêque de Périgueux à la Première Croisade: Raynaud de Thiviers', *Bulletin de la Société Historique et Archéologique du Périgord*, 87 (1960), 66–9.

—— 'Les Chanoines réguliers en Limousin aux XIᵉ et XIIᵉ siècles', *Analecta Praemonstratensia*, 36 (1960), 193–235.

—— 'Les Chanoines réguliers du Chalard (Haute-Vienne)', *BSAHL* 98 (1971), 153–72.

—— 'Les Chanoines réguliers de Lesterps, Bénévent et Aureil en Limousin aux XIᵉ et XIIᵉ siècles', *BSAHL* 99 (1972), 80–135.

—— 'Le Bullaire du Limousin', *BSAHL* 100 (1973), 111–49; 109 (1982), 53–69.

BECQUET, J., 'Chanoines réguliers en Limousin au XIIe siècle: sanctuaires régularisés et dépendances étrangères', *BSAHL* 101 (1974), 67–111.

——— 'Les Évêques de Limoges aux Xe, XIe et XIIe siècles', *BSAHL* 104 (1977), 63–90; 105 (1978), 79–104; 106 (1979), 85–114; 107 (1980), 109–41; 108 (1981), 98–116.

——— 'Un acte de l'évêque Pierre de Limoges (1101)', *BSAHL* 112 (1985), 14–19.

BEECH, G. T., 'Contemporary Views of William the Troubadour, IXth Duke of Aquitaine (1086–1126)', in N. Bulst and J.-P. Genet (edd.), *Medieval Lives and the Historian: Studies in Medieval Prosopography* (Kalamazoo, Mich., 1986), 73–89.

BEITSCHER, J. K., 'Monastic Reform at Beaulieu 1031–1095', *Viator*, 5 (1974), 199–210.

BISHKO, C. J., 'Fernando I and the Origins of the Leonese-Castilian Alliance with Cluny', *Studies in Medieval Spanish Frontier History* (London, 1980).

BISSON, T. N., 'The Organized Peace in Southern France and Catalonia, ca.1140–ca.1233', *AHR* 82 (1977), 290–311.

BLANC, C., 'Les Pratiques de piété des laïcs dans les pays du Bas-Rhône aux XIe et XIIe siècles', *AM* 72 (1960), 137–47.

BLUMENTHAL, U.-R., *The Early Councils of Pope Paschal II 1100–1110* (Pontifical Institute of Mediaeval Studies, Studies and Texts, 43; Toronto, 1978).

BOEHM, L., 'Gedanken zum Frankreich-Bewußtsein im frühen 12. Jahrhundert', *Historisches Jahrbuch*, 74 (1955), 681–7.

——— 'Die "Gesta Tancredi" des Radulf von Caen: Ein Beitrag zur Geschichtsschreibung der Normannen um 1100', *Historisches Jahrbuch*, 75 (1956), 47–72.

BOISSONNADE, P., *Du nouveau sur la Chanson de Roland* (Paris, 1923).

BONNASSIE, P., *La Catalogne du milieu du Xe à la fin du XIe siècle: croissance et mutations d'une société*, 1 vol. in 2 (Publications de l'Université de Toulouse-Le Mirail^A, 23 and 29; Toulouse, 1975–6).

BONNAUD-DELAMARE, R., 'La Paix en Flandre pendant la Première Croisade', *Revue du Nord*, 39 (1957), 147–52.

——— 'Les Institutions de paix en Aquitaine au XIe siècle', *Recueils de la Société Jean Bodin*, 14 (1961), 415–87.

——— 'La Paix de Touraine pendant la Première Croisade', *Revue d'histoire ecclésiastique*, 70 (1975), 749–57.

BOSCH VILÁ, J., 'Al-Bakri: dos fragmentos sobre Barbastro en el "Bayan Al-Mugrib"', *EEMCA* 3 (1947–8), 242–61.

BOÜARD, A. DE, *Manuel de diplomatique française et pontificale*, 2 vols. (Paris, 1929–48).

BOUCHARD, C. B., *Sword, Miter, and Cloister: Nobility and the Church in Burgundy 980–1198* (Ithaca, NY, 1987).

BOUILLET, A., and SERVIÈRES, L., *Sainte Foy, vierge et martyre*, 1 vol. in 2 (Rodez, 1900).

BREDERO, A. H., 'Le Moyen Âge et le Purgatoire', *Revue d'histoire ecclésiastique*, 78 (1983), 429–52.

BREUILS, A., *Saint Austinde archevêque d'Auch (1000–1068) et la Gascogne au XIe siècle* (Auch, 1895).

BRÜCKNER, W., 'Sterben im Mönchsgewand: Zum Funktionswandel einer Totenkleidsitte', in H. F. Foltin *et al.* (edd.), *Kontakte und Grenzen: Probleme der Volks-, Kultur- und Sozialforschung. Festschrift für Gerhard Heilfurth zum 60. Geburtstag* (Göttingen, 1969), 259–77.

BRUNDAGE, J. A., *Medieval Canon Law and the Crusader* (Madison, Wis., 1969).

—— *Law, Sex, and Christian Society in Medieval Europe* (Chicago, 1987).

CALLAHAN, D. F., 'The Sermons of Adémar of Chabannes and the Cult of St. Martial of Limoges', *RB* 86 (1976), 251–95.

—— 'Adémar de Chabannes et la paix de Dieu', *AM* 89 (1977), 21–43.

—— 'William the Great and the Monasteries of Aquitaine', *Studia Monastica*, 19 (1977), 321–42.

—— 'The Peace of God and the Cult of the Saints in Aquitaine in the Tenth and Eleventh Centuries', *HR* 14 (1987), 445–66.

CALZADA, L. DE LA, 'La proyección del pensamiento de Gregorio VII en los reinos de Castilla y León', *Studi Gregoriani*, 3 (1948), 1–87.

CASTAING-SICARD, M., 'Donations toulousaines du Xe au XIIIe siècle', *AM* 70 (1958), 27–64.

CATE, J. L., 'A Gay Crusader', *Byzantion*, 16 (1942–3), 503–26.

CHANDLER, V., 'Politics and Piety: Influences on Charitable Donations during the Anglo-Norman Period', *RB* 90 (1980), 63–71.

CHAPEAU, G., 'Un pèlerinage noble à Charroux au XIe siècle', *Bulletin de la Société des Antiquaires de l'Ouest³*, 13 (1943), 250–71.

CHAULIAC, A., *Histoire de l'abbaye Sainte-Croix de Bordeaux* (Archives de la France monastique, 9; Ligugé, 1910).

COENS, M., 'La *Scriptura de sancto Fronto nova*, attribuée au chorévêque Gauzbert', *AB* 75 (1957), 340–65.

COLLINS, R., *The Basques* (Oxford, 1986).

CONRAD, H., 'Gottesfrieden und Heeresverfassung in der Zeit der Kreuzzüge: Ein Beitrag zur Geschichte des Heeresstrafrechts im Mittelalter', *Zeitschrift der Savigny-Stiftung für Rechtsgeschichte: Germanistische Abteilung*, 61 (1941), 71–126.

CONSTABLE, G., *Monastic Tithes from their Origins to the Twelfth Century* (Cambridge, 1964).

—— '"Famuli" and "Conversi" at Cluny: A Note on Statute 24 of Peter the Venerable', *RB* 83 (1973), 326–50.

—— 'The Financing of the Crusades in the Twelfth Century', in Kedar, Mayer, and Smail (edd.), *Outremer*, 64–88.

—— 'Medieval Charters as a Source for the History of the Crusades', in Edbury (ed.), *Crusade and Settlement*, 73–89.

CONTAMINE, P., *War in the Middle Ages*, trans. M. Jones (Oxford, 1984).

COWDREY, H. E. J., 'Unions and Confraternity with Cluny', *JEH* 16 (1965), 152–62.

—— 'Bishop Ermenfrid of Sion and the Penitential Ordinance following the Battle of Hastings', *JEH* 20 (1969), 225–42.

—— *The Cluniacs and the Gregorian Reform* (Oxford, 1970).

—— 'The Peace and the Truce of God in the Eleventh Century', *Past and Present*, 46 (1970), 42–67.

—— 'Pope Urban II's Preaching of the First Crusade', *History*, 55 (1970), 177–88.

—— 'Cluny and the First Crusade', *RB* 83 (1973), 285–311.

—— 'The Origin of the Idea of Crusade', *International History Review*, 1 (1979), 121–5.

—— 'Pope Gregory VII's "Crusading" Plan of 1074', in Kedar, Mayer, and Smail (edd.), *Outremer*, 27–40.

—— *The Age of Abbot Desiderius: Montecassino, the Papacy, and the Normans in the Eleventh and Early Twelfth Centuries* (Oxford, 1983).

—— 'Martyrdom and the First Crusade', in Edbury (ed.), *Crusade and Settlement*, 46–56.

CROZET, R., 'Le Voyage d'Urbain II en France (1095–1096) et son importance au point de vue archéologique', *AM* 49 (1937), 42–69.

—— 'Le Voyage d'Urbain II et ses négociations avec le clergé de France (1095–1096)', *Revue historique*, 179 (1937), 271–310.

DAUPHIN, H., *Le Bienheureux Richard abbé de Saint-Vanne de Verdun* (Bibliothèque de la Revue d'Histoire Ecclésiastique, 24; Louvain, 1946).

DAVID, P., 'Études sur le livre de Saint-Jacques attribué au Pape Calixte II', *Bulletin des études portugaises et de l'Institut Français en Portugal*, 10 (1945), 1–41; 11 (1947), 113–85; 12 (1948), 70–223; 13 (1949), 52–104.

—— *Études historiques sur la Galice et le Portugal du VIᵉ au XIIᵉ siècle* (Collection portugaise, 7; Lisbon, 1947).

DAVIS, R. H. C., *The Normans and their Myth* (London, 1976).

—— *The Medieval Warhorse: Origin, Development and Redevelopment* (London, 1989).

DEBORD, A., *La Société laïque dans les pays de la Charente Xᵉ–XIIᵉ siècles* (Paris, 1984).

DEFOURNEAUX, M., *Les Français en Espagne aux XIᵉ et XIIᵉ siècles* (Paris, 1949).

DELARUELLE, É., 'Essai sur la formation de l'idée de Croisade', *Bulletin de littérature ecclésiastique*, 42 (1941), 24–45, 86–103; 45 (1944), 13–46, 73–90; 54 (1953), 226–39; 55 (1954), 50–63.

—— 'La Culture religieuse des laïcs en France aux XIᵉ et XIIᵉ siècles', in *I laici nella 'Societas Christiana' dei secoli XI e XII: atti della terza settimana internazionale di studio Mendola, 21–27 agosto 1965* (Miscellanea del Centro di Studi Medioevali, 5; Milan, 1968), 548–81.

DELISLE, L., 'Notice sur les manuscrits disparus de la Bibliothèque de Tours pendant la première moitié du XIXᵉ siècle', *Notices et extraits des manuscrits de la Bibliothèque Nationale et autres bibliothèques*, 31 (1884), 157–356.

—— 'Notice sur les manuscrits originaux d'Adémar de Chabannes', *Notices et extraits des manuscrits de la Bibliothèque Nationale et autres bibliothèques*, 35 (1896), 241–358.

DEVAILLY, G., *Le Berry du Xe siècle au milieu du XIIIe: étude politique, religieuse, sociale et économique* (Civilisations et sociétés, 19; Paris, 1973).

DIENER, H., 'Das Itinerar des Abtes Hugo von Cluny', in G. Tellenbach (ed.), *Neue Forschungen über Cluny und die Cluniacenser* (Freiburg, 1959), 353–426.

DINZELBACHER, P., *Vision und Visionsliteratur im Mittelalter* (Monographien zur Geschichte des Mittelalters, 23; Stuttgart, 1981).

DORMEIER, H., *Montecassino und die Laien im 11. und 12. Jahrhundert* (MGH Schriften, 27; Stuttgart, 1979).

DOUGLAS, D. C., *The Norman Achievement 1050–1100* (London, 1969).

DOZY, R., *Recherches sur l'histoire et la littérature de l'Espagne pendant le Moyen Âge*, 3rd edn., 2 vols. (Paris, 1881).

DUBY, G., *La Société aux XIe et XIIe siècles dans la région mâconnaise*, 2nd edn. (Paris, 1971).

—— 'Laity and the Peace of God', *The Chivalrous Society*, trans. C. Postan (London, 1977), 123–33.

—— 'The Origins of Knighthood', *The Chivalrous Society*, trans. C. Postan (London, 1977), 158–70.

—— *The Three Orders: Feudal Society Imagined*, trans. A. Goldhammer (Chicago, 1980).

—— *The Knight, the Lady and the Priest: The Making of Modern Marriage in Medieval France*, trans. B. Bray (Harmondsworth, 1985).

DU CANGE, C. DU F., *Glossarium Mediae et Infimae Latinitatis*, rev. G. A. L. Henschel, 7 vols. (Paris, 1840–50).

DUCHESNE, L., 'Saint Martial de Limoges', *AM* 4 (1892), 289–330.

—— *Fastes épiscopaux de l'ancienne Gaule*, 2nd edn., 3 vols. (Paris, 1907–15).

DUNBABIN, J., *France in the Making 843–1180* (Oxford, 1985).

DURÁN GUDIOL, A., *El hospital de Somport entre Aragón y Bearn (siglos XII y XIII)* (Saragossa, 1986).

EDBURY, P. W. (ed.), *Crusade and Settlement: Papers read at the First Conference of the Society for the Study of the Crusades and the Latin East and presented to R. C. Smail* (Cardiff, 1985).

ENGELS, O., 'Papsttum, Reconquista und spanisches Landeskonzil im Hochmittelalter', *Annuarium Historiae Conciliorum*, 1 (1969), 37–49, 241–87.

—— 'Vorstufen der Staatwerdung im Mittelalter: Zum Kontext der Gottesfriedensbewegung', *Historisches Jahrbuch*, 97/8 (1978), 71–86.

ERDMANN, C., *Die Entstehung des Kreuzzugsgedankens* (Forschungen zur Kirchen- und Geistesgeschichte, 6; Stuttgart, 1935).

—— *The Origin of the Idea of Crusade*, trans. M. W. Baldwin and W. Goffart (Princeton, NJ, 1977).

EWALD, P., *Walram von Naumburg: Zur Geschichte der publicistischen Literatur des XI. Jahrhunderts* (Bonn, 1874).

EWALD, P., 'Die Papstbriefe der Brittischen Sammlung', *Neues Archiv der Gesellschaft für ältere deutsche Geschichtskunde*, 5 (1880), 275–414.

FERREIRO, A., 'The Siege of Barbastro 1064–1065: A Reassessment', *Journal of Medieval History*, 9 (1983), 129–44.

FICHTENAU, H., *Arenga: Spätantike und Mittelalter im Spiegel von Urkundenformeln* (Mitteilungen des Instituts für Österreichische Geschichtsforschung, 18; Graz, 1957).

FIGUERAS, C. M., 'Acerco del rito de la profesión monástica medieval "ad succurrendum"', *Liturgica*, 2 (1958), 359–400.

FITA, F., 'El concilio nacional de Palencia en el año 1100 y el de Gerona en 1101', *BRAH* 24 (1894), 215–35.

FLANDRIN, J.-L., *Un temps pour embrasser: aux origines de la morale sexuelle occidentale (VIᵉ–XIᵉ siècle)* (Paris, 1983).

FLETCHER, R. A., *Saint James's Catapult: The Life and Times of Diego Gelmírez of Santiago de Compostela* (Oxford, 1984).

—— 'Reconquest and Crusade in Spain c.1050–1150', *TRHS*⁵, 37 (1987), 31–47.

—— *The Quest for El Cid* (London, 1989).

FLORI, J., 'Sémantique et société médiévale: le verbe adouber et son évolution au XIIᵉ siècle', *AESC* 31 (1976), 915–40.

—— 'Les Origines de l'adoubement chevaleresque: étude des remises d'armes et du vocabulaire qui les exprime dans les sources historiques latines jusqu'au début du XIIIᵉ siècle', *Traditio*, 35 (1979), 209–72.

—— *L'Idéologie du glaive: préhistoire de la chevalerie* (Travaux d'histoire éthico-politique, 43; Geneva, 1983).

—— *L'Essor de la chevalerie XIᵉ–XIIᵉ siècles* (Travaux d'histoire éthico-politique, 46; Geneva, 1986).

—— 'Encore l'usage de la lance . . . la technique du combat chevaleresque vers l'an 1100', *CCM* 31 (1988), 213–40.

—— 'Mort et martyre des guerriers vers 1100: l'exemple de la Première Croisade', *CCM* 34 (1991), 121–39.

FONTETTE, F. DE, 'Évêques de Limoges et Comtes de Poitou au XIᵉ siècle', in *Études d'histoire du droit canonique dédiées à Gabriel Le Bras*, 1 vol. in 2 (Paris, 1965), i. 553–8.

FOREY, A. J., *The Templars in the Corona de Aragón* (London, 1973).

—— 'The Military Orders and the Spanish Reconquest in the Twelfth and Thirteenth Centuries', *Traditio*, 40 (1984), 197–234.

FORSYTH, I. H., *The Throne of Wisdom: Wood Sculptures of the Madonna in Romanesque France* (Princeton, NJ, 1972).

FOURNIER, G., *Le Peuplement rural en Basse Auvergne durant le Haut Moyen Âge* (Publications de la Faculté des Lettres et Sciences Humaines de Clermont-Ferrand², 12; Aurillac, 1962).

—— *Le Château dans la France médiévale: essai de sociologie monumentale* (Paris, 1978).

FOURNIER, P., and LE BRAS, G., *Histoire des collections canoniques en Occident depuis les Fausses Décrétales jusqu'au Décret de Gratien*, 2 vols. (Paris, 1931–2).

FRANÇOIS, M. (ed.), *Le Livre des miracles de Notre-Dame de Rocamadour* (2ᵉ Colloque de Rocamadour; Rocamadour, 1973).

FRIED, J., *Der päpstliche Schutz für Laienfürsten: Die politische Geschichte des päpstlichen Schutzprivilegs für Laien (11.–13. Jahrhundert)* (Abhandlungen der Heidelberger Akadamie der Wissenschaften, Phil.-hist. Kl. 1980, 1; Heidelberg, 1980).

FROLOW, A., *La Relique de la Vraie Croix: recherches sur le développement d'un culte* (Archives de l'Orient Chrétien, 7; Paris, 1961).

GALLAIS, P., and RIOU, Y.-J. (edd.), *Mélanges offerts à René Crozet*, 1 vol. in 2 (Poitiers, 1966).

GARAUD, M., *Les Châtelains de Poitou et l'avènement du régime féodal XIᵉ et XIIᵉ siècles* (Mémoires de la Société des Antiquaires de l'Ouest⁴, 8; Poitiers, 1964).

GEARY, P. J., *Furta Sacra: Thefts of Relics in the Central Middle Ages*, rev. edn. (Princeton, NJ, 1990).

GENICOT, L., *Les Généalogies* (Typologie des sources du Moyen Âge occidental, 15; Turnhout, 1975).

GIEYSZTOR, A., 'The Genesis of the Crusades: The Encyclical of Sergius IV (1009–1012)', *Medievalia et Humanistica*, 5 (1948), 3–23; 6 (1950), 3–34.

GILCHRIST, J. T., 'The Erdmann Thesis and the Canon Law 1083–1141', in Edbury (ed.), *Crusade and Settlement*, 37–45.

GIRY, A., *Manuel de diplomatique*, 2nd edn., 2 vols. (Paris, 1925).

GOETZ, H.-W., 'Kirchenschutz, Rechtswahrung und Reform: Zu den Zielen und zum Wesen der frühen Gottesfriedensbewegung in Frankreich', *Francia*, 11 (1983), 193–239.

GOLDING, B., 'Anglo-Norman Knightly Burials', in Harper-Bill and Harvey (edd.), *Ideals and Practice of Medieval Knighthood from the First and Second Conferences*, 35–48.

GOÑI GAZTAMBIDE, J., *Historia de la Bula de la Cruzada en España* (Victoriensia, 4; Vitoria, 1958).

GONZÁLEZ MIRANDA, M., 'La condesa doña Sancha y el monasterio de Santa Cruz de la Serós', *EEMCA* 6 (1956), 185–202.

GOTTLOB, A., *Kreuzablass und Almosenablass: Eine Studie über die Frühzeit des Ablasswesens* (Kirchenrechtliche Abhandlungen, 30–1; Stuttgart, 1906).

GOUGAUD, L., *Devotional and Ascetic Practices in the Middle Ages*, trans. G. C. Bateman (London, 1927).

GRUNDMANN, H., 'Adelsbekehrungen im Hochmittelalter: *Conversi* und *nutriti* im Kloster', in J. Fleckenstein and K. Schmid (edd.), *Adel und Kirche: Gerd Tellenbach zum 65. Geburtstag dargebracht von Freunden und Schülern* (Freiburg, 1968), 325–45.

GUILLOT, O., *Le Comte d'Anjou et son entourage au XIᵉ siècle*, 2 vols. (Paris, 1972).

GUREVICH, A. J., 'Au Moyen Âge: conscience individuelle et image de l'au-delà', *AESC* 37 (1982), 255–75.

GUREVICH, A. J., 'Popular and Scholarly Medieval Cultural Traditions: Notes in the Margin of Jacques Le Goff's Book', *Journal of Medieval History*, 9 (1983), 71–90.

—— *Medieval Popular Culture: Problems of Belief and Perception*, trans. J. M. Bak and P. A. Hollingsworth (Cambridge, 1988).

HALPHEN, L., *Le Comté d'Anjou au XIe siècle* (Paris, 1906).

HARPER-BILL, C., 'The Piety of the Anglo-Norman Knightly Class', in R. A. Brown (ed.), *Proceedings of the Battle Conference on Anglo-Norman Studies, ii. 1979* (Woodbridge, 1980), 63–77.

—— and HARVEY, R. (edd.), *The Ideals and Practice of Medieval Knighthood: Papers from the First and Second Strawberry Hill Conferences* (Woodbridge, 1986).

—— and HARVEY, R. (edd.), *The Ideals and Practice of Medieval Knighthood II. Papers from the Third Strawberry Hill Conference 1986* (Woodbridge, 1988).

HEAD, T., 'Andrew of Fleury and the Peace League of Bourges', *HR* 14 (1987), 513–29.

—— *Hagiography and the Cult of Saints: The Diocese of Orléans 800–1200* (Cambridge, 1990).

HERBERS, K., *Der Jakobuskult des 12. Jahrhunderts und der 'Liber Sancti Jacobi'. Studien über das Verhältnis zwischen Religion und Gesellschaft im hohen Mittelalter* (Historische Forschungen, 7; Wiesbaden, 1984).

HIGOUNET, C., 'Les Chemins de Saint-Jacques et les sauvetés de Gascogne', *AM* 63 (1951), 293–304.

—— 'En Bordelais: "Principes castella tenentes"', in P. Contamine (ed.), *La Noblesse au Moyen Âge XIe–XVe siècles: essais à la mémoire de Robert Boutruche* (Paris, 1976), 97–104.

HILL, J. H., and HILL, L. L., *Raymond IV Count of Toulouse* (Syracuse, NY, 1962).

HOFFMANN, H., *Gottesfriede und Treuga Dei* (MGH Schriften, 20; Stuttgart, 1964).

HOHLER, C., 'A Note on *Jacobus*', *Journal of the Warburg and Courtauld Institutes*, 35 (1972), 31–80.

HOUSLEY, N., *The Italian Crusades: The Papal–Angevin Alliance and the Crusades against Christian Lay Powers 1254–1343* (Oxford, 1982).

—— *The Avignon Papacy and the Crusades 1305–1378* (Oxford, 1986).

HUICI MIRANDA, A., *Historia musulmana de Valencia y su región: novedades y rectificaciones*, 3 vols. (Valencia, 1969–70).

HÜLS, R., *Kardinäle, Klerus und Kirchen Roms 1049–1130* (Bibliothek des Deutschen Historischen Instituts in Rom, 48; Tübingen, 1977).

HUNT, N., *Cluny under Saint Hugh 1049–1109* (London, 1967).

JAURGAIN, J. DE, *La Vasconie*, 2 vols. (Pau, 1898–1902).

JOHNSON, P. D., *Prayer, Patronage, and Power: The Abbey of la Trinité, Vendôme, 1032–1187* (New York, 1981).

JOHRENDT, J., '"Milites" und "Militia" im 11. Jahrhundert in Deutschland', in A. Borst (ed.), *Das Rittertum im Mittelalter* (Wege der Forschung, 349; Darmstadt, 1976), 419–36.

JOMBART, E., 'Indulgences', *DDC* 5 (1953), 1331–52.

JORANSON, E., 'The Great German Pilgrimage of 1064–1065', in L. J. Paetow (ed.), *The Crusades and Other Historical Essays Presented to Dana C. Munro by his Former Students* (New York, 1928), 3–43.

JORDEN, W., *Das cluniazensische Totengedächtniswesen* (Münsterische Beiträge zur Theologie, 15; Münster, 1930).

JUILLET, J., 'Lieux et chemins', in François (ed.), *Le Livre des miracles*, 25–43.

KEDAR, B. Z., MAYER, H. E., and SMAIL, R. C. (edd.), *Outremer: Studies in the History of the Crusading Kingdom of Jerusalem Presented to Joshua Prawer* (Jerusalem, 1982).

KEEN, M. H., *Chivalry* (New Haven, Conn., 1984).

KEHR, P., 'Cómo y cuándo se hizo Aragón feudatario de la Santa Sede', *EEMCA* 1 (1945), 285–326.

—— 'El Papado y los reinos de Navarra y Aragón hasta mediados del siglo XII', *EEMCA* 2 (1946), 74–179.

KENDRICK, T. D., *St James in Spain* (London, 1960).

KIENAST, W., *Der Herzogstitel in Frankreich und Deutschland (9. bis 12. Jahrhundert)* (Munich, 1968).

KÖRNER, T., *Iuramentum und frühe Friedensbewegung (10.–12. Jahrhundert)* (Münchener Universitätsschriften: Juristische Fakultät, Abhandlungen zur rechtswissenschaftlichen Grundlagenforschung, 26; Berlin, 1977).

LABANDE, E.-R., 'Recherches sur les pèlerins dans l'Europe des XIe et XIIe siècles', *CCM* 1 (1958), 159–69, 339–47.

—— 'Situation de l'Aquitaine en 1066', *Bulletin de la Société des Antiquaires de l'Ouest*⁴, 8 (1966), 339–63.

LACARRA, J. M., 'La conquista de Zaragoza por Alfonso I (18 diciembre 1118)', *Al-Andalus*, 12 (1947), 65–96.

—— 'Gastón de Bearn y Zaragoza', *Pirineos*, 8 (1952), 127–36.

—— 'A propos de la colonisation "franca" en Navarre et en Aragon', *AM* 65 (1953), 331–42.

—— 'Aspectos económicos de la sumisión de los reinos de taifas (1010–1102)', in J. Malaquer de Motes (ed.), *Homenaje a Jaime Vicens Vivens*, 2 vols. (Barcelona, 1965–7), i. 255–77.

—— 'Los franceses en la reconquista y repoblación del valle del Ebro en tiempos de Alfonso el Batallador', *Cuadernos de Historia*, 2 (1968), 65–80.

—— *Vida de Alfonso el Batallador* (Saragossa, 1971).

LAHARIE, M., 'Évêques et société en Périgord du Xe au milieu du XIIe siècle', *AM* 94 (1982), 343–68.

LAIR, J., *Études critiques sur divers textes des Xe et XIe siècles*, ii. *Historia d'Adémar de Chabannes* (Paris, 1899). (Includes parallel editions of the various recensions of Adhemar's chronicle, bk. III, chs. 16–70.)

LANDES, R., 'The Dynamics of Heresy and Reform in Limoges: A Study of Popular Participation in the "Peace of God" (994–1033)', *HR* 14 (1987), 467–511.

LASTEYRIE, C. DE, *L'Abbaye de Saint-Martial de Limoges: étude historique, économique et archéologique, précédée de recherches nouvelles sur la vie du saint* (Paris, 1901).

LAURANSON-ROSAZ, C., *L'Auvergne et ses marges (Velay, Gévaudan) du VIIIᵉ au XIᵉ siècle* (Le Puy, 1987).

LAWRENCE, C. H., *Medieval Monasticism: Forms of Religious Life in Western Europe in the Middle Ages* (London, 1984).

LECLER, A., *Dictionnaire topographique, archéologique et historique de la Creuse* (Limoges, 1902).

LECLERCQ, J., 'La Vêture "ad succurrendum" d'après le moine Raoul', *Analecta Monastica*³ (Studia Anselmiana, 37; Rome, 1955), 158–68.

LECOUTEUX, C., *Geschichte der Gespenster und Wiedergänger im Mittelalter* (Cologne, 1987).

LE GOFF, J., *The Birth of Purgatory*, trans. A. Goldhammer (London, 1984).

LEMAÎTRE, J.-L., 'Les Confraternités de La Sauve-Majeure', *Revue historique de Bordeaux et du Département de la Gironde*, NS 28 (1981), 5–34.

—— 'Note sur le texte de la *Chronique* d'Étienne Maleu, chanoine de Saint-Junien', *RM* 60 (1982), 175–91.

—— *Mourir à Saint-Martial: la commémoration des morts et les obituaires à Saint-Martial de Limoges du XIᵉ au XIIIᵉ siècle* (Paris, 1989).

LEVILLAIN, L., 'Adémar de Chabannes, généalogiste', *Bulletin de la Société des Antiquaires de l'Ouest*³, 10 (1934), 237–63.

LEWIS, A. R., 'Count Gerald of Aurillac and Feudalism in South Central France in the Early Tenth Century', *Traditio*, 20 (1964), 41–58.

LEWIS, P. A., 'Mortgages in the Bordelais and Bazadais', *Viator*, 10 (1979), 23–38.

LINEHAN, P., 'Religion, Nationalism and National Identity in Medieval Spain and Portugal', in S. Mews (ed.), *Religion and National Identity* (Studies in Church History, 18; Oxford, 1982), 161–99.

LITTLE, L. K., 'Pride goes before Avarice: Social Change and the Vices in Latin Christendom', *AHR* 76 (1971), 16–49.

—— 'Formules monastiques de malédiction aux IXᵉ et Xᵉ siècles', *RM* 58 (1975), 377–99.

—— *Religious Poverty and the Profit Economy in Medieval Europe* (London, 1978).

—— 'La Morphologie des malédictions monastiques', *AESC* 34 (1979), 43–60.

LOMAX, D. W., 'The First English Pilgrims to Santiago de Compostela', in Mayr-Harting and Moore (edd.), *Studies in Medieval History*, 165–75.

LOTTER, F., 'Das Idealbild adliger Laienfrömmigkeit in den Anfängen Clunys: Odos Vita des Grafen Gerald von Aurillac', in W. Lourdaux and D. Verhelst (edd.), *Benedictine Culture 750–1050* (Mediaevalia Lovaniensia¹, 11; Louvain, 1983), 76–95.

LUGGE, M., *'Gallia' und 'Frankreich' im Mittelalter: Untersuchungen über den Zusammenhang zwischen geographisch-historischer Terminologie und politischem Denken vom 6.–15. Jahrhundert* (Bonner Historische Forschungen, 15; Bonn, 1960).

LYNCH, J. H., 'Monastic Recruitment in the Eleventh and Twelfth Centuries:

Some Social and Economic Considerations', *American Benedictine Review*, 26 (1975), 425–47.

—— *Simoniacal Entry into Religious Life from 1000 to 1260* (Columbus, Ohio, 1976).

McGINN, B., 'Iter Sancti Sepulchri: The Piety of the First Crusaders', in B. K. Lackner and K. R. Philp (edd.), *Essays on Medieval Civilization* (The Walter Prescott Webb Memorial Lectures, 12; Austin, Tex., 1978), 33–71.

McGUIRE, B. P., 'Purgatory, the Communion of Saints, and Medieval Change', *Viator*, 20 (1989), 61–84.

MACKINNEY, L. C., 'The People and Public Opinion in the Eleventh-Century Peace Movement', *Speculum*, 5 (1930), 181–206.

MARTINDALE, J., 'The Origins of the Duchy of Aquitaine and the Government of the Counts of Poitou (902–1137)', D. Phil. thesis (Oxford, 1965).

—— 'Conventum inter Guillelmum Aquitanorum comes et Hugonem Chiliarchum', *EHR* 84 (1969), 528–48.

—— ' "Cavalaria et Orgueill": Duke William IX of Aquitaine and the Historian', in Harper-Bill and Harvey (edd.), *Ideals and Practice of Medieval Knighthood II*, 87–116.

MAYER, H. E., *The Crusades*, trans. J. Gillingham, 2nd edn. (Oxford, 1988).

MAYR-HARTING, H., and MOORE, R. I. (edd.), *Studies in Medieval History Presented to R. H. C. Davis* (London, 1985).

MÉGIER, E., 'Deux exemples de "prépurgatoire" chez les historiens: à propos de *La Naissance du Purgatoire* de Jacques Le Goff', *CCM* 28 (1985), 45–62.

MENÉNDEZ PIDAL, R., *La España del Cid*, 5th edn., 1 vol. in 2 (Madrid, 1956).

MOISAN, A., 'Aimeri Picaud de Parthenay et le "Liber Sancti Jacobi" ', *Bibliothèque de l'École des Chartes*, 143 (1985), 5–52.

MOORE, R. I., 'Family, Community and Cult on the Eve of the Gregorian Reform', *TRHS*[5], 30 (1980), 49–69.

MORISON, P. R., 'The Miraculous and French Society, circa 950–1100', D. Phil. thesis (Oxford, 1984).

MORTIMER, R., 'Religious and Secular Motives for some English Monastic Foundations', in D. Baker (ed.), *Religious Motivation: Biographical and Sociological Problems for the Church Historian* (Studies in Church History, 15; Oxford, 1978), 77–85.

MÜLLER, K., 'Der Umschwung in der Lehre von der Busse während des 12. Jahrhunderts', in *Theologische Abhandlungen Carl von Weizsäcker zu seinem siebzigsten Geburtstage 11. December 1892 gewidmet* (Freiburg, 1892) 287–320.

MURATORI, E., 'L'assedio di Barbastro, prima crociata di Spagna, e la canzone di gesta omonima: occasioni della storia e scarto retorico', *Francofonia*, 8 (1985), 23–35.

MURRAY, A., *Reason and Society in the Middle Ages*, rev. edn. (Oxford, 1985).

MUSSET, L., 'Recherches sur les pèlerins et les pèlerinages en Normandie jusqu'à la Première Croisade', *Annales de Normandie*, 12 (1962), 127–50.

MÜSSIGBROD, A., *Die Abtei Moissac 1050–1150: Zu einem Zentrum cluniacensischen*

Mönchtums in Südwestfrankreich (Münstersche Mittelalter-Schriften, 58; Munich, 1988).

MUSSOT-GOULARD, R., *Les Princes de Gascogne* (Marsolan, 1982).

NELSON, L. H., 'Rotrou of Perche and the Aragonese Reconquest', *Traditio*, 26 (1970), 113–33.

NOTH, A., *Heiliger Krieg und Heiliger Kampf in Islam und Christentum: Beiträge zur Vorgeschichte und Geschichte der Kreuzzüge* (Bonner Historische Forschungen, 28; Bonn, 1966).

OURY, G. M., 'Gérard de Corbie avant son arrivée à la Sauve-Majeure', *RB* 90 (1980), 306–14.

—— 'La Spiritualité du fondateur de La Sauve-Majeure, saint Gérard (v. 1020–1095)', *Revue historique de Bordeaux et du Département de la Gironde*, NS 29 (1982), 5–19.

PAUL, J., *L'Église et la culture en Occident IXe–XIIe siècles*, 1 vol. in 2 (Nouvelle Clio, 15; Paris, 1986).

PAULUS, N., *Geschichte des Ablasses im Mittelalter vom Ursprunge bis zur Mitte des 14. Jahrhunderts*, 2 vols. (Paderborn, 1922–3).

PAXTON, F. S., 'The Peace of God in Modern Historiography: Perspectives and Trends', *HR* 14 (1987), 385–404.

PEIRCE, I., 'The Knight, his Arms and Armour in the Eleventh and Twelfth Centuries', in Harper-Bill and Harvey (edd.), *Ideals and Practice from the First and Second Conferences*, 152–64.

PÉREZ DE URBEL, J., *Sancho el Mayor de Navarra* (Madrid, 1950).

PERNOUD, R., 'Le Livre des miracles de Notre-Dame de Rocamadour: étude des manuscrits de 1172', in François (ed.), *Le Livre des miracles*, 9–23.

PHILLIPS, J. P., 'A Note on the Origins of Raymond of Poitiers', *EHR* 106 (1991), 66–7.

POECK, D., 'Laienbegräbnisse in Cluny', *Frühmittelalterliche Studien*, 15 (1981), 68–179.

POLY, J.-P., *La Provence et la société féodale (879–1166): contribution à l'étude des structures dites féodales dans le Midi* (Paris, 1976).

PONCELET, A., 'La Plus Ancienne Vie de S. Géraud d'Aurillac', *AB* 14 (1895), 89–107.

—— 'Boémond et S. Léonard', *AB* 31 (1912), 24–44.

PORGES, W., 'The Clergy, the Poor, and the Non-combatants on the First Crusade', *Speculum*, 21 (1946), 1–23.

POSCHMANN, B., *Die abendländische Kirchenbuße im frühen Mittelalter* (Breslauer Studien zur historischen Theologie, 16; Breslau, 1930).

—— *Der Ablass im Licht der Bussgeschichte* (Theophaneia, 4; Bonn, 1948).

POULIN, J.-C., *L'Idéal de sainteté dans l'Aquitaine carolingienne d'après les sources hagiographiques (750–950)* (Travaux du laboratoire d'histoire religieuse de l'université Laval, 1; Quebec, 1975).

RASSOW, P., 'La Cofradía de Belchite', *Anuario de historia del derecho español*, 3 (1926), 200–26.

REILLY, B. F., *The Kingdom of León-Castilla under Queen Urraca 1109–1126* (Princeton, NJ, 1982).

—— *The Kingdom of León-Castilla under King Alfonso VI 1065–1109* (Princeton, NJ, 1988).

RICHARD, A., *Histoire des comtes de Poitou 778–1204*, 2 vols. (Paris, 1903).

RICHÉ, P., 'Les Moines bénédictins, maîtres d'école (VIIIe–XIe siècles), in W. Lourdaux and D. Verhelst (edd.), *Benedictine Culture 750–1050* (Mediaevalia Lovaniensia[1], 11; Louvain, 1983), 96–113.

RICKARD, P., *A History of the French Language*, 2nd edn. (London, 1989).

RILEY-SMITH, J. S. C., 'Crusading as an Act of Love', *History*, 65 (1980), 177–92.

—— 'The First Crusade and St. Peter', in Kedar, Mayer, and Smail (edd.), *Outremer*, 41–63.

—— *The First Crusade and the Idea of Crusading* (London, 1986).

—— *The Crusades: A Short History* (London, 1987).

—— 'Family Traditions and Participation in the Second Crusade', in M. Gervers (ed.), *The Second Crusade and the Cistercians* (New York, 1992), 101–8.

—— (ed.), *The Atlas of the Crusades* (London, 1991).

ROBINSON, I. S., *Authority and Resistance in the Investiture Contest: The Polemical Literature of the Late Eleventh Century* (Manchester, 1978).

—— *The Papacy 1073–1198: Continuity and Innovation* (Cambridge, 1990).

ROCACHER, J., *Rocamadour et son pèlerinage: étude historique et archéologique*, 2 vols. (Toulouse, 1979).

ROLLASON, D. W., 'The Miracles of St Benedict: A Window on Early Medieval France', in Mayr-Harting and Moore (edd.), *Studies in Medieval History*, 73–90.

ROSENWEIN, B. H., *To be the Neighbor of Saint Peter: The Social Meaning of Cluny's Property 909–1049* (Ithaca, NY, 1989).

ROUSSET, P., *Les Origines et les caractères de la Première Croisade* (Neuchâtel, 1945).

—— 'L'Idéal chevaleresque dans deux *Vitae* clunisiennes', in *Études de civilisation médiévale (IXe–XIIe siècles): mélanges offerts à Edmond-René Labande* (Poitiers, 1974), 623–33.

—— *Histoire d'une idéologie: la Croisade* (Lausanne, 1983).

RUPIN, E., *Roc-Amadour: étude historique et archéologique* (Paris, 1904).

SCHMITT, J.-C., 'Les Revenants dans la société féodale', *Temps de la réflexion*, 3 (1982), 285–306.

SCHNEIDMÜLLER, B., *Nomen Patriae: Die Entstehung Frankreichs in der politisch-geographischen Terminologie (10.–13. Jahrhundert)* (Nationes, 7; Sigmaringen, 1987).

SEGL, P., *Königtum und Klosterreform in Spanien: Untersuchungen über die Cluniacenserklöster in Kastilien-León vom Beginn des 11. bis zur Mitte des 12. Jahrhunderts* (Kallmünz, 1974).

SENAC, R.-A., 'Essai de géographie et d'histoire de l'évêché de Gascogne (977–1059)', *Bulletin philologique et historique* (1983), 11–25.

SIBERRY, E., *Criticism of Crusading 1095–1274* (Oxford, 1985).

SIGAL, P.-A., *Les Marcheurs de Dieu: pèlerinages et pèlerins au Moyen Âge* (Paris, 1974).

—— *L'Homme et le miracle dans la France médiévale (XIᵉ–XIIᵉ siècle)* (Paris, 1985).

SMAIL, R. C., *Crusading Warfare (1097–1193)* (Cambridge, 1956).

SOHN, A., *Der Abbatiat Ademars von Saint-Martial de Limoges (1063–1114): Ein Beitrag zur Geschichte des cluniacensischen Klösterverbandes* (Beiträge zur Geschichte des alten Mönchtums und des Benediktinertums, 37; Münster, 1989).

SOMERVILLE, R., 'The Council of Clermont (1095), and Latin Christian Society', *Archivum Historiae Pontificiae*, 12 (1974), 55–90.

SPRANDEL, R., *Ivo von Chartres und seine Stellung in der Kirchengeschichte* (Pariser Historische Studien, 1; Stuttgart, 1962).

SUMPTION, J., *Pilgrimage: An Image of Mediaeval Religion* (London, 1975).

TENANT DE LA TOUR, G., *L'Homme et la terre de Charlemagne à Saint Louis: essai sur les origines et les caractères d'une féodalité* (Paris, 1943).

TESKE, W., 'Laien, Laienmönche und Laienbrüder in der Abtei Cluny: Ein Beitrag zum "Konversen-Problem"', *Frühmittelalterliche Studien*, 10 (1976), 248–322; 11 (1977), 288–339.

THOMAS, G., 'Les Comtes de la Marche de la Maison de Charroux (Xᵉ siècle–1177)', *Mémoires de la Société des Sciences Naturelles et Archéologiques de la Creuse*, 23 (1925–7), 561–700.

TÖPFER, B., 'Reliquienkult und Pilgerbewegung zur Zeit der Klosterreform im burgundisch-aquitanischen Gebiet', in H. Kretzschmar (ed.), *Vom Mittelalter zur Neuzeit: Zum 65. Geburtstag von Heinrich Sproemberg* (Forschungen zur Mittelalterlichen Geschichte, 1; Berlin, 1956), 420–39.

—— *Volk und Kirche zur Zeit der beginnenden Gottesfriedensbewegung in Frankreich* (Berlin, 1957).

TROTTER, D. A., *Medieval French Literature and the Crusades (1100–1300)* (Histoire des idées et critique littéraire, 256; Geneva, 1987).

TUCOO-CHALA, P., *La Vicomté de Béarn et le problème de sa souveraineté des origines à 1620* (Bordeaux, 1961).

UBIETO ARTETA, AGUSTÍN, *Los 'tenentes' en Aragón y Navarra en los siglos XI y XII* (Valencia, 1973).

UBIETO ARTETA, ANTONIO, 'La participación navarro-aragonesa en la primera Cruzada', *Príncipe de Viana*, 8 (1947), 357–84.

—— 'La creación de la Cofradía militar de Belchite', *EEMCA* 5 (1952), 427–34.

—— *Historia de Aragón: la formación territorial* (Saragossa, 1981).

VALOUS, G. DE, *Le Monachisme clunisien des origines au XVᵉ siècle*, 2 vols. (Archives de la France monastique, 39–40; Ligugé, 1935).

VAUCHEZ, A., *La Spiritualité du Moyen Âge occidental VIIIᵉ–XIIᵉ siècles* (Paris, 1975).

—— *La Sainteté en Occident aux derniers siècles du Moyen Âge d'après les procès de canonisation et les documents hagiographiques* (Bibliothèque des Écoles Françaises d'Athènes et de Rome, 241; Rome, 1981).

VÁZQUEZ DE PARGA, L., LACARRA, J. M., and URÍA RÍU, J., *Las peregrinaciones a Santiago de Compostela*, 3 vols. (Madrid, 1948–9).

VERDON, J., 'Une source de la reconquête chrétienne en Espagne: la Chronique de Saint-Maixent', in Gallais and Riou (edd.), *Mélanges*, i. 273–82.

VIDIER, A., *L'Historiographie à Saint-Benoît-sur-Loire et les Miracles de saint Benoît* (Paris, 1965).

VILLEY, M., *La Croisade: essai sur la formation d'une théorie juridique* (L'Église et l'État au Moyen Âge, 6; Paris, 1942).

VOGEL, C., 'Le Pèlerinage pénitentiel', in *Pellegrinaggi e culto dei Santi in Europa fino alla I^A Crociata* (Convegni del Centro di Studi sulla Spiritualità Medievale, 4; Todi, 1963), 37–94.

—— 'Les Rites de la pénitence publique aux xe et xie siècles', in Gallais and Riou (edd.), *Mélanges*, i. 137–44.

—— *Le Pécheur et la pénitence au Moyen Âge* (Paris, 1969).

—— *Les 'libri paenitentiales'* (Typologie des sources du Moyen Âge occidental, 27; Turnhout, 1978).

VONES, L., *Die 'Historia Compostellana' und die Kirchenpolitik des nordwestspanischen Raumes 1070–1130* (Kölner historische Abhandlungen, 29; Cologne, 1980).

WAAS, A., *Geschichte der Kreuzzüge*, 2 vols. (Freiburg, 1956).

WALKER, D., 'The Organization of Material in Medieval Cartularies', in D. A. Bullough and R. L. Storey (edd.), *The Study of Medieval Records: Essays in Honour of Kathleen Major* (Oxford, 1971), 132–50.

WARD, B., *Miracles and the Medieval Mind: Theory, Record and Event 1000–1215*, rev. edn. (Aldershot, 1987).

WEINSTEIN, D., and BELL, R. M., *Saints and Society: The Two Worlds of Western Christendom 1000–1700* (Chicago, 1982).

WERNER, K. F., 'Les Nations et le sentiment national dans l'Europe médiévale', *Revue historique*, 244 (1970), 285–304.

WHITE, S. D., *Custom, Kinship, and Gifts to Saints: The Laudatio Parentum in Western France 1050–1150* (Chapel Hill, NC, 1988).

WILMART, A., 'Les Ouvrages d'un moine du Bec: un débat sur la profession monastique au xiie siècle', *RB* 44 (1932), 21–46.

WINTER, J. M. VAN, '*Cingulum militiae*, Schwertleite en *miles*-terminologie als spiegel van veranderend menselijk gedrag', *Tijdschrift voor Rechtsgeschiedenis*, 44 (1976), 1–92.

WOLFF, R. L., 'How the News was brought from Byzantium to Angoulême; or, The Pursuit of a Hare in an Ox Cart', *Byzantine and Modern Greek Studies*, 4 (1978), 139–89.

ZAFARANA, Z., 'Bosone', *Dizionario biografico degli Italiani*, 13 (Rome, 1971), 267–70.

Index